Lecture Notes in Mechanical Engineering

Lecture Notes in Mechanical Engineering (LNME) publishes the latest developments in Mechanical Engineering - quickly, informally and with high quality. Original research reported in proceedings and post-proceedings represents the core of LNME. Volumes published in LNME embrace all aspects, subfields and new challenges of mechanical engineering. Topics in the series include:

- Engineering Design
- Machinery and Machine Elements
- Mechanical Structures and Stress Analysis
- Automotive Engineering
- Engine Technology
- Aerospace Technology and Astronautics
- Nanotechnology and Microengineering
- Control, Robotics, Mechatronics
- MEMS
- Theoretical and Applied Mechanics
- Dynamical Systems, Control
- Fluid Mechanics
- Engineering Thermodynamics, Heat and Mass Transfer
- Manufacturing
- Precision Engineering, Instrumentation, Measurement
- Materials Engineering
- Tribology and Surface Technology

To submit a proposal or request further information, please contact the Springer Editor in your country:

China: Li Shen at li.shen@springer.com
India: Dr. Akash Chakraborty at akash.chakraborty@springernature.com
Rest of Asia, Australia, New Zealand: Swati Meherishi at swati.meherishi@springer.com
All other countries: Dr. Leontina Di Cecco at Leontina.dicecco@springer.com

To submit a proposal for a monograph, please check our Springer Tracts in Mechanical Engineering at http://www.springer.com/series/11693 or contact Leontina.dicecco@springer.com

Indexed by SCOPUS. The books of the series are submitted for indexing to Web of Science.

More information about this series at http://www.springer.com/series/11236

Dean Vucinic · Fabiana Rodrigues Leta ·
Sheeja Janardhanan
Editors

Advances in Visualization and Optimization Techniques for Multidisciplinary Research

Trends in Modelling and Simulations for Engineering Applications

Springer

Editors
Dean Vucinic
Vesalius College (VeCo)
Vrije Universiteit Brussel (VUB)
Brussels, Belgium

Fabiana Rodrigues Leta
Mechanical Engineering Department
Universidade Federal Fluminense
Niterói—Rio de Janeiro, Brazil

Sheeja Janardhanan
Department of Mechanical Engineering
SCMS School of Engineering and Technology
Karukutty, Ernakulam, Kerala, India

ISSN 2195-4356 ISSN 2195-4364 (electronic)
Lecture Notes in Mechanical Engineering
ISBN 978-981-13-9808-7 ISBN 978-981-13-9806-3 (eBook)
https://doi.org/10.1007/978-981-13-9806-3

This Springer imprint is published by the registered company Springer Nature Singapore Pte Ltd.
The registered company address is: 152 Beach Road, #21-01/04 Gateway East, Singapore 189721, Singapore

Preface

"Seeing is Believing" as the famous saying goes....

It is well known that the visualization and optimization techniques are essential parts in today's complex modelling and simulation applications performed in many multidisciplinary research and development projects, aiming at excellence in engineering results.

The challenge in producing this book was to assemble the best papers and lectures given during the last several years at the International Conference on Advanced Computational Engineering and Experimenting (ACE-X), presented in the Special Session 11 "Advance Scientific Visualisation for Multidisciplinary Engineering", addressing the following topics:

- Advanced Scientific Visualization
- Information Visualization and Visual Analytics
- Big Data Visualization, Exploration and Analysis
- Computational Models and Human-Computer Interaction
- Image Analysis and Pattern Recognition
- Multidisciplinary Simulation and Modelling
- Engineering Design Optimization
- Computer Vision and Robotics
- Astronomy and Geology
- Biomedical Imaging
- Sensor Networks and Cybersecurity
- Nano Technology Visualization

The Special Session 11 objectives are to promote and put forward the latest advances in scientific visualization techniques and their usage when exploring and analysing engineering data and their respective solutions—being critical to establish a more direct and efficient way of communication between engineers, scientists and general public, especially valuable when seeking new innovative solutions and presenting their expected benefits. Today, the continuous and rapid technological changes are demanding new ways in exploring "big data" information sources underpinning the knowledge discovery, interpretation and understanding of the results coming from—more and more—complex computations and experiments, as being essential part of the common engineering practice. These large data sets represent an extremely important information base, which enables trustable multidisciplinary design optimization and prototyping, following tightly validated and coupled computational and experimental procedures. Such well-known setups are motivating the seeking for new innovations and improvements in finding the "best" engineering solutions fulfilling high-quality modern standards, by bringing together the best practices, from academia, science, industry and beyond.

The book is organized in 12 chapters covering multidisciplinary scientific domains such as space, aerospace, shipbuilding, turbomachinery, green transportation, renewable energy.

Chapter "Advanced in Mach-Zehnder Interferometer Using Windowed Fourier Transform to Analyse Coupled Heat and Mass Transfer" gives an insight into optical digital interferometry as a superior tool over the conventional FFT techniques in the studies pertaining to heat and mass transfer.

Metaheuristic algorithms for optimizing flight trajectory addressing the reduction in flight cost and pollution are discussed in Chapter "Commercial Aircraft Trajectory Optimization to Reduce Flight Costs and Pollution: Metaheuristic Algorithms".

Chapter "Uncertainty Quantification and Robust Optimization in Engineering" presents a novel method known as polynomial chaos, incorporating uncertainties into the design based on the gradient-based robust optimization, presenting an example of shape optimization for transonic aerofoil.

Chapter "Determination of Linear and Non-linear Hydrodynamic Derivatives of a Surface Ship in Manoeuvring Using CFD Method" finds its importance in replicating the towing tank model tests by using the CFD simulations, which results are analysed using Fourier series for obtaining the hydrodynamic derivatives of the container ship hull model.

The application of pattern recognition method for estimating the wind load on ships and marine objects is discussed in Chapter "Application of Pattern Recognition Method for Estimating Wind Loads on Ships and Marine Objects". The chapter also brings to light the need for series of experiments in order to achieve a successful prediction of wind loads.

Chapter "Tensairity, an Extra-Light Weight Structure for Airships" is devoted to airships wherein the concept of tensairity, a lightweight inflatable reinforced beam that can be used as a structural element for airship keel, as novel technology that has large potential for building new airships of any size and geometry.

It is no more a problem watching television programs of different languages, as Chapter "Realization of Subtitle Support in Hybrid Digital TV Applications" addresses this very well, describing the subtitles support in the hybrid digital TV applications with relatively simple hardware equipment.

The computational hemodynamics is gaining importance as an engineering application, which is becoming more and more important in the medical domain. Chapter "Human Heart Blood Flow Numerical Modelling and Simulations" presents a fluid structure interaction approach to simulate the human heart blood flow, also with the artificial valve implant for the treatment of the heart problems.

In Chapter "A Review on Computer Vision Applied to Mechanical Tests in Search for Better Accuracy", a review of over 30 years in the application of computer vision in material testing is presented. Computer vision systems are helping better prediction of material properties, as discussed against the conventional destructive and conventional nondestructive testing.

Yet another application of CFD as a powerful tool in visualizing complex flows around the blades in turbomachinery is presented in Chapter "CFD—A Powerful Visualization Tool in Turbomachinery Applications", where additional codes for high resolution of the boundary layer and better convergence are discussed.

Chapter "Application of Simulation for Emulating and Visualizing Nanotechnology" presents a literature review of simulation and visualization and discusses the applications of various online simulation tools, laboratory instruments and remote access instruments used for visualizing nanotechnology devices and processes.

Chapter "Hybrid Optimization Method and Algorithms for Monochrome Images Tone Approximation with Implementation" considers the relevant problem of monochrome multi-tone images tone approximation and presents the methodology of developing advanced hybrid optimization algorithms.

As the reader will notice the authors are coming from almost every continent of the world, making this book a very special and unique reference for international research, by empowering engineering community to share the latest advances coming out of such multidisciplinary best practices experienced in Europe, Asia, North and South America.

With big wish that you will enjoy the reading, and that you might possibly be the next author of a chapter, as it is envisaged that there will be a follow-up book coming in the near future.

Brussels, Belgium
Rio de Janeiro, Brazil
Kochi, India
June 2019

Dean Vucinic
Fabiana Rodrigues Leta
Sheeja Janardhanan

Contents

About the Editors

Prof. Dr. Ir. Dean Vucinic has joined Vesalius College (VeCo), which is affiliated to the Vrije Universiteit Brussel (VUB), as senior scientist and advisor in 2017, and became the adjunct professor for its two educational programs (Foundation/European Business Communication). He has been affiliated to VUB since 1988. Before joining VeCo, he was guest professor and senior research scientist at the VUB Faculty of Engineering Sciences (IR), as member of its two departments: Mechanical Engineering (MECH) and Electronics and Informatics (ETRO).

He is also the visiting associate professor at the Faculty of Electrical Engineering, Computer Science and Information Technology (FERIT), University of Osijek, Croatia, at the Department of Software Engineering holding the chair of visual computing.

His work is mostly related to research and development (R&D) projects, and his interest covers the topics of scientific visualization, modeling and simulation, optimization methodologies and techniques, which are very often found together in solving complex problems within the multidisciplinary engineering and computer science domains.

His Ph.D. thesis became a book in 2010, ISBN 978-3-8383-3500-1. In the early 1990s, he developed "CFView—Computational Field Visualization System", the first-time-ever interactive visualization software adapted to numerical simulation solvers, completely based on the object-oriented technology and fully implemented in C++. It has to be noted that VUB spin-off NUMECA is still using his software after more than 20 years, and in addition, more than 20 VUB Ph.D.s applied CFView in their visualization and data analysis tasks.

During almost 30 years at VUB, he successfully participated in more than 20 European projects under the European frameworks, EUREKA/ITEA and Tempus educational programs.

He is author of more than 60 scientific papers in international reviewed journals and conference proceedings and, in addition, author of several chapters in various books.

He is member of International Advisory Boards and Scientific Committees of Journals and Conferences, acting as chair, session organizer, reviewer and editor of respective paper proceedings.

Due to his international presence, he is promoting and encouraging international cooperation in research and development, and education, just mentioning some of the realized initiatives with USA, Canada, Russia, Brazil, India, Egypt, Japan and among others.

He is the European Commission expert in H2020 and member of international organizations: AIAA, IEEE, ACM, SAE and ASME.

Fabiana Rodrigues Leta is professor of mechanical engineering at School of Engineering in Federal Fluminense University (UFF), Rio de Janeiro, Brazil, since 1989. She is full professor at UFF since 2015. She has a bachelor's degree in mechanical engineering from Federal Fluminense University (1989) and Doctor of Science degree in mechanical engineering from Pontifical Catholic University of Rio de Janeiro, Brazil (1998).

She was head of the Mechanical Engineering Department (2007–2013), coordinator of Scientific and Technological Research (2006–2007), director of the Innovation Agency (2009–2014), deputy director of the School of Engineering (2011–2018) and became dean of the School of Engineering in November 2018 at Federal Fluminense University, she is the first women in this position.

She was responsible for the creation of the Innovation Agency of the Federal Fluminense University and she also created and coordinated the Integrated Center for Technology and Innovation of the School of Engineering in 2014. She is editor of the journal Engevista and the magazine Engenharia and Design. She coordinates the Laboratory of Dimensional and Computational Metrology, created by her in 1992.

She is a member of the Brazilian Association of Mechanical Engineering Sciences and of the Brazilian Society of Metrology. She took part of the Municipal Council of Science and Technological Innovation of Niterói (2006–2008 and 2013–2015). She is tutor of the Tutorial Education Program (Brazilian program supported by Education Ministry). She was a founding member of the Forum of Directors of Schools of Engineering in Brazil.

She develops research mainly in the following areas: visual computing, metrology by image technological innovation, oil and gas industry and engineering education. She is part of the scientific committee of IWSSIP International Conference on Systems, Signals and Image Processing and of ACE-X—International Conference on Advanced Computational Engineering and Experimenting since 2010.

She published more than 260 papers in conferences and journals and edited five books concerning her research specialty. She advised numerous undergraduate and graduate students and has coordinated and participated in several projects, financed by development agencies and industry. Some of them can be highlighted: Morphing for Generation of Spoken Portraits; Metrology based on Computational Vision; Implementation of the Innovation Agency of the UFF; Production of Didactic Material for Industrial Metrology; Computing System for Automatic Visual Inspection of Printed Circuit Boards; Rock Geochemical Analysis System; Automatic Interpreter System for Graphics Images; Basaltic Petrogenesis Geodetic Scenarios; System to Identify and Estimate the Stress Orientation from Oil Well; Corrosion Estimation by Image Analysis; Deep Learning Algorithms for Monitoring Oil Well Construction.

Sheeja Janardhanan is professor of mechanical engineering at SCMS School of Engineering and Technology, Ernakulam, India (from 2013). She has a bachelor's degree in mechanical engineering from Andhra University (2002). She was awarded a master's degree in computer aided structural analysis and design from Cochin University of Science and Technology, Kochi, India, with first rank (2005) and Ph.D. in numerical ship hydrodynamics from Department of Ocean Engineering, Indian Institute of Technology Madras, Chennai, India (2010).

She formerly worked as a researcher in the research and development wing of Indian Register of Shipping, Mumbai, India (2010–2013) where she shouldered the responsibility of leading a team of young scientists in developing numerical towing tank. She is among the pioneers to apply computational fluid dynamics (CFD) method in ship maneuvering in India. She was the organizing chair for the International Conference on Computing in Mechanical Engineering (ICCME 2015 and 2017). She is also the coordinator of Research and Consultancy cell of SCMS School of Engineering and Technology, single point of contact of national level online educational program, National Programme on Technology Enhanced Learning, India.

She is actively involved in research pertaining to CFD applications in ship and marine hydrodynamics, hydroelasticity, vortex-induced vibrations, bio-inspired propulsion of AUVs and metal casting processes using CFD. She has been supervising many graduate and doctoral theses and also has been reviewing articles in peer-reviewed journals with about thirty-five publications to her credit.

Advanced in Mach-Zehnder Interferometer Using Windowed Fourier Transform to Analyse Coupled Heat and Mass Transfer

Amirhossein Ahadi and M. Ziad Saghir

Abstract Series of thermodiffusion experiments using optical digital interferometry (ODI) have been conducted onboard the International Space Station. Conventionally, the two-dimensional (2D) fast Fourier transform (FFT) fringe analysis technique has been applied as a fast and reliable technique to extract data. In this study, for the first time, the windowed Fourier transform (WFT) method is used to analyse the same experiments. In this method, a Fourier transformation is applied on the fringes at two different stages: initially, during the filtration of the non-zero peaks and then on the wrapped phase image. We provide a detailed comparison between FFT and WFT results of binary and ternary mixtures for ODI thermodiffusion experiments. The substantial enhancements of this method are presented and discussed for different experiments conducted for both binary and ternary mixtures. We show that while disturbances in the phase fringe pattern can cause significant error in FFT techniques, if the windowed Fourier filtration (WFF) parameters are properly chosen this type of noise can be eliminated during WFF analysis. The importance of replacing the FFT method becomes more pronounced for the ternary system, as this method fails to reconcile reliable concentration profiles. The results of this chapter can show that the application of the windowed Fourier transform in optical digital interferometry investigations show improved results over the same experiments analysed using FFT methods, especially for experiments involving very small heat and mass fluxes such as the Soret effect in multicomponent mixtures.

Keywords Fourier image processing · Interferometry · Thermodiffusion · Ternary mixtures

A. Ahadi (✉) · M. Z. Saghir
350 Victoria Street, Toronto, ON, Canada
e-mail: aahadi@ryerson.ca

D. Vucinic et al. (eds.), *Advances in Visualization and Optimization Techniques for Multidisciplinary Research*, Lecture Notes in Mechanical Engineering, https://doi.org/10.1007/978-981-13-9806-3_1

Nomenclature

Change in refractive index	Δn
Coordinate index of pixel	i, j
Coordinate system	x, y
Fourier transform of coordinate	u, v, ξ, η
Initial concentration	c_O
Molecular diffusion coefficient	D (m^2 s^{-1})
Maximum concentration difference	ΔC
Non zero picks, Fourier transform	C, C*
Number of pixels in x, y direction	m, n
Optical length of the cell	L (mm)
Phase distribution	$\Delta\phi$
Soret coefficient	S_T (K^{-1})
Temperature difference	ΔT (K)
Time	t (s)
Thermodiffusion coefficient	D_T (m^2 s^{-1} K^{-1})
Vibration amplitude	A (mm)
WFT spectrum	*Sf*

Greek Symbol

Concentration contrast factor	$(\partial n \partial c)_{p.T}(-)$
Density	$\rho\left(\mathrm{kg/m^3}\right)$
Interference phase	ϕ
Laser wavelength	λ (nm)
Mixture viscosity	λ (nm)
Relaxation time	τ (s)
Temperature contrast factor	$(\partial n \partial T)_{p.c}$ (1/K)
Thermal diffusivity	$\chi\,(\mathrm{m^2/s})$
Standard deviations	δ_x, δ_y

Subscript

Experimental	exp
Hour	hr
Minute	min
Reference	ref
Thermal	th
Threshold	thr
Steady	st

1 Introduction

Image processing is a well-established and highly developed research field [1, 2] in which two-dimensional transform techniques are applied to optical images in order to extract useful information. Precision optical equipment, very fast computers, and reliable image processing software has increased the total reliability and accuracy of the measurement system, making optical metrology the first choice across many engineering and scientific disciplines, such as medicine, engineering and industry, robot and machine vision, among others.

Interferometry is a method that measures the displacement between regions of different refractive indices. It therefore has many applications in the measurement of fluid flows, photo elastic stresses, surfaces and thin films, and vibrations. By extracting the phase information from the fringe pattern, a quantitative height map is produced. The height map may represent a surface profile, a displacement, or the optical path difference through a phase object. In practice, obtaining this height map is difficult due to ambiguities in the fringe data, speckle noise, and the resolution of the imaging system [3–5]. There are many different transform techniques used for image processing; however, these techniques all have limitations and no single method is suitable for all types of problems [6]. There are continuous research efforts into expanding, adapting, and improving the available transform techniques for different applications of image processing. A 2D domain such as an image or a 2D audio representation allows the application of transform techniques in two dimensions. The use of 2D transform techniques for image processing has been explored and it can produce interesting and useful results [2, 7, 8].

Most of the optical methods for measuring the phase are based on spectral shearing interferometry with two spectrally sheared driving pulses [9]. Ishii et al. proposed a demodulation technique for the interferometric phase using discrete Hilbert transform and discrete high pass filtering [10]. This algorithm can be applied to interference signals, which have a range in spatial frequencies. They demonstrated a useful demodulation scheme on signals with a wide spectral bandwidth. A two-dimensional phase profile was calculated by using one-dimensional discrete signal processing with the raster scanning procedure [10].

Although nonlinear methods can provide only the amplitude and the phase of an isolated ultra-short pulse, linear techniques can possess a much better sensitivity and reliability when a reference pulse is available. These techniques are simple to implement, very sensitive, and provide a complete measurement of the complex electric field as a continuous function of frequency [11]. In the late 19th century, Joseph Fourier proposed a theorem that any periodic signal can be decomposed into a series of harmonically related sinusoidal functions with specified amplitudes, frequencies, and phase offsets; this is known as the Fourier series [8, 11, 12]. The Fourier transform uses this concept for decomposition of a continuous-time signal into sinusoidal components; any signal of infinite extent and bandwidth can be perfectly reconstructed using an infinite series of sinusoidal waves representing all frequencies. Fourier analysis is extremely important in many areas of signal processing. The phase information is programmed into the interferogram in the form of bending of the fringes. By extracting this fringe frequency from the dominant carrier frequency, the required phase information can be isolated. For many years, the two-dimensional fast Fourier transform fringe analysis technique has been regarded as a fast and reliable technique for the analysis of fringe patterns projected onto static objects.

Many different software programs have been developed to focus on high-resolution digital fringe evaluation including phase stepping, Fourier domain evaluation, and unwrapping techniques for both regular and irregular fringe patterns [13, 14]. Since the harmonic modes used in the expansions are periodic in nature, Fourier series are ideal for analysing periodic signals. As a result of having a very noisy refractive index pattern caused by FFT, it was necessary to develop and implement another technique to enhance the quality of the results. Optical digital interferometry (ODI) has been used extensively as a method to extract temperature, temperature gradient, and concentration of species [14–17]. In this chapter, we examine the implementation of image processing methods on thermodiffusion experiments.

Due to local temperature differences, components of a mixture tend to move to either the hot or cold sides of a chamber in a process known as thermodiffusion [18]. Based on different applications of this phenomenon, it is called thermal diffusion, the Ludwig-Soret effect, or the Soret effect [19–21]. The experimental approaches to investigate the Soret effect are defined in two main categories: classical methods [22] and recent laser-based measurements techniques such as laser beam deflection [23, 24] and holographic interferometry [25].

The Soret coefficient, which is a materials property, is recognized to be difficult to determine at ground conditions because of technical constraints in the control of liquid convection and convective instabilities due to gravity [18, 26]. Measurements conducted in controlled microgravity environments such as the International Space Station (ISS) minimize the perturbation effects of gravity and allow purely diffusive conditions to be achieved. The best example of such measurements is the experiments run on the SODI (Selectable Optical Diagnostics Instrument). The two major detailed experiments that have been performed under the SODI mission are IVIDIL and DSC. The IVIDIL (Influence of VIbrations on DIffusion of Liquids) experiment investigated the effect of vibrations on the thermal separation of binary mixtures [14, 15, 20, 26, 27]. The main purpose of DSC (Diffusion Soret Coefficient) was to measure the diffusion and Soret coefficients of selected ternary mixtures, a process that is fundamentally more complex and sensitive than the measurements on binary mixtures [28]. The outcome of these experiments can be employed to test thermodiffusion theories and develop physical and mathematical models for the estimation of thermodiffusion and diffusion coefficients [15, 17]. Two binary mixtures of water-isopropanol (IPA) and the ternary mixture of tetrahydronaphthalene-isobutylbenzene-dodecane (THN-IBB-C_{12}) with various compositions were hosted in the cell arrays for the IVIDIL and DSC experiments, respectively.

The experimental data for both of these experiments were obtained using the interferometry image processing technique [15, 17], which is one of the most effective non-contact optical techniques for these systems. The samples were monitored by a Mach-Zehnder interferometer at two wavelengths. Because of small scale of the variations in concentration, the processed results can be affected by the inherent noise that exists in the system due to many reasons such as presence of dust particles on the front glass disturbing a clear view of the cell.

In other studies, it was shown that using more complex image processing techniques improved the contrast of the reconstructed images and reduced the noise produced by parasitic reflections reaching the hologram plane at angles of incidence angle different from that of the object wave. The FFT method has thus been improved further and windowed Fourier transform processing was developed. This new proposed algorithm analyses the sequence of fringe patterns with two continuous steps of Fourier transformation: first during the filtration of one of the non-zero peaks and the other on the wrapped phase image. This has the advantage of making the algorithm more robust, accurate, and suitable for the measurement of phase differences between two images. Moreover, the noise that may occur in the phase image as result of the FFT method can be accurately eliminated. This method filters out the spatial frequencies associated with undesired terms in the computed Fourier transform of the hologram [29]. Moreover, by use of the two complementary outputs of the interferometer, it is possible to remove the zero-order terms and thus to increase the resolution.

In this investigation, for the first time we have implemented a windowed Fourier transform algorithm to study the thermodiffusion phenomenon to improve the contrast of the reconstructed images and reduce the noise. First, we discuss different image processing techniques used to process optical digital interferometry (ODI) experiments that deal with coupling between the sensitive measurements of heat and mass transport in the domain. Section 2 briefly presents the experimental setup and the experimental procedure that was performed on board the ISS. The detailed information and mathematical formulas are discussed in Sect. 3. The principles of the fast Fourier transform and the windowed Fourier transform methods for fringe pattern processing of ODI images are introduced using references [30, 31]. Section 4 presents a short mathematical description of extracting the temperature and concentration from the phase images for binary and ternary mixtures. Section 5 introduces the software tool developed during this study with MATLAB to process the thermodiffusion ODI fringes and validate the software against the benchmark values of Soret coefficient of one of the test mixtures. The results and discussion are provided in Sect. 6, and they focus on a detailed comparison between the FFT and WFT results in studying the binary and ternary ODI experiments. Finally, a summary of this chapter is provided in the conclusion section.

2 Procedure of the Experiment

Fringe analysis techniques [7, 12] are considered effective, reliable, and robust optical non-contact methods for measuring refractive index variation. These methods have been employed in both the DSC and IVIDIL experiments. In this method, a structured lighting pattern was initially projected onto the experimental cell. As shown in the Fig. 1, the generated laser beam was propagated into an optical fibre cable until reaching an optical coupler. This optical coupler divides the laser beam into two beams. The sample was monitored using a Mach-Zehnder interferometer at two wavelengths, depending on whether the sample was a binary or ternary mixture. Figure 1 shows the two different paths inside the cell holder of the split beam. One of the paths moved the beam through the front of the cell (in the direction perpendicular to the thermal gradient, as shown in Fig. 2) and the other one passed the beam through the void area to be used as the reference beam. After reflection from the mirrors, both beams interfered with each other at the second beam splitter. In this way, interference fringes were generated and captured by a CCD camera.

The first step of the thermodiffusion experiments that will be analysed in this chapter was thermalization of the sample at the mean temperature to reach the initial state of uniform concentration and temperature distribution inside the cavity. At this stage, the top and bottom plates of the cell were maintained at the mean temperature. Then, the temperature gradient was built up by application of heat to both sides of the cavity. ODI images and the corresponding temperature differential between the hot and cold sides were acquired during the experiment. After completion of the designated experimental time span, the temperature difference was removed and the cell was brought back to the mean temperature. If needed, the run was repeated at a different temperature difference. Afterwards, the moveable optic was relocated to another sample and the cycle was repeated.

Thermal equilibrium in the sample mixture is established at a characteristic time that is a function of the length of the experimental cell and the thermal diffusivity of

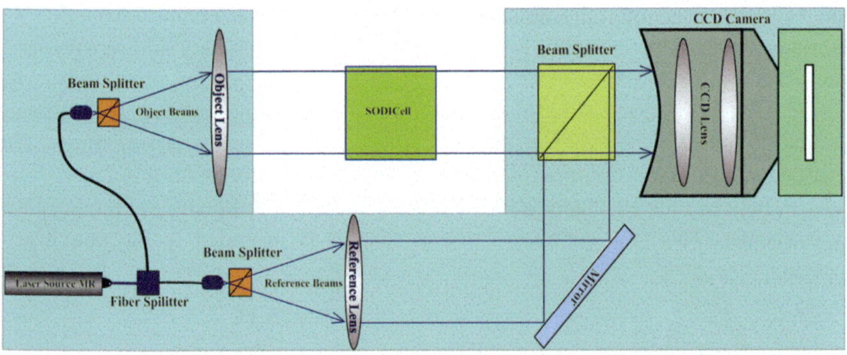

Fig. 1 Scheme of the interferometer setup, it represents the moving bridge f SODI apparatus

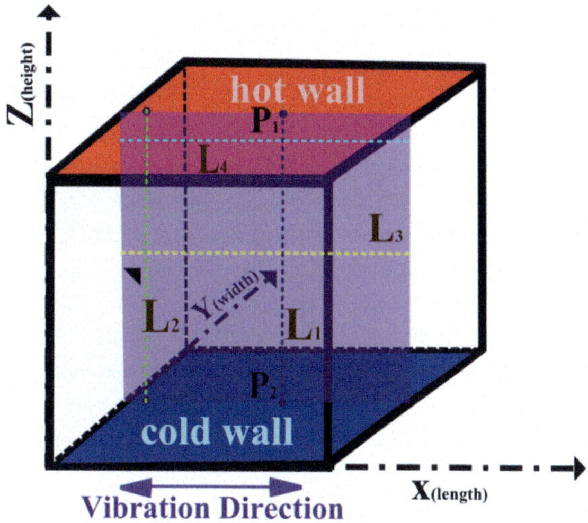

Fig. 2 Geometry of the SODI-IVIDIL cell. The same cell with an aspect ratio of 1:1:0.5 is used in DSC respect to the x, y and z directions

the liquid mixture: $\tau_{th} \propto L^2/\chi$. Any change in the refractive index is assumed to be due to temperature in this period [31]. Eventually, a concentration gradient is established as a result of the Soret effect. The concentration profile is slowly generated in the initially homogeneous mixture. In addition, there is a characteristic diffusion time that is similar to the thermal time as it is also a function of the cell length: $\tau_C \propto L^2/D$, where D is the smallest eigenvalue of the diffusion matrix. By definition, at the end of the diffusion time the linear separation of species has been reached [17]. In general, the thermal time is about 100 times smaller than the diffusion time. As a result, the temperature field reaches the steady state a few minutes after applying the temperature difference, after which time the changes in the refractive index are purely due to the variation in species concentration.

3 Transform Methods of Fringe Analysis

The output of the optical experiment and methods were in the form of a fringe pattern, which requires further processing to be analysed [1, 7, 32]. Two typical techniques for phase retrieval are the importer technique with a Fourier transform [33–35], and the phase-shifting technique [36]. The phase-shifting technique evaluates the fringe patterns pixel by pixel so there is no effect of any pixel on any other. It should be noted that noise in the image could have a large effect on the final output from this technique. On the other hand, the importer technique with a Fourier transform processes the whole frame of a fringe pattern at the same time. It is more tolerant to noise, but neighbouring pixels affect each other. In local processing methods such as phase-shifting this behaviour can be detected.

3.1 Fast Fourier Transform

The procedure established for image processing of thermodiffusion experiment is as follows: after downloading the raw data and converting them to digital images, each interferogram was subjected to resizing or cropping to remove the affected pixels close to the vicinity of the walls, known as crashed pixels. Then, each image was reconstructed by performing a two-dimensional fast Fourier transform, filtering a selected band of the spectrum that includes one of the non-zero peaks, placing this band at the middle of the spectrum, performing an inverse 2D fast Fourier transform of the filtered result, and phase unwrapping. After these steps, the gradients of composition inside of the fluid were calculated from gradients of the refractive index and local temperature for both the steady and dynamic regimes.

Three unknowns exist in Eq. (1): $a(x, y)$, $b(x, y)$, and $\varphi(x, y)$. The phase $\varphi(x, y)$ is the information of interest and the other two terms $a(x, y)$ and $b(x, y)$ need to be eliminated. The mathematical description of above steps is shown in this section. The measured intensity distribution $i(x, y)$ can be written in the form [15, 33, 37]:

$$i(x,y) = a(x,y) + \frac{1}{2}b(x,y)\left(e^{j\phi(x,y)} + e^{-j\phi(x,y)}\right) \tag{1}$$

where $a(x, y)$ is the background noise and $b(x, y)$ the modulation noise or the local contrast of the pattern. $\phi(x, y)$ is the phase of the intensity $i(x, y)$. Equation (1), can be rewritten as one zero peak and two non-zero peaks:

$$i(x,y) = a(x,y) + c(x,y) + c^*(x,y) \tag{2}$$

Applying the FFT to Eq. (2) with respect to x and y gives

$$I(u,v) = A(u,v) + C(u,v) + C^*(u,v) \tag{3}$$

in which the zero peak, C, and C^* are placed symmetrically to the origin. The next step was to filter out either of the two spectra on the carrier. The unwanted background variation had been filtered out at this stage using a filter mask. Once more translating the intensity function to the origin by the carrier frequency and applying an inverse fast Fourier transform (iFFT) algorithm to $C(u, v)$, the phase pattern can be found as follows [38]:

$$\log[c(x,y)] = \log\left[\left(\frac{1}{2}\right)b(x,y)\right] + i\phi(x,y) \tag{4}$$

3.2 *Windowed Fourier Transform*

The WFT which is also known as short-time Fourier transform establishes the use of two algorithms, the windowed Fourier filtering (WFF) and the windowed Fourier ridges (WFR) [39–42]. A smoothing filter, which is used in this study as a local processor, assumes that the intensity values in a small block around each pixel are the same and averages the values of that block. The use of such an averaging technique is not suitable for a fringe pattern since its intensity undulates as a cosine function. Because of this, more advanced and effective techniques, such as regularized phase tracking [43, 44], wavelet transform [45, 46], Wigner-Ville distribution [47], and windowed Fourier transform (WFT) were established. In this investigation, the principle of the WFT is considered, used, and compared with the FFT. While the entire procedure of the WFT method can be found in the literature [2, 31, 48], in this study, for the first time, the windowed Fourier transform method was used to analyse thermodiffusion interferometry experiments. Prior to this study, FFT was solely used for this goal [14, 15, 27, 49], despite its various limitations and inaccuracies that will be discussed later on in this chapter.

3.2.1 The Windowed Fourier Transform Mathematical Formulation

The WFT and its inverse are a pair of transforms as given below [32, 40, 42, 50]:

$$Sf(u,v,\xi,\eta) = \int_{-\infty}^{\infty}\int_{-\infty}^{\infty} f(x,y)g^*(x,y)dxdy \tag{5}$$

$$f(x,y) = \frac{1}{4\pi^2}\int_{-\infty}^{\infty}\int_{-\infty}^{\infty}\int_{-\infty}^{\infty}\int_{-\infty}^{\infty} Sf(u,v,\xi,\eta)g^*_{u,v,\xi,\eta}(x,y)d\xi\, d\eta\, du\, dv \tag{6}$$

where the symbol * denotes the complex conjugate. In Eq. (5), the analysis of a 2D image *f*(*x*, *y*) into the WFT basis $g_{u,v,\xi,\eta}(x,y)$, results in the 4D coefficients (or the WFT spectrum) $Sf(u,v,\xi,\eta)$. Then, Eq. (6) reconstructs the image. The WFT basis consists of a series of windowed Fourier elements as described below [32, 40, 42, 50]:

$$g_{u,v,\xi,\eta}(x,y) = g(x-u, y-v)e^{(j\xi x + j\eta y)} \tag{7}$$

Associated with the Fourier basis $\exp(j\xi x + j\eta y)$, which has an infinite spatial extension [32], the WFT elements have an incomplete spatial extension due to the windowed function $g(x-u,\, y-v)$.Accordingly, the WFT spectrum, $Sf(u,v,\xi,\eta)$ gives the frequency information at each pixel in the image, which is difficult for the Fourier transform [32]. The cost is that WFT computation requires more computer power since its basis is dismissed and not orthogonal [41, 51]. The WFT method is also named the Gabor transform if $g(x, y)$ is a Gaussian function given as follows:

$$g(x,y) = e^{\left(-\frac{x^2}{2\delta_x^2}-\frac{y^2}{2\delta_y^2}\right)} \tag{8}$$

where δ_x, δ_y are the standard deviations of the Gaussian function in the x and y directions, respectively, which control the spatial extension of $g(x, y)$. The Gaussian window is set throughout this research, although a simple square window also works well. The Gaussian window function is divided by $\sqrt{\pi\delta_x\delta_y}$ for normalization such that $\|g(x, y)\| = 1$.

3.2.2 Windowed Fourier Filtering

The similarity of WFT and FFT reminds us that a fringe pattern can be filtered by processing its WFT spectrum [40]. Commonly, noise in the original image permeates the entire spectrum domain with small coefficients due to its randomness and incoherence with the WFT basis. Consequently, most noise can be suppressed by discarding spectrum coefficients if their amplitudes are smaller than a preset threshold. The scheme can be expressed as [32, 40, 42, 50];

$$f(x,y) = \frac{1}{4\pi^2}\int_{-\infty}^{\infty}\int_{-\infty}^{\infty}\int_{\eta_1}^{\eta_2}\int_{\xi_1}^{\xi_2}\overline{Sf}(u,v,\xi,\eta)g_{u,v,\xi,\eta}^{*}(x,y)d\xi\, d\eta\, du\, dv \tag{9}$$

$$\overline{Sf}(u,v,\xi,\eta) = \begin{cases} Sf(u,v,\xi,\eta) & if\ |Sf(u,v,\xi,\eta)| \geq thr \\ 0 & if\ |Sf(u,v,\xi,\eta)| < thr \end{cases} \tag{10}$$

where $\overline{Sf}(u,v,\xi,\eta)$ denotes the threshold spectrum and $\bar{f}(x,y)$ is the filtered fringe pattern; *thr* denotes the threshold value. A windowed Fourier filter (WFF) can provide an exponential field, from which phase can be extracted. This is similar to the traditional Fourier transform technique for demodulation of carrier fringe patterns. Although, if the spectrum of all the possible frequencies is selected, the WFF provides a filtered fringe pattern. The output is frequently a complex field and its real part should be used. If either ξ or η is chosen to be positive, then the WFF gives an exponential phase field. Its angle gives an ambiguous phase distribution [52].

Note that the integration limits in Eq. (9) are set to be from ξ_l to ξ_h and from η_l to η_h for ξ and η, respectively, instead of from $-\infty$ to ∞. This represents that only a particular spectrum should be computed if the frequencies of the fringe pattern are within this set integration range. It is not necessary to compute the other coefficients, and therefore the computational cost is reduced. The range can be estimated by analysing the concentration or Fourier spectrum of a fringe pattern. A smooth image is shaped after an iWFT that represents the phase map, which is utilized for the rest of the post-processing in this experiment.

4 Theoretical Equations to Study MZI Soret Experiments

The procedure explained in the previous section was applied to the reference image to evaluate $\phi_{\text{ref}}(\text{x}, \text{y})$. The reference image can be any image during the experiment that was able to satisfy the initial conditions of the experiment, or even another image during the process that possessed predefined reference conditions. This reference was compared with other images to track the changes from the chosen reference image. The required phase difference, $\Delta\phi$, or phase distribution, was given by the difference between the phase of ith image and the reference image [53]:

$$\Delta\phi(x, y, t_i) = \phi(x, y, t_i) - \phi_{ref}(x, y) \tag{11}$$

The phase difference that was obtained by the unwrapping process was used to calculate the refractive index variation in the sample. This distribution led to the calculation of the temperature and concentration variations inside the cell. The relationship between the phase difference $\Delta\phi$ and Δn is also a function of optical length (L) and the wavelength of the laser:

$$\Delta n(x, y) = n(x, y) - n(x_0, y_0) = \frac{\lambda}{2\pi L}\Delta\phi(x, y) \tag{12}$$

The diffusive mass transport in a multicomponent mixture in the presence of a temperature gradient is described by more than two fluxes: one as a result of the temperature difference and other independent mass fluxes. There is no need to define the last mass flux for the ith component because the final closing relationship describes the last mass flux completely. The mass fluxes along the direction of the thermal gradient are given by Eq. (13):

$$\vec{J}_i = -\rho\left(\sum_{j=1}^{n-1} D_{ij}\nabla c_i + D_{T,i}c_i(1 - c_0)\nabla T\right) \tag{13}$$

where ρ is the density of the mixture, c_i is the mass fraction of the ith component, D_{ii} is the pure diffusion coefficient with $i \neq j$ for the cross diffusion coefficients in porous media, and $D_{T,i}$ represents the thermodiffusion coefficients of the components in the ternary mixture. The ODI is set up to test thermodiffusion phenomena in multicomponent mixtures that include n various components. This process theoretically requires $n - 1$ laser sources with different wavelengths. According to interferometry principles, each laser can provide one equation, as below, that has

$n - 1$ unknowns during the diffusion time: ΔC_i where $i = 1, 2, \ldots, n - 1$. For a binary mixture and in a convectionless regime the analytical transient solution is given by Eq. (14):

$$\Delta C(t) = C(t, L) - C(t, 0) = -S_T C_0 (1 - C_0) \Delta T \left[1 - \frac{8}{\pi^2} \sum_{n,odd}^{\infty} \frac{1}{n^2} \exp\left(-n^2 \frac{t}{\tau_D} \right) \right] \tag{14}$$

In the above equation, S_T (Soret coefficient) and τ_D (a function of mass diffusion coefficient) are unknown coefficients that were calculated using the temporal concentration difference between hot and cold sides of the cell. Here Eq. (14) was used for validation using the benchmark value of S_T and τ_D coefficients for water/IPA mixtures. In addition, the refractive index variation in the ODI interferometry experiment can be formulated as

$$\Delta n_i(x, y) = \left(\frac{\partial n_i}{\partial T} \right)_{T_0, \lambda_i} \Delta T(x, y) + \sum_{j=1}^{n-1} \left(\left(\frac{\partial n_i}{\partial c_j} \right)_{T_0, \lambda_i} \Delta C_j(x, y) \right) \tag{15}$$

where $\Delta T(x, y)$ and $\Delta C_i(x, y)$ are respectively the temperature and concentration changes in an arbitrary point (x, y); $(\partial n / \partial T)$ and $(\partial n / \partial c)$ are known contrast factors defined as the variation of the refractive index due to temperature and concentration, respectively, while keeping the other parameter constant. The contrast factors are functions of composition and temperature; however, due to the small variations of these parameters in the current study, these coefficients are assumed constant for each mixture. The values for the contrast factors are adopted from reference [17, 54] for the mean temperature of the mixtures (25 °C) for binary and ternary mixtures.

Because thermal time was noticeably smaller than diffusion time, ignoring a few minutes at the beginning of the experiment will not affect the final concentration profile at the end of diffusion time. Consequently, during the first thermal time Eq. (15) can be rewritten as follows:

$$\Delta T(x, y) = \frac{\Delta n_i(x, y)}{\left(\frac{\partial n_i}{\partial T} \right)_{T_0, C_0, \lambda_i}} \tag{16}$$

According to the theory of heat transfer in a liquid system, the temperature profile will not change after the first thermal time. Consequently, after that time any derivative of temperature is equal to zero and Eq. (15) can be rewritten as follows when there are two lasers for a ternary mixture:

$$\Delta n_1(x, y) = \left(\frac{\partial n_1}{\partial c_1} \right)_{T_0, C_2, \lambda_1} \Delta C_1(x, y) + \left(\frac{\partial n_1}{\partial c_2} \right)_{T_0, C_1, \lambda_1} \Delta C_2(x, y) \tag{17}$$

$$\Delta n_2(x,y) = \left(\frac{\partial n_2}{\partial c_1}\right)_{T_0,C_2,\lambda_2} \Delta C_1(x,y) + \left(\frac{\partial n_2}{\partial c_2}\right)_{T_0,C_1,\lambda_2} \Delta C_2(x,y) \tag{18}$$

So, the mass fraction of the first two components can be calculated as follows:

$$\begin{pmatrix} \Delta C_1(x,y) \\ \\ \Delta C_2(x,y) \end{pmatrix} = \begin{pmatrix} \left(\frac{\partial n_1}{\partial c_1}\right)_{T_0,C_2,\lambda_1} & \left(\frac{\partial n_1}{\partial c_2}\right)_{T_0,C_1,\lambda_1} \\ \left(\frac{\partial n_2}{\partial c_1}\right)_{T_0,C_2,\lambda_2} & \left(\frac{\partial n_2}{\partial c_2}\right)_{T_0,C1,\lambda_2} \end{pmatrix}^{-1} \begin{pmatrix} \Delta n_1(x,y) \\ \\ \Delta n_2(x,y) \end{pmatrix} \tag{19}$$

The concentration of the third component can be found based on species mass conservation [16]. It is worth to mention that for a binary mixture, Eq. (19) can be simplified and a similar equation as Eq. (16) can be determined for measurement of concentration [15]. The entire image processing procedure of the current work is illustrated in Fig. 3. The main difference between using this new method and the simple FFT is the implementation of steps 5 and 6 that represent the windowed Fourier filtration technique. The addition of these two steps increases the total processing time by approximately 40 times. Nevertheless, the outcome of the analysis justifies the importance of using this method for such ODI experiments that deal with simultaneous heat and mass transfer.

The optical properties of the liquids at various laser wavelengths used in this study are mentioned in Table 1 and 2.

5 Software Development

In order to implement the theoretical framework described earlier, a software program was developed and used to investigate thermodiffusion interferometry experiments for different binary and ternary mixtures, including the SODI project that has been performed on board the ISS as shown in Fig. 4.

The FFT, WFT, and WFF algorithms were integrated into a MATLAB® function for the image processing. Values of sigma corresponding to σ_x and σ_y for the Gaussian window function and, and also the choices of ξ_l, ξ_0, ξ_h, η_l, η_0 and η_h are recommended by literature [40, 55]. WFF are explained briefly in this report and in detail in reference [41]. While less than 4 s are needed for processing fringes via FFT, about 2 or 3 min are required for processing the image using WFF for a fringe pattern of 714 × 706 by a core i7 2.9 GHz PC. This software is integrated with a graphical user interface. The GUI for the main window is displayed in Fig. 4. Finally, an evolutionary optimized method was integrated into the software that provides an accurate estimation of S_T for binary and ternary mixtures according to the temporal concentration data.

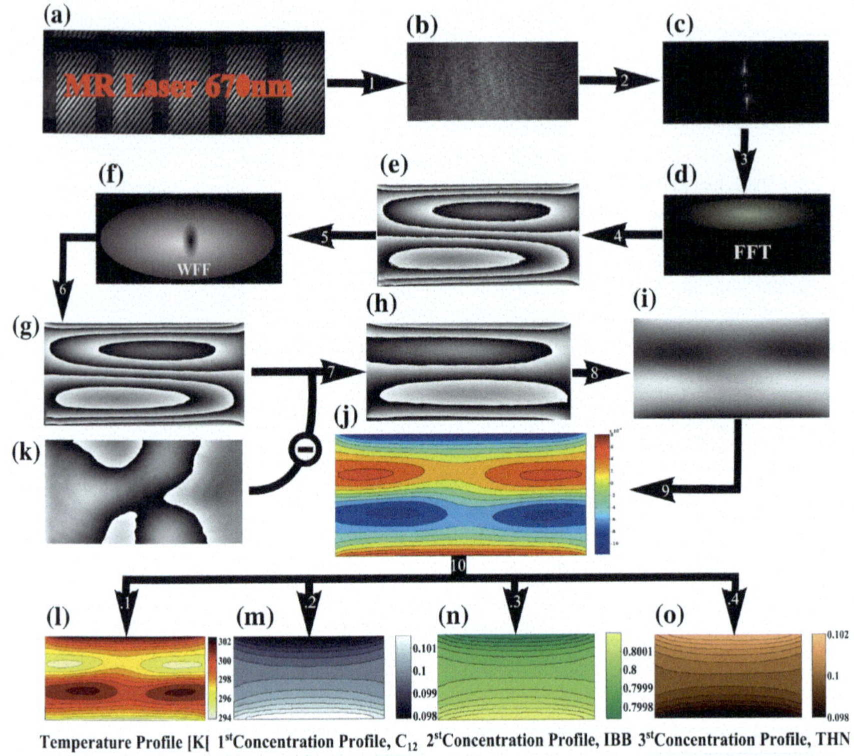

Fig. 3 Principle scheme of Fourier image processing in this study (**1**: Converting and Cropping, **2**: Fast Fourier Transform, **3**: Filter out the non-zero peaks using band filter, **4**: Inverse 2D FFT, **5**: Windowed Fourier Filter, **6**: Inverse WFT, **7**: Subtracting the image from the reference image (k), **8**: Unwrapping, **9**:Calculationg of Refractive index, and **10**: Concentration and Temperature)

5.1 Software Validation

In total, six runs of the SODI-IVIDIL and SODI-DSC experiments were analysed in this study (see Table 3): one run with a forced vibration and positive Soret coefficient, and the other five without an external vibration. Four of IVIDIL runs with no forced vibration were considered in this section, and they have been selected herein for validation. Run 2 and Run2R were used for comparison to runs with negative Soret coefficients with 10% initial mass fractions of IPA and a 5 K temperature differential; in addition, Run 33 and Run33R were included and possess (50% initial IPA mass fraction) positive Soret coefficients and a ΔT = 15 K. The 'R' suffix in the title of the runs instants for "repeated". In order to validate the software for the different conditions of the Soret effect, two major benchmark cases with positive or negative Soret coefficients were chosen to prove the accuracy of the software for stable and unstable conditions of Soret. The properties of these

Table 1 Optical properties of the water-IPA mixture measured at T = 298 K for laser with the wavelength of 670 nm (MR) and IVIDIL cases [17, 56]

$c_{o,IPA}$ (wt−) (%)	$(\partial n/\partial T)_{p.c} \times 10^{-4}$ (K^{-1})	$(\partial n/\partial c)_{p.T} \times 10^{-2}$ (−)	$S_T \times 10^{-3}$ (K^{-1})	$D \times 10^{-10}$ $(m^2 s^{-1})$	$D_T \times 10^{-13}$ $(m^2 s^{-1} K^{-1})$
10	−1.3427	−9.23	−8.47 ± 1.5	7.11 ± 0.4	−60.2 ± 10
50	−3.474	−3.64	5.68 ± 0.3	1.60 ± 0.1	9.57 ± 0.5

Table 2 Optical properties of the THN-IBB-C12 mixture measured at T = 298 K for SODI-DSC case Run12 [54]

Laser type	$c_{o,THN}$ (wt−) (%)	$c_{o,IBB}$ (wt−) (%)	$c_{o,C_{12}}$ (wt−) (%)	λ (nm)	$(\partial n/\partial T)_{p,IBB,THN} \times 10^{-4}$ (K^{-1})	$(\partial n/\partial c_{C_{12}})_{p.T,THN} \times 10^{-2}$ (−)	$(\partial n/\partial c_{THN})_{p.T,IBB} \times 10^{-1}$ (−)	$(\partial n/\partial c_{IBB})_{p.T,C_{12}} \times 10^{-2}$ (−)
MR	10	80	10	670	−4.812	−5.72	1.213	−7.57
MN	10	80	10	935	−4.763	−5.38	1.240	−6.96

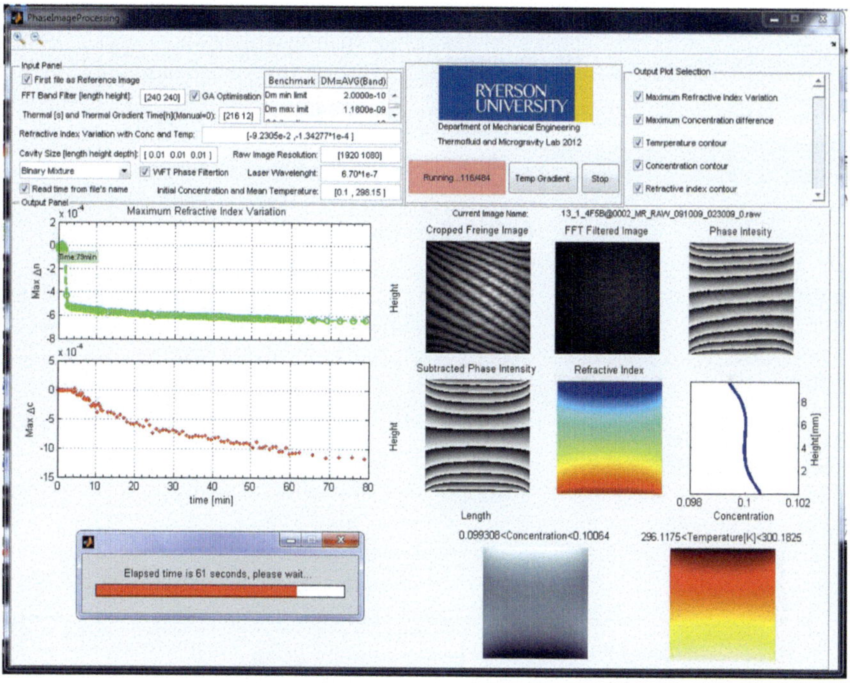

Fig. 4 Main GUI window of the software, processing one on of IVIDIL runs, different plots illustrate various Fourier image processing, heat and mass transfer parameters inside the cell and also concentration along the line at the cavity centre

mixtures, including the benchmark Soret and diffusion coefficients (adapted from [46, 47]), are included in Table 3.

The temporal separation of the isopropanol is shown in Fig. 5 for runs 2, 2R, 33, and 33R. The maximum separation was calculated based on the average concentration of the 100 pixels along the x-direction and close to points P_1 and P_2. Moreover, the best fit curves to the runs with positive and negative Soret coefficients are plotted in this Figure. The first 12 h illustrates the thermodiffusion separation, while the last 6 h is the diffusion time of the experiment. The results show the reproducibility the system as the repeated runs are fit closely. The curve of the separation rate is fairly smooth as a result of using the accurate image processing method (WFT). Moreover, the results are in agreement with the theoretical trends obtained by using the benchmark Soret and diffusion coefficients. For a quantitative comparison, the Soret and diffusion coefficients were determined for each experimental run based on the concentration difference presented in Fig. 5 and included in Table 3. All of the measured coefficients are within the error band of the benchmark studies and the relative error is maintained at less than 8% with respect to the mean benchmark values of the diffusion coefficients.

Table 3 Details of considered SODI runs in this study

SODI Run #	$C_{0,IPA}$	f (Hz)	A (mm)	ΔT (K)	$S_T \times 10^3$ (K^{-1}) (this study)	$D \times 10^{10}$ (m^2s^{-1}) (this study)
IVIDIL Run 2	10%	0	0	5	−8.81	6.59
IVIDIL Run 2R	10%	0	0	5	−8.58	6.85
IVIDIL Run 33	50%	0	0	15	5.16	1.61
IVIDIL Run 33R	50%	0	0	15	4.96	1.60
IVIDIL Run 39	50%	0.05	70	10	–	–
DSC Run12	80%IBB, 10% THN, 10%C_{12}	–	–	–	–	–

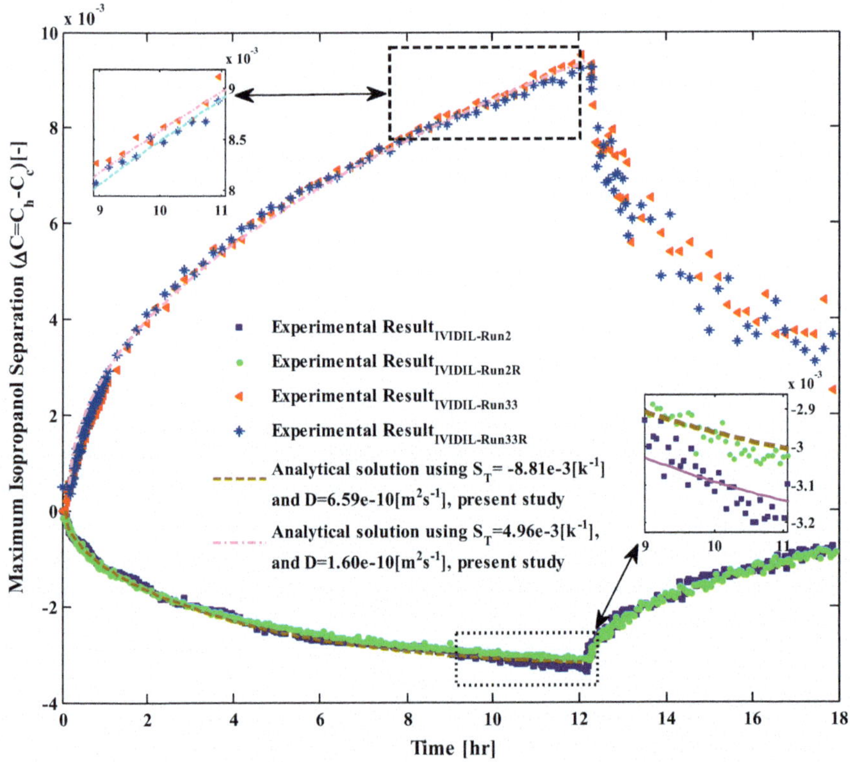

Fig. 5 Maximum concentration variation between hot and cold walls for negative and positive Soret effect for the binary mixture

6 Results and Discussion

In this section, the results obtained by windowed Fourier transform (WFT) for a typical ODI thermodiffusion experiment were compared with the basic Fourier transforms or phase shifting methods. Several parameters, such as the refractive index, temperatures, and concentration fields, for the binary and ternary mixtures were investigated. The concentration and temperature profiles were captured on a plane at the middle of the cavity as shown in Fig. 2. The variation of the considered parameters along lines L_1, L_2, L_3, and L_4 at the middle and close to the walls on this plane are studied. Finally, the separation between points P_1 and P_2 was also used for comparison purposes.

Before starting the main dissection of the FFT method, it must be noted that in generally the phase extracted using other methods such as a phase-shifting algorithm produces more noise than the FFT method. In addition, the phase shifter must be calibrated to establish the relationship between the input of the phase shifter and the resulting phase shift between the consecutive fringe patterns [13, 57, 58].

Many studies have used the FFT method to analyse the images obtained by the ODI technique for thermodiffusion experiments, such as the extensive works of Shevtsova et al. [14, 27, 59, 60]. The results obtained by this method usually contain a noisy concentration (or refractive index) field, especially at the interface of the contour bands. These massive fluctuations are not related to the physical characteristics of the system, such as the size of the cavity, but rather are a consequence of the image processing technique.

A sample fringe pattern of a SODI-IVIDIL experiment is shown in Fig. 6a. Many imperfections can be observed in this sample fringe pattern, some of which are indicated by a circle with white dashed lines, and the most dominant one is specified by a red circle in the vicinity of the upper wall. Other images that were taken during the experiment suffer from the same or similar sources of error. These imperfections form due to reflection off the walls for pixels close to boundaries or unwanted particles attached to the cell and the beam splitters and mirrors, among other causes. These sources of error are unavoidable due to the complexity of the integrated apparatus on board the ISS. In the following sections we will discuss the effects of the dominant flaw indicated by the red circle in Fig. 6a on the main parameters.

The influence of the dominant flaw on the wrapped phase map is shown in Fig. 6b, d. The magnified picture in Fig. 6e contrasts with the error in the wrapped phase. In Fig. 6d, a bridge between two fringe bands was formed. This bridge broke the discontinuity between the phases and resulted in the removal of one the fringe bands from the phase map. A comparison with the results in Fig. 6 illustrates the fact that the FFT method is rather sensitive to noise since it is a pixel-wise fringe pattern processing method. This is also true for the entire1D local and global methods. Among the transform-based methods, the 2D WFR and 2D windowed transform seem to be the most robust algorithms; nevertheless, they are time-consuming and not suitable for real-time processing. If parameters are

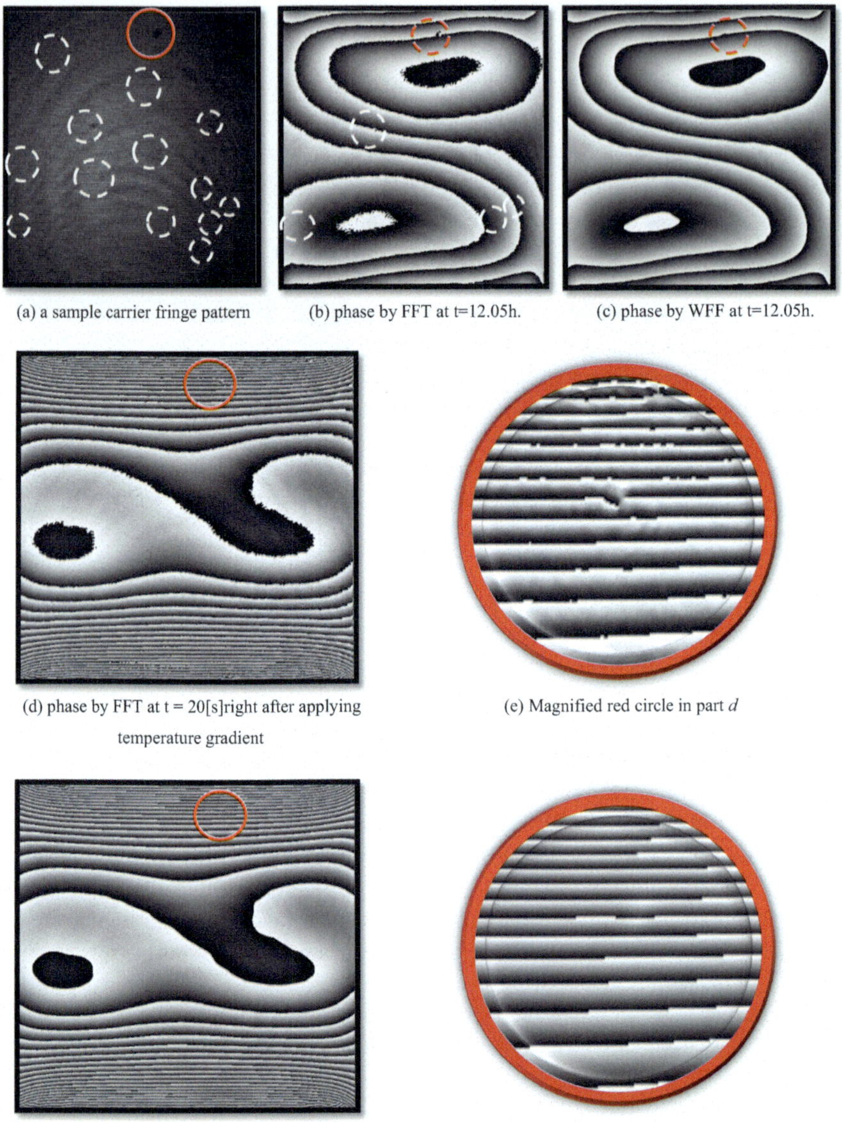

(a) a sample carrier fringe pattern (b) phase by FFT at t=12.05h. (c) phase by WFF at t=12.05h.

(d) phase by FFT at t = 20[s]right after applying temperature gradient (e) Magnified red circle in part *d*

(f) phase by WFF at t = 20[s]right after applying temperature gradient (g) Magnified red circle in part *f*

Fig. 6 Input is a noisy fringe; output is a filtered fringe pattern, phase map using FFT and WFF

properly chosen, 2D WFF and 2D WFR can reduce the noise in the phase caused by the fringe pattern. The results of the phase image processing by windowed Fourier transform is shown in Fig. 6c, f. The noise in the FFT method was caused by the 1D algorithms that compute the phase distribution line by line. Consequently, the phase images were more exposed to noise than in 2D methods [31]. Figure 6g shows the same region as Fig. 6d but for a 2D FFT processed image. Not only is there no bridge between the two phase bands, but the noise in the phase map is completely removed.

Phase unwrapping is frequently desirable to construct a continuous phase map from a wrapped phase map [37]. The phase can be unwrapped by scanning the phase map pixel by pixel and then line by line and compensating for any jumps in the data. However, this method is usually unsuccessful when applied to a noisy wrapped phase map. Two widely used strategies to overcome the unsuccessful unwrapping process are circumventing the "bad" pixels and approximating the phase in a least-squares sense [48, 61]. One strategy is to remove the noise before phase unwrapping, which can be effectively fulfilled by WFF. The wrapped phase map by FFT is illustrated in Fig. 7a and the corresponding unwrapped image is shown in Fig. 7c. Performing the unwrapping procedure with a noisy boundary causes all neighbouring pixels to be affected by the noise. Despite this, the wrapped phase filtered by WFF becomes clear and smooth, as shown in Fig. 7d. The corresponding plot using FFT is shown in Fig. 7c. Sometimes it is not only "bad" pixels but also "bad" regions that require smoothing. This problem can again be solved by WFF. Since the spectrum in the "bad" region is usually broader than in other places, more energy will be removed in WFF, and the amplitude of the filtered exponential field becomes smaller [37]. In this case, the strongest spot that exists in the fringe images may cause 5 or 25% error in the calculation of the refractive index of Fig. 6b, d, respectively, because it combines two phase bands together during the unwrapping process.

According to Eq. (12), the refractive index can be calculated from the unwrapped phase map. However, when FFT was used, fluctuations in the refractive index contours were created, as shown in Fig. 8a, especially close to the hot and cold walls. The pattern obtained by the WFF method (Fig. 8b), however, was smooth and all of the local noise and fluctuations were removed.

These observations were also obtained for the temperature profile calculated according to Eq. (15). A linear temperature variation was attained in the cavity at the end of the thermal time (t = 12 h) at the middle of the cell. Figure 9 displays the contours of the temperature field inside the IVIDIL cell during Run33. The obtained temperature field was much smoother for the case of WFT (Fig. 9b) over that of the FFT method (Fig. 9a). WFF provided the exact and precise temperature profile inside the cell; however, the FFT measurements were also acceptable for the temperature field and the errors were locally limited. Thus, the FFT method was reliable in measuring the temperature in the range of the current study, that is, with a 5 K temperature difference in the system. The nonlinearity in the temperature variation observed at the corners was due to a small lateral gap between the quartz glass and both the cold and hot walls of the cell. The same observation was recorded in other studies [17].

Studying the variation of the concentration in the entire domain is the main key to understand the double diffusive interaction phenomenon, which was one of the fundamental motivations of the IVIDIL experiment. Figure 10 illustrates the

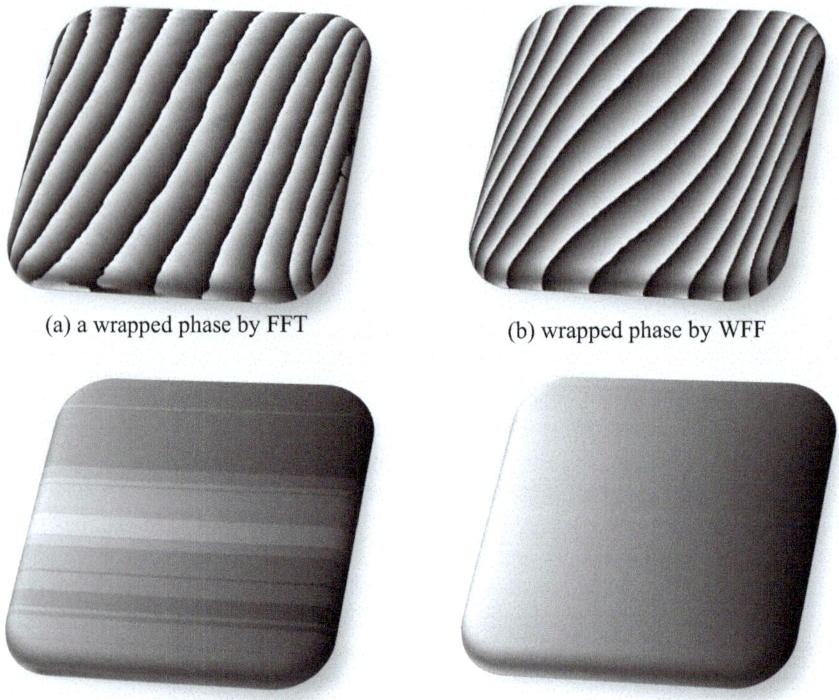

(a) a wrapped phase by FFT (b) wrapped phase by WFF

(c) unwrapped phase of (a) (d) unwrapped phase of (b)

Fig. 7 General results of the unwrap a phase map **a** with and **b** without badpixels or bad region close to boundaries showed in part **c** and **d** respectively

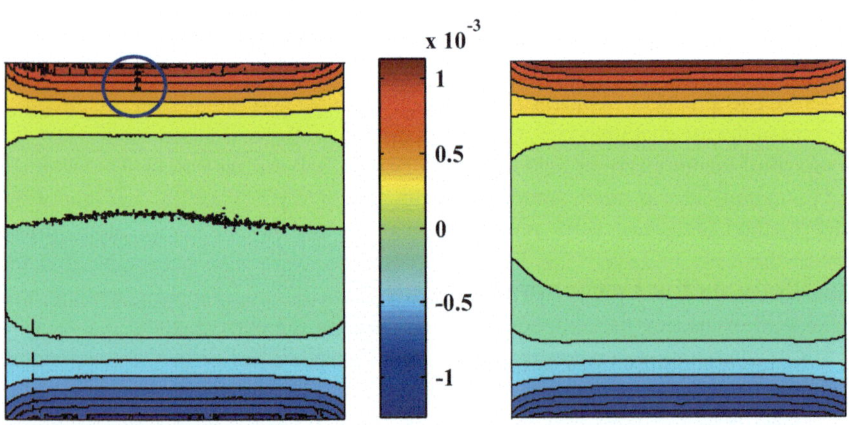

(a) Refractive indexby FFT (b) Refractive indexby WFT

Fig. 8 Refractive index contour from a carrier fringes (t = 125 s), noisy domain in evident in the first part of the figure

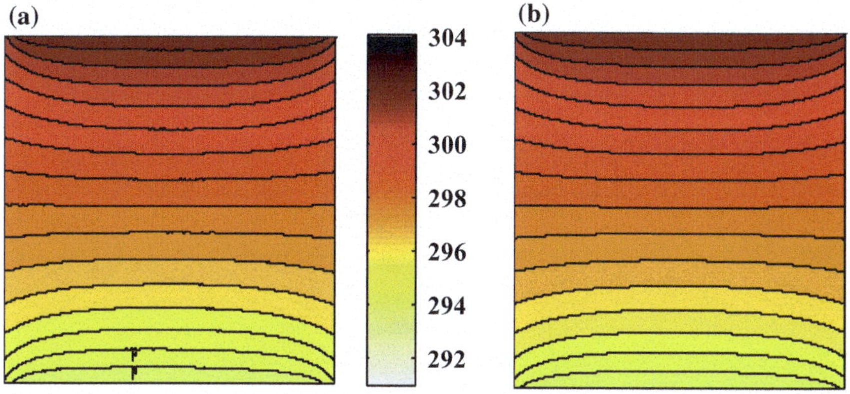

Fig. 9 Temperature contours by **a** FFT, and **b** WFF after two thermal times for IVIDIL Run2R

concentration contours inside the primary IVIDIL cell (from the front view of the cell that is shown in Fig. 2 during Run39 at the end of the thermodiffusion phase (t = 12 h) obtained by both the FFT and WFF methods. The last two parts of this figure demonstrate the same field one hour after the termination of the applied temperature difference, as the species were diffusing to establish a homogeneous mixture. The FFT results illustrate a noisy pattern with stronger noise close to the walls, while the WFF results show isolines that are clear and smooth even at the corners of the cell. That the isolines can be distinguished trivially by WFF is an important result, since it is usually at the boundaries that the noise most affects the results. Lastly, it is worth mentioning that the nonlinear temperature profile affects the concentration pattern in the cavity close to each corner at the end of thermodiffusion phase.

Despite the better result of WFF, it might be argued that the concentration map calculated by FFT provides a good enough agreement that reduces the calculation cost remarkably. It must be noted that the most important parameter to be extracted from the thermodiffusion experiment is the maximum separation of the components, and according to this parameter the Soret coefficient can be measured. In short, while having a clear view is helpful to analyse the pure diffusion in the system; it is essential to have the most accurate measurement of the concentration near both hot and cold walls in order to precisely measure S_T. The maximum separation must be calculated using the extracted data inside the noisier region (close to walls).

Thus, Fig. 11a shows concentration variations along the thermal gradient direction close to the lateral wall of the cavity (line L_2 in Fig. 2), and Fig. 11b illustrates the same but near the hot wall (line L_4 in Fig. 2). The main oscillatory pattern is due to the external forced vibration imposed during Run33, which is traced by both methods. However, Fig. 11a shows continuous fluctuations extracted using the FFT method, which do not exist in the real behaviour that is extracted by WFF. It seems that the noise of the FFT method have a particular frequency and

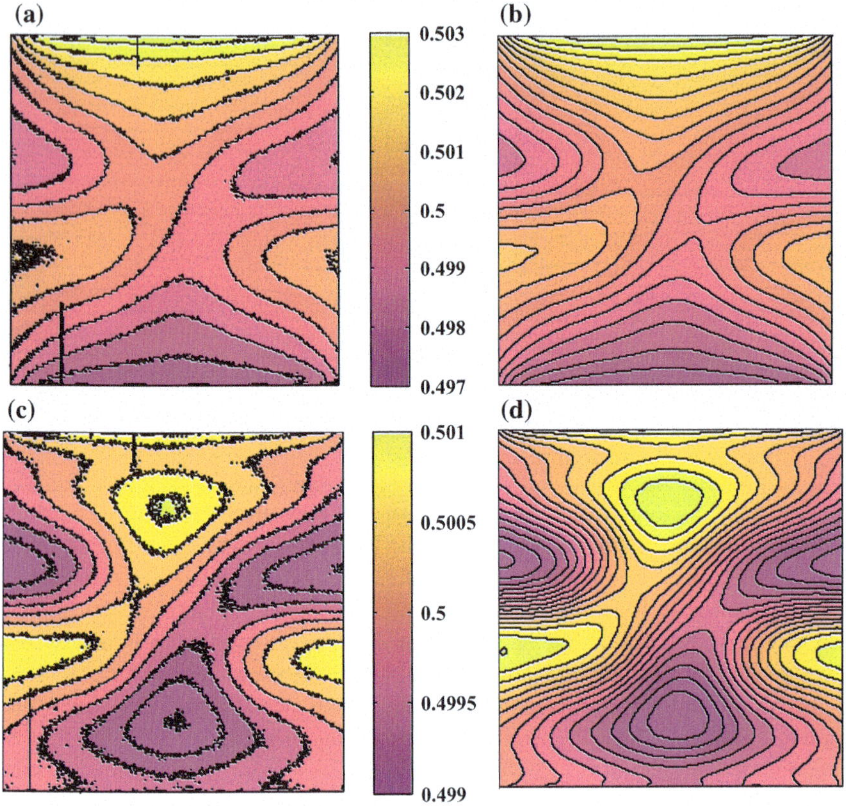

Fig. 10 IPA concentration contours in side cell for IVIDIL-Run33 using **a** FFT at t = 12 h, **b** WFF at t = 12 h, **c** FFT at t = 13 h, and **d** WFF at t = 13 h

amplitude in this plot that is completely removed by WFF. The concentration near the hot and cold walls are the most important variable, which results in the quality of the FFT method being lower near these walls, as shown in Fig. 11b. The closest row of the pixels to the hot wall (FFT_1) shows dominant fluctuations. This failure in the FFT method may be improved by analysing the pixels at a larger distance from the wall.

The mass fractions on the tenth pixel row (out of more than 800 row pixels) is also plotted in Fig. 11b in green (captioned as FFT_2), which displays a smoother and better performance at the middle region of the line L_4. Nevertheless, strong fluctuations close to both ends are still evident, which are caused by strong noise close to the walls that can crash the unwrapping over many pixels. We observed that WFF provides a noticeable improvement of the concentration curve along the length of line L_4. Although the concentration profile of the binary system extracted from the FFT method illustrated the noisy domain, the profile pattern can still

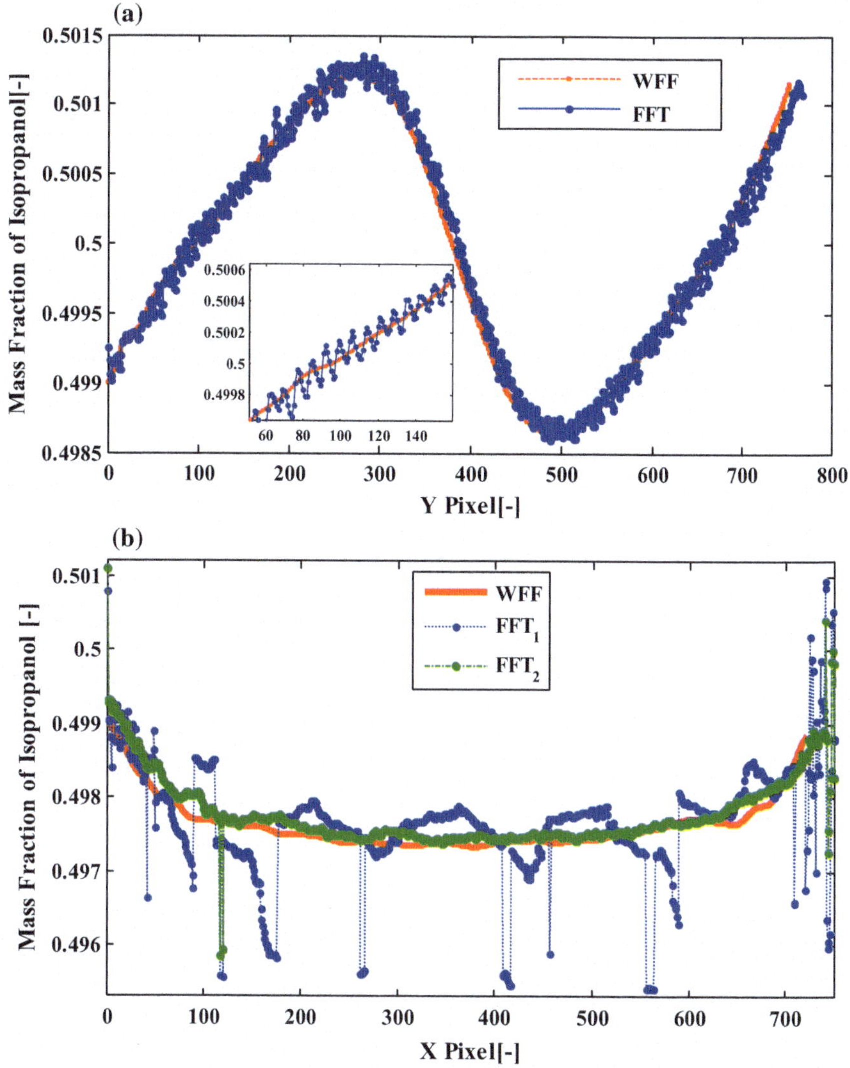

Fig. 11 Results of mass fraction of IPA **a** along lines "L_2": comparison between FFT and WFT **b** along lines "L_4": Wavy lines on the first row (close to wall) for FFT method (FFT_1)

imitate the real one. However, for the case of the ternary mixtures the measurement conditions became more critical.

The impact of the WFF on the concentration profiles proved more essential for ternary systems. Solving for the flow variables is a more complex process in case of a ternary mixture over a binary mixture, and the error is more pronounced as a result of the linked system of equations. The concentration distribution was completely different when comparing the FFT and WFF methods as demonstrated by

Run12 from the SODI-DSC shown in Fig. 12. The variation of the species concentration was very small for this ternary mixture (less than 0.003). The contours of the concentration obtained by the FFT method (Fig. 12a, c) are not distinguishable from each other. Due to the lack of considerable separation of C_{12} in comparison with THN or IBB, these observations were even worse for the C_{12} contour. When WFF was employed (Fig. 12b, d), the concentration gradient became discernable, but the errors were still not eliminated completely. However, the enhancement of WFF, especially for the case of THN, was substantial and was used to extract the needed parameters. The results show when the maximum separation in the domain is small [<0.003 (wt)], the FFT method cannot be employed as a reliable tool.

As mentioned earlier, the concentration profile at the centre of the cavity is less affected by noise in the system. The concentration of THN along lines L_1 and L_3 (Fig. 2) are plotted in Fig. 13a. The black solid and dashed lines correspond to the results of the WFF method and the FFT method, respectively. The fluctuation along the x-direction is stronger than the y-direction for both methods. The variation along the x-direction is plotted at the centre of the cell, which must display the initial concentration of the components along L_3 during the experiment. WFF extracts a more accurate result that satisfies the requirement of the initial concentration at the centre of the cell [$\sim$0.1 $\pm$ 0.0003 (wt)]. The FFT result does not satisfy this condition as it deviates $\pm$0.0012 from the required initial condition. In this case, the maximum separation in the system was less than 0.0065 (wt), which means that there was more than 18% variation in the intensity at any point in the domain using FFT. It is worth to mention that this error for the C_{12} is more than 40%, because the maximum separation of this component is about 0.0014 (wt), which is in the same order as error bar of the processing.

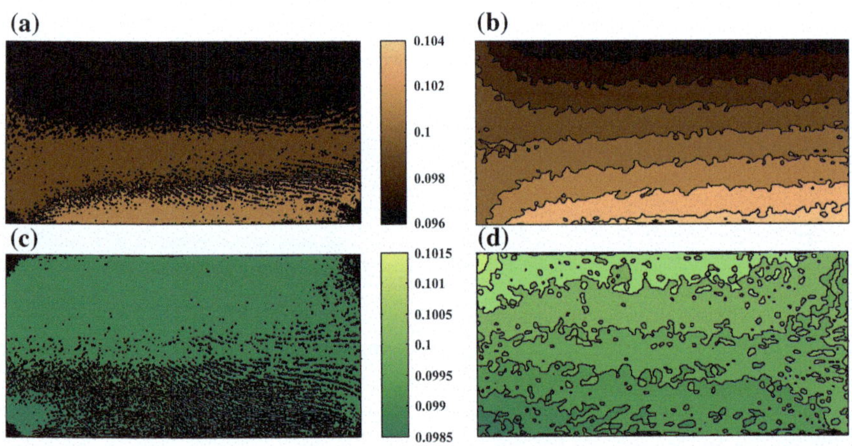

Fig. 12 Concentration variation for DCS Run 12 at the end of thermodiffusion phase that shows the maximum separation in the system **a** THN by FFT, **b** THN by WFF, **c** IBB by FFT, and **d** IBB by WFF

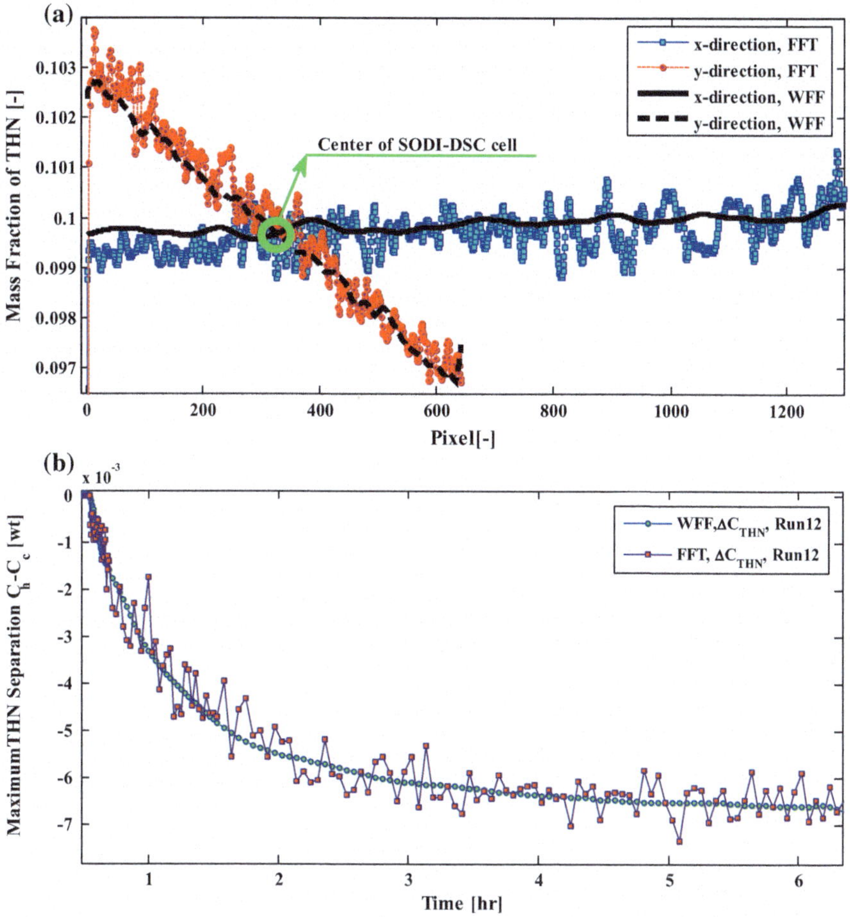

Fig. 13 Comparison results of FFT and WFF for a ternary case **a** along two perpendicular lines at the centre of the cell and **b** maximum separation between hot and cold walls versus time

The similar deflection along the y-direction was also noticeable using FFT, where the maximum concentration fluctuation compared to linear variation between the hot and cold walls along line L_3 was 0.0016. The ODI method provided the advantage of live tracing of the concentration distribution in the cell, a feature which is not accessible while using other experimental methods. Accurate capture of this capability is important for many different techniques of Soret effect measurements. Figure 13b illustrates the maximum temporal separation of THN in the ternary system during the thermodiffusion phase. Despite the more accurate values of Soret coefficient extracted with WFF, it must be noted that the FFT plot may represent the real trend of the maximum separation with a specific of fluctuation from reference of 0.0015 (wt) for the mixture in the normal condition. However, this error for the ternary mixtures at the critical conditions, such as choosing the points closer to walls to get the maximum possible separation or utilizing two lasers with close wavelengths in the experiment that results in more sensitive measurements, is more important and produces a deviation of 0.003 (wt) based on Figs. 12 and 13a. This deviation in the ternary case may cause more than 30–50% error in measurement of the Soret coefficient if FFT is used.

7 Conclusion

The application of the two-dimensional windowed Fourier transform and its use for filtering during image processing of thermodiffusion interferometry experiments were investigated for the first time. A MATLAB program was developed that is facilitates the use of both FFT and WFT methods. In order to determine the impact the FFT and WFF methods on ODI thermodiffusion experiments, six different runs that were performed on board the ISS were analysed according to both FFT and WFT. The overall requirements and different aspects of both FFT and WFT methods were discussed. It is shown that if WFT parameters are properly chosen, it can significantly reduce the noise from the fringe pattern of an ODI thermodiffusion experiment. The FFT method gives a reliable measurement of temperature for such a problem when a large thermal gradient ($\sim$5 K) exists between the cold and hot walls and when it is not essential to record precisely the temperature variations at different points in the domain. The comparison of the results shows that the WFT provides a noticeable improvement on the measurement of concentration. This improvement is more pronounced when dealing with very small variations in ternary systems. The FFT fails to extract a reasonable concentration profile for the ternary system. It is concluded that when the maximum separation in the domain is less than 0.003 (wt), the results of the FFT method is not reliable and may introduce more than 40% error in the measurements. This work provides the necessity of using the windowed Fourier transform method to study optical digital interferometry experiments involving both heat and mass transfer. The vital role of WFT method is indispensable in studying thermodiffusion in ternary systems.

Acknowledgements Authors would like to acknowledge the financial support of the Canadian Space Agency (CSA) and the Natural Sciences and Engineering Research Council of Canada (NSERC) in funding this work.

References

1. Heise B (2010) Image processing for phase-sensitive optical coherence tomography, 3838120582, Paperback
2. Bahri M, Hitzer ESM, Ashino R, Vaillancourt R (2010) Windowed Fourier transform of two-dimensional quaternionic signals. Appl Math Comput 216:2366

3. Judge TR (1996) Automatic analysis of interferometric data—FRAN (description and user guide)
4. Judge TR, Bryanston-Cross PJ (1994) A review of phase unwrapping techniques in fringe analysis. Opt Lasers Eng 21:199–239
5. Bryanston-Cross PJ, Quan C, Judge TR (1994) Application of the FFT method for the quantitative extraction of information from high-resolution interferometric and photoelastic data. Opt Laser Technol 26:147
6. Pike C (2008) Two-dimensional Fourier processing of rasterised audio. Abstract A comprehensive software tool has been de
7. Vander R, Lipson SG, Leizerson I (2003) Fourier fringe analysis with improved spatial resolution. Appl Opt 42:6830–6837
8. Robinson EA (1982) A historical perspective of spectrum estimation. Proc IEEE 70:885–907
9. Cormier E, Corner L, Kosik EM, Walmsley IA, Wyatt AS (2005) Spectral phase interferometry for complete reconstruction of Attosecond pulses, Attosecond Science and technology. Laser Phys 15:909–915
10. Onodera R, Watanabe H, Ishii Y (2005) Interferometric phase-measurement using a one-dimensional discrete hilbert transform. Opt Rev 12:29–36
11. Lepetit L, Chériaux G, Joffre M (1995) Linear techniques of phase measurement by femtosecond spectral interferometry for applications in spectroscopy. J Opt Soc Am B 12:2467–2474
12. Macy WW Jr (1983) Two-dimensional fringe-pattern analysis. Appl Opt 22:3898–3901
13. Kreis T (1986) Digital holographic interference-phase measurement using the Fourier-transform method. J Opt Soc Am A 3:847–855
14. Shevtsova V, Lyubimova T, Saghir Z, Melnikov D, Gaponenko Y, Sechenyh V, et al (2011) IVIDIL: on-board g-jitters and diffusion controlled phenomena. In: International symposium on physical sciences in space, vol 327, p 12031
15. Kianian A, Ahadi A, Saghir MZ (2012) Experimental evidence of low Rayleigh vibration on mixture during thermodiffusion experiment. Can J Chem Eng 91(9):1568–1574
16. Ahadi A, Van Varenbergh S, Saghir MZ (in press) Measurement of Soret coefficients for a ternary hydrocarbon mixture in low gravity environment. J Chem Phys
17. Ahadi A, Kianian A, Saghir MZ (2013) Heat and mass transport phenomena under influence of vibration using a new aided image processing approach. Int J Therm Sci, 1–12
18. Ahadi A, Saghir MZ (2013) Transient effect of micro vibration from two space vehicles on mixture during thermodiffusion experiment. Micro Sci Technol 9999:12
19. Costeseque P, Mojtabi A, Platten JK (2011) Thermodiffusion phenomena. CR Mec 339:275–279
20. Ahadi A, Saghir MZ (2012) Quasi steady state effect of micro vibration from two space vehicles on mixture during thermodiffusion experiment. FDMP Fluid Dyn Mater Process, 287
21. Blanco P, Bou-Ali MM, Platten JK, Urteaga P, Madariaga JA, Santamaria C (2008) Determination of thermal diffusion coefficient in equimolar n-alkane mixtures: empirical correlations. J Chem Phys 129:174504
22. Ahadi A, Yousefi T, Saghir MZ (2013) Double diffusive convection and thermodiffusion of fullerene–toluene nanofluid in a porous cavity. Can J Chem Eng 91(12):1918–1927
23. Koniger A, Wunderlich H, Koehler W (2010) Measurement of diffusion and thermal diffusion in ternary fluid mixtures using a two-color optical beam deflection technique. J Chem Phys 132:174506
24. Koehler W, Koniger A, Meier B (2009) Measurement of the Soret, diffusion, and thermal diffusion coefficients of three binary organic benchmark mixtures and of ethanol water mixtures using a beam deflection technique. Phil Mag 89:907
25. Zhong X (2006) Methods for removal of spurious reflection effect on phase-shifting interferometry. J Opt A: Pure Appl Opt 8:617
26. Khoshnevis A, Ahadi A, Saghir MZ (2012) Influence of static and oscillatory gravity fields on thermodiffusion of a binary mixture. Int J Therm Sci, 1–12

27. Ahadi A, Saghir MZ (2013) Experimental study of the impacts of forced vibration on thermodiffusion phenomenon in microgravity environment. Appl Therm Eng 60(1):348–358
28. Mazzoni S (2011) DSC as part of the DCMIX project experiment scientific requirements
29. Cuche E, Marquet P, Depeursinge C (2000) Spatial filtering for zero-order and twin-image elimination in digital off-axis holography. Appl Opt 39:4070–4075
30. Zhong J, Weng J (2004) Spatial carrier-fringe pattern analysis by means of wavelet transform: wavelet transform profilometry. Appl Opt 43:4993–4998
31. Zhang Z, Jing Z, Wang Z, Kuang D (2012) Comparison of Fourier transform, windowed Fourier transform, and wavelet transform methods for phase calculation at discontinuities in fringe projection profilometry. Opt Lasers Eng 50:1152
32. González RC, Woods RE (2008) Digital image processing. Prentice Hall, Pearson
33. Abdul-Rahman H (2007) Three dimensional Fourier fringe analysis and phase unwrapping, Liverpool John Moores University for the degree of Doctor of Philosophy
34. Goldberg KA, Bokor J (2001) Fourier-transform method of phase-shift determination. Appl Opt 40:2886–2894
35. Deck LL (2003) Fourier-transform phase-shifting interferometry. Appl Opt 42:2354–2365
36. Creath K (1998) In: Wolf E (ed),V phase-measurement interferometry techniques. Elsevier, p 349
37. Ghiglia DC, Pritt MD (1998) Two-dimensional phase unwrapping: theory, algorithms, and software. Wiley
38. Hipp M, Woisetschlager J, Reiterer P, Neger T (2004) Digital evaluation of interferograms. Measurement 36:53–66
39. Qian K, Seah HS, Asundi A (2005) Fault detection by interferometric fringe pattern analysis using windowed Fourier transform. Meas Sci Technol 16:1582
40. Kemao Q (2007) Two-dimensional windowed Fourier transform for fringe pattern analysis: principles, applications and implementations. Opt Lasers Eng 45:304
41. Kemao Q, Soon SH, Asundi A (2003) Smoothing filters in phase-shifting interferometry. Opt Laser Technol 35:649
42. Chen W (2008) Retrieval of instantaneous frequency from digital holograms based on adaptive windows. Opt Eng 47:065801
43. Servin M, Marroquin JL, Cuevas FJ (1997) Demodulation of a single interferogram by use of a two-dimensional regularized phase-tracking technique. Appl Opt 36:4540–4548
44. Villa J, Servin M (1999) Robust profilometer for the measurement of 3-D object shapes based on a regularized phase tracker. Opt Lasers Eng 31:279
45. Cai LZ, Liu Q, Yang XL (2003) Phase-shift extraction and wave-front reconstruction in phase-shifting interferometry with arbitrary phase steps. Opt Lett 28:1808–1810
46. Fang J, Xiong CY, Yang ZL (2001) Digital transform processing of carrier fringe patterns from speckle-shearing interferometry. J Mod Opt 48:507–520
47. Federico A, Kaufmann GH (2003) Phase retrieval in digital speckle pattern interferometry by use of a smoothed space-frequency distribution. Appl Opt 42:7066–7071
48. Kemao Q (2008) A simple phase unwrapping approach based on filtering by windowed Fourier transform: a note on the threshold selection. Opt Laser Technol 40:1091
49. Shevtsova V, Melnikov D, Legros JC, Yan Y, Saghir Z, Lyubimova T et al (2007) Influence of vibrations on thermodiffusion in binary mixture: a benchmark of numerical solutions. Phys Fluids 19:17111
50. Quan C, Niu H, Tay CJ (2010) An improved windowed Fourier transform for fringe demodulation. Opt Laser Technol 42:126–131
51. Kemao Q, Soon SH, Asundi A (2003) Instantaneous frequency and its application to strain extraction in Moire interferometry. Appl Opt 42:6504–6513
52. Kemao Q, Wang H, Gao W (2008) Windowed Fourier transform for fringe pattern analysis: theoretical analyses. Appl Opt 47:5408–5419
53. Takeda M, Ina H, Kobayashi S (1982) Fourier transform methods of fringe-pattern analysis for computer-based topography and interferometry. J Opt Soc Am 72:156–160

54. Galand Q (2012) Experimental investigation of the diffusive properties of ternary liquid systems. Ph.D. thesis, University De Bruxelles, ECOLE Polytechnique De Bruxelles, vol 1
55. Kemao Q (2004) Windowed Fourier transform for fringe pattern analysis. Appl Opt 43:2695
56. Mialdun A, Yasnou V, Shevtsova V, Königer A, Köhler W, Alonso de Mezquia D et al (2012) A comprehensive study of diffusion, thermodiffusion, and Soret coefficients of water-isopropanol mixtures. J Chem Phys 136:244512
57. Servin M, Marroquin JL, Cuevas FJ (2001) Fringe-follower regularized phase tracker for demodulation of closed-fringe interferograms. J Opt Soc Am A 18:689–695
58. Servin M, Rodriguez-Vera R, Marroquin JL, Malacara D (1998) Phase-shifting interferometry using a two-dimensional regularized phase-tracking technique. J Mod Opt 45:1809–1819
59. Shevtsova V (2010) IVIDIL experiment onboard the ISS. Adv Space Res 46:672–679
60. Mialdun A, Shevtsova VM (2008) Development of optical digital interferometry technique for measurement of thermodiffusion coefficients. Int J Heat Mass Transf 51:3164
61. Kemao Q (2007) A simple phase unwrapping approach based on filtering by windowed Fourier transform: the phase near edges. Opt Laser Technol 39:1364

Commercial Aircraft Trajectory Optimization to Reduce Flight Costs and Pollution: Metaheuristic Algorithms

Alejandro Murrieta-Mendoza and Ruxandra Mihaela Botez

Abstract Aircraft require significant quantities of fuel in order to generate the power required to sustain a flight. Burning this fuel causes the release of polluting particles to the atmosphere and constitutes a direct cost attributed to fuel consumption. The optimization of various aircraft operations in different flight phases such as cruise and descent, as well as terminal area movements, have been identified as a way to reduce fuel requirements, thus reducing pollution. The goal of this chapter is to briefly explain and apply different metaheuristic optimization algorithms to improve the cruise flight phase cost in terms of fuel burn. Another goal is to present an overview of the most popular commercial aircraft models. The algorithms implemented for different optimization strategies are genetic algorithms, the artificial bee colony, and the ant colony algorithm. The fuel burn aircraft model used here is in the form of a Performance Database. A methodology to create this model using a Level D aircraft research flight simulator is briefly explained. Weather plays an important role in flight optimization, and so this work explains a method for incorporating open source weather. The results obtained for the optimization algorithms show that every optimization algorithm was able to reduce the flight consumption, thereby reducing the pollution emissions and contributing to airlines' profit margins.

Keywords Algorithms · Cost · Optimization · Pollution · Trajectory · Aircraft · Aeronautics

A. Murrieta-Mendoza · R. M. Botez (✉)
Laboratory of Research in Active Controls, Avioncis and Aeroservoelasticity (LARCASE), Université Du Québec/École de Technologie Supérieure, Montreal H2J 3L6, Canada
e-mail: ruxandra.botez@etsmtl.ca

A. Murrieta-Mendoza
e-mail: alejandro.murrieta-mendoza.1@etsmtl.net

D. Vucinic et al. (eds.), *Advances in Visualization and Optimization Techniques for Multidisciplinary Research*, Lecture Notes in Mechanical Engineering, https://doi.org/10.1007/978-981-13-9806-3_2

Nomenclature

Most of the units are expressed under the International System of Units. However, because of aviation conventions, feet from the imperial system is used to express altitude

h (ft)	Altitude
Pij	Ants decision parameter
Bg (ft, Mach or degrees)	Best global trajectory's position
P	Bilinear interpolation weight
va (kts or Kelvin)	Bilinear value to interpolate
Dr (N)	Drag
Rm (m)	Earth radio
e_{ij} (hrs)	Flight Time difference
F (kg)	Fuel burn
Fij (kg)	Fuel burn for a given segment ij
Cons (kg)	Fuel consumption
ff (kg/hr)	Fuel flow
g (m/s^2)	Gravity
GS (kts)	Ground Speed
L (N)	Lift
m (kg)	Mass
c_1	Particle influence towards the global leader
c_2	Particle influence towards the local leader
D (Mach, degrees, ft)	Particle's displacement
X (Mach, degrees, ft)	Particle's position
C	Pheromone
R_2	Random parameter related to the global leader
R_1	Random parameter related to the local leader
$R\omega$	Random parameter related to the particle's inertia
v (kts)	South wind
Bl (ft, Mach or degrees)	Best local trajectory's position
Th (N)	Thrust
VTAS (kts)	True Air Speed
w (kts)	Vertical wind
u (kts)	West wind
w_d (kts)	Wind rates parallel to the aircraft

Greek Symbol

χ_{HDG} (deg)	Aircraft azimuth
α	Algorithm convergence parameter
β	Algorithm convergence parameter
ϕ (deg)	Latitude
λ (deg)	Longitude
ω	Particle's inertia

γ	Pheromone evaporation rate
γ_{TAS} (deg)	Pitch
μ_{TAS} (deg)	Roll
η	Thrust-specific fuel consumption

1 Introduction

The development of the aeronautical industry has provided important benefits in terms of economic, cultural, and knowledge exchange, mainly because it is the fastest means of transportation. To put this in perspective, in the ages before the maturity of aviation, it would normally take between 4 to 5 weeks to cross the Atlantic, nowadays, a London—New York flight takes roughly 8 h, and the Concord was able to cross the Atlantic in 3 h. The Air Transportation Air Group (ATAG) [1] has estimated that in the year 2014, 3.3 billon passengers were transported by aircraft (of which 54% were international tourists) and cargo worth US$ 6.4 trillion were transported by aircraft (50.4 million tons in freight, 35% of world trade). This key industry has led to the creation of millions of jobs, contributing to growing economies.

1.1 Pollution

In spite the enormous benefits of aviation, there is one main drawback: Pollution. Aircraft require quite large quantities of fossil fuel in order to power their engines. The ATAG [1] estimated that in 2015, 294 billion litres of jet fuel were used by commercial operators alone. For a single example, a Montreal to Paris commercial flight, the consumption of 30 tons of fuel is a typical value. It is well known that combustion ejects polluting particles into the atmosphere. The emissions that have attracted the attention of researches are mainly carbon dioxide (CO_2), nitrogen oxides (NOx), and hydrocarbons. Aviation also produces other detrimental effects, such as noise and the formation of clouds in the form of contrails in zones where there would not otherwise be any clouds.

Vapour contrails and CO_2 are known for their contribution to global warming. NOx is detrimental to the ozone layer [2], high hydrocarbon concentration at low altitudes near populated areas is harmful to humans, contributing to respiratory problems [3], and noise is linked to discomfort, stress, sleep deprivation and even cancer [4]. Noise regulations and the continuous descent approach procedure (descending with the motor in the IDLE setting) have reduced pollution and noise to acceptable levels [5, 6]. New engine developments have reduced NOx emissions.

The current aeronautical goal is to reduce CO_2 emissions by 2050 to 50% of those recorded in 2005 [7].

1.2 Alternatives for Pollution Reduction in Aircraft Operations

There are various ways to reduce pollution in aircraft operations. Logically, improving engines so they require less fuel to generate the same or even more thrust is desirable. Changing from fossil fuel to bio-fuel is a promising approach as it could be a more sustainable solution [8]. Improving aerodynamics to reduce drag, and thus the required thrust to power a flight has been a part of each generation aircraft. Reducing weight with new lighter materials and changing old bulky and heavy avionics to the lighter new options is also desirable. However, it will take years before old fleets can be converted to new generation, more efficient aircraft; a similar situation exists in terms of the overhaul of avionics. However, there are also operational improvements that can be implemented on every aircraft regardless of their generation, such as engine washing, assisted taxiing, reducing the Auxiliary Power Unit (APU) use and trajectory optimization [9]. This last option is especially promising as it has been established that aircraft do not fly at the speeds and altitudes that provide the most economical trajectory in terms of fuel burn [10–13]. By providing efficient trajectories, the amount of fuel needed to fly the required distance is reduced, thereby reducing the pollution released to the atmosphere. Optimizing trajectories would also reduce the need for fuel-intensive and thus high-polluting procedures such as go-around (or missed approach) [14].

Trajectory optimization is in line with the aeronautical desire to improve the use of airspace. Flights flying at low-efficiency trajectories, combined with the depletion of the airspace, especially in the eastern United States and in Europe, have motivated nations to develop and implement infrastructure to increase airspace capacity. In this context the Next Generation of ATM System (NextGen) has been developing in the United States, and Europe has been working with the Single European Sky ATM Research (SESAR).

These programs propose what is known as Trajectory/Intended Based Operations (TBO/IBO). Under this operations concept, aircraft will be able to dynamically negotiate their optimal reference trajectory with air traffic control instead of following their imposed trajectory. This procedure may also reduce the vocal communication between crew and traffic controllers, reducing the possibility of errors. Future systems would use TBO/IBO, also known as 4D trajectories, which would require each aircraft's position (lat./long.), speed, altitude and their Required Time of Arrival (RTA) to a given (or to a set) waypoint. This RTA is of special importance, especially during the last part of cruise, as it is intended that aircraft should arrive at their Top of Descent at a given time to allow the correct flow of traffic arriving to the terminal area and to allow a successful Continuous

Descent Approach (CDA). For this process to work, aircraft should be able to provide their location at all times. This is being worked out by programs encouraging the installation of infrastructure capable of supporting aircraft equipped with an automatic dependence surveillance broadcast (ADS-B). This device constantly emits an aircraft's current position to air traffic control and to aircraft in the vicinity. Installation of ADS-B is also being encouraged by different agencies around the world.

However, 3D trajectories (aircraft position, altitude, and speed) are still used in current systems, and represent the basis of 4D trajectories. Reference trajectory optimization can be classified into two main different types: lateral navigation reference trajectory and vertical navigation reference trajectory. Lateral reference trajectory, also known as ground-track, is a set of points that an aircraft should follow from the initial point to the destination. Vertical reference trajectory refers to the altitude and the speed that an aircraft should attain at each waypoint in a pre-defined ground-track trajectory. The lateral reference trajectory is mostly dependent upon zones with sufficient air traffic management coverage (in order to guarantee safety), avoiding forbidden zones (No-flight zones), and following advantageous wind patterns. As current regulation normally require aircraft to fly at around constant Mach numbers, the vertical reference trajectory consists mostly of finding that Mach number and the waypoints where to perform changes in altitude to reduce fuel consumption [15]. These changes in altitude are called "step-climbs". In practice, Air Traffic Control must authorize the execution of "step-climbs" to guarantee enough clearance and separation between aircraft. One key advantage of TBO/IBO trajectories is that aircraft would automatically negotiate this change of altitudes with Air Traffic Control.

Computing the most efficient trajectory in terms of fuel burn is complicated because aircraft fuel burn is affected by multiple parameters such as weather, the change of mass due to fuel consumption, the diverse combinations in speeds and altitudes available to perform the flight, and the many positions where the step-climb could be executed.

To compute the most economical trajectory, different algorithms employing optimal control have been used, as shown in [16–21]. Classical optimization techniques such as those related to finding the shortest path have been implemented to find the most efficient route while avoiding obstacles in the lateral reference trajectory [22, 23]. The vertical reference trajectory for all flight phases has been optimized by using the Golden Search Section algorithm and step climbs [24], search space reduction algorithms [25, 26], flight cost estimators [27], and the beam search algorithm [28]. The latter two algorithms have been coupled to reduce the computation time [29]. The vertical and the lateral reference trajectories have been optimized by determining the optimal ground track and then finding the most economical waypoints to execute step climbs [30]. The vertical and the lateral reference trajectories have been coupled using dynamic programming for a 4D flight in [31, 32].

Commercial aircraft utilize a device called the Flight Management System (FMS) that, among many different tasks, manages the flight plan and optimizes the flight trajectory. Most of the current FMSs only manage 3D trajectory optimization, as the infrastructure required for the 4D has not yet been fully deployed.

Flight trajectory optimization is of interest not only for airborne equipment, but also for ground-based algorithms that compute the flight trajectory to be proposed and submitted for approval by traffic control authorities.

The objective of this chapter is to explain the implementation of different metaheuristic optimization algorithms utilized to find the optimal trajectory of reference for a commercial aircraft. Through this work, the word "optimal" refers to the trajectory that provides the most economical fuel consumption. Two methods for creating a fuel consumption model are explained, followed by a methodology to compute the flight cost using that model. An open weather data source is introduced, as well as how to retrieve its information, and the effects of weather on fuel flow are explained. This chapter also shows how to model air space as a graph, a concept used in several algorithms. Genetic Algorithms are explained in detail, as they are used here to find the most economical set of altitudes during the cruise phase. An algorithm able to compute the combination of Mach numbers to fulfil the RTA constraint using the Artificial Colony Optimization (ACO) taking into account weather is shown. The last algorithm explored in this chapter is the artificial bee colony. This chapter ends with some conclusions and recommendations for future work.

2 Aircraft and Weather Models for Reference Trajectory Optimization

One particularity of the aeronautical industry is that aircraft models are very complex, in the sense that the parameters that compose the well-known equations of motion describing the aircraft dynamics for a wide flight envelope are not publicly available. The same is true for engine fuel burn computations.

Nevertheless, the two main models used to compute the fuel burn are the 3DoF point mass equations of motion and the use of a performance database. Other models include energy formulations or use Dubins's curve. However, the latter is mostly used to define the trajectory of any particle; an additional model for fuel burn is required in order to compute the flight cost.

2.1 DOF Equations of Motion

The more complex equations of motion can be reduced for commercial trajectory optimization as described by Bronsvoort [33]. For these equations to work, the following assumptions must be made: Non-rotating earth, constant gravity, aircraft mass is not a state, thrust force vector is aligned with airspeed, no side slip, and the path angle changes are instantaneous. The resulting equations are given by Eqs. (1)–(6):

$$\dot{V}_{TAS} = \frac{Th - Dr - m \cdot g \cdot \sin(\gamma_{TAS})}{m} - \dot{w}_{Xd} \tag{1}$$

$$\dot{\chi}_{HDG} = \frac{1}{V_{TAS} \cdot \cos(\gamma_{TAS})}\left(\frac{L \cdot \sin(\mu_{TAS})}{m} - \dot{w}_{Yd}\right) \tag{2}$$

$$L \cdot \cos(\mu_{TAS}) - m \cdot g \cdot \cos(\gamma_{TAS}) + m \cdot \dot{w}_{Zd} = 0 \tag{3}$$

$$\dot{\varphi} = \frac{V_{TAS} \cdot \cos(\gamma_{TAS}) \cdot \sin(\chi_{HDG}) + \upsilon}{R_M + h} \tag{4}$$

$$\dot{\lambda} = \frac{V_{TAS} \cdot \cos(\gamma_{TAS}) \cdot \cos(\chi_{HDG}) + u}{(R_N + h) \cdot \cos\varphi} \tag{5}$$

$$\dot{h} = V_{TAS} \cdot \sin\gamma_{TAS} + w \tag{6}$$

$$\dot{m} = -F \tag{7}$$

where λ and ϕ represent the aircraft longitude and latitude, respectively, h is the altitude, Th, Dr and L are the thrust, drag and lift, respectively, χ is the azimuth, γ is the pitch, μ is the roll, and $\dot{w}_{Xd}, \dot{w}_{Yd}$ and $\dot{w}_{Zd}$ are the wind rates parallel to the aircraft, parallel to the earth surface and perpendicular to the aircraft, respectively. In addition, u, v and w indicate the west, south and vertical wind, m is the aircraft mass, and F represents the fuel burn. The fuel burn is computed with the fuel as with Eq (8).

$$ff = \eta(V_{TAS}) \cdot Th \tag{8}$$

where η is the thrust-specific fuel consumption (as a function of V_{TAS}). Fuel burn is computed simply by multiplying ff times the time spent under that flight condition.

It is common to find in the literature that Th, D, L, and 2 η coefficients are computed using the Base of Aircraft Data (BADA) developed by Eurocontrol. Wind information can be pre-defined data, or it can be given in the form of a forecast.

By solving Eqs. (1–8), the fuel required to power a flight for a given trajectory can be computed.

2.2 Numerical Performance Database

An alternative way to compute the fuel burn to accomplish a given flight is by means of a performance database (PDB). This type of model can be used by the FMS to compute fuel burn. However, PDBs are not easy to obtain since, as noted earlier, aircraft performance information is restricted by aircraft manufacturers. Some avionics companies perform flight tests, either in real aircraft or in high quality flight simulators, to experimentally develop their aircraft's PDB.

A PDB is a set of tables that provide a certain output for some given inputs. PDB inputs mostly contain constant Indicated Air Speed (IAS), acceleration in IAS during climb, constant indicated Mach numbers, Cruise at constant Mach numbers, Cruise at constant IAS, Descent at a constant Mach number, Descent at a constant IAS, and Deceleration during IAS descent.

The required inputs for a PDB are: Aircraft Weight, Speed (Mach or IAS), Altitude, and the Temperature Standard Deviation (DEV_TMP). The DEV_TEMP is the difference between the real temperature and the temperature provided by the standard atmosphere.

2.3 PDB Creation Using a Level D Flight Simulator

This section summarizes the methodology proposed in [34] to develop a PDB using a Level D flight simulator for the cruise phase. Level D represents the maximum accuracy available with a flight simulator. The climb phase generation can be found in [35].

During the cruise phase, the fuel information is normally provided in terms of fuel flow (kg/hr).

To develop a PDB, fuel flow samples for different flight conditions are obtained for the aircraft in flight. These samples are taken in such small steps (in terms of altitude, Mach, and weight) that the PDB outputs between two samples are considered to be linear.

Once the flight envelope of interest has been defined, the "sampling process" for each flight condition is carried out according to the following methodology: First, the aircraft mass is set to its desired value. Next, the aircraft is flown to its desired altitude while the desired Mach number is fixed manually. Once the aircraft is stable at the desired flight conditions, the test pilot should wait approximately three minutes to let the fuel flow stabilize [36], and then take the first sample. This sampling is done roughly every two seconds. Five different measures are taken. The average of the sampled values is considered to be the fuel flow for the evaluated flight condition. Many different measurements were taken to obtain Fig. 1, which represents the average PDB for the cruise phase for the Citation X under ISA conditions.

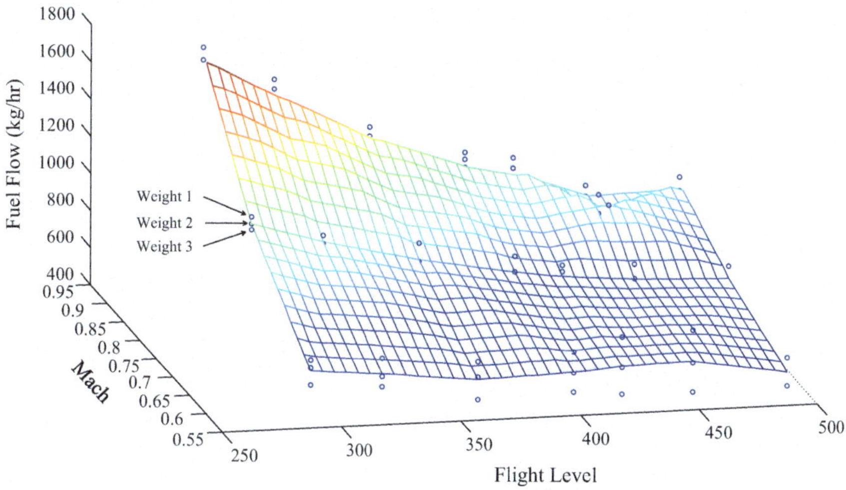

Fig. 1 Average grid of the PDB for the cruise phase taking 3 weights

Different flights for distances of 75 and 150 nautical miles at different Mach numbers and different altitudes were evaluated to validate the performance model quality. The results showed that a maximal difference between the flight provided by the flight simulator and the flight evaluated with the created PDB was about 3%, and the lowest difference was close to 0.4%.

In [35, 37] an identification algorithm was developed to compute the same outputs provided by the PDB using the same inputs required for a PDB. This identification method is valid for the climb phase, and its results can be used to generate a PDB. The main benefit of the algorithm developed in [35, 37] is that it reduces the number of flight tests required; for example, the parameters for the Equations of Motion for the climb regime can be identified using only 9 flight tests. The fuel burn of the algorithm developed in [35, 37] was compared against the fuel burn from 61 different flights (every flight was flown under the same conditions). Fuel burn results showed a fuel burn difference of maximal 4% and as low as 0.04%.

2.4 Fuel Burn Computation Using a PDB

The algorithms presented in this chapter are based on a PDB. A complete methodology to compute the overall flight cost from a PDB by taking into account all flight phases such as climb, acceleration, cruise, descent, and deceleration was discussed in [38, 39]. The basic aspects of this methodology are described next. Please notice that due to aviation conventions, the altitude units are provided in feet (ft), deviation from the International Systems Units used for all the other units.

2.5 *Obtaining the Required Data from the PDB*

PDBs are constructed for a limited number of inputs; in this case these are aircraft mass, speed, altitude and temperature. The available values for each input are also limited in the number of elements. Let's suppose that the weight is only separated in 10,000 kg steps, Mach number is separated in 0.05 M steps, altitude in 1000 ft steps, and temperature in 5° steps. Under these circumstances, it is easy to obtain the fuel flow from an aircraft with a mass of 150,000 kg flying at an altitude of 34,000 ft at a Mach number of 0.785 and at an ISA temperature deviation of 5 °C. However, if the values change to be 153,135 kg, 34,500 ft, 0.793 M, and an ISA temperature deviation of 3 °C, it becomes impossible to directly obtain the required values. This example is an extreme one. Normally, PDBs are created such that the available altitudes and the available Mach numbers are those that will always be selected. For example, altitudes are always separated in 1000 ft in order to follow the International Civil Aviation Organization (ICAO) convention to always fly at altitudes that are multiples of 1000 ft.

Following the example given above, it can be concluded that not every exact required input value is available to provide the required output (i.e. fuel flow for the cruise phase). The PDB sampling allows the use of linear interpolations to compute the exact required output when the provided inputs are not among the available PDB inputs. The process followed to compute the fuel flow for an aircraft with a mass of 153,135 kg, an ISA temperature deviation of 3 °C at a Mach number of 0.8, and an altitude of 34,000 ft would be as follows:

The first step is to identify the lower and the upper limits surrounding the required value. In our example, for an aircraft mass of 153,135 kg, the lower limit (W_l) would be 150,000 kg, and the upper limit (W_u) would be 155,000 kg, and for the ISA temperature deviation, the upper limit (*ISADevu*) would be 5 °C and the lower limit (*ISADevl*) would be 0 °C.

The second step is to determine the fuel flow for each combination as follows.

$$ff_{11} = PDBOutput(Wl, M, Alt, ISADevl) \tag{9}$$

$$ff_{12} = PDB\ Output(Wl, M, Alt, ISADevu) \tag{10}$$

$$ff_{21} = PDB\ Output(Wu, M, Alt, ISADevl) \tag{11}$$

$$ff_{22} = PDB\ Output(Wu, M, Alt, ISADevu) \tag{12}$$

The third step is to interpolate for *ISADev:*

$$ff_1 = Interpolation(ff_{11}, ff_{12}, ISADev) \tag{13}$$

$$ff_2 = Interpolation(ff_{21}, ff_{22}, ISADev) \tag{14}$$

The fourth step is to interpolate in terms of the weight in order to obtain the required output (the fuel flow in this case)

$$ff = Interpolation(ff_1, ff_2, weight) \tag{15}$$

This process is graphically shown in Fig. 2.

2.6 Computing the Required Fuel Burn from the Fuel Flow

The fuel flow is normally given in units such as kg/hr. Therefore, it is important to determine the amount of time an aircraft would be at those conditions. As temperature and weight are continuously changing, these values are assumed to remain constant during a pre-defined distance. This way the Flight Time required to cover that distance (where weight and temperature are assumed constant) can be computed by computing the GS (ground distance), defined as the aircraft's speed in relation with the ground. This GS is influenced by the wind, as indicated in Eq. (16):

$$GS = V_{TAS} \pm Wind\,Speed \tag{16}$$

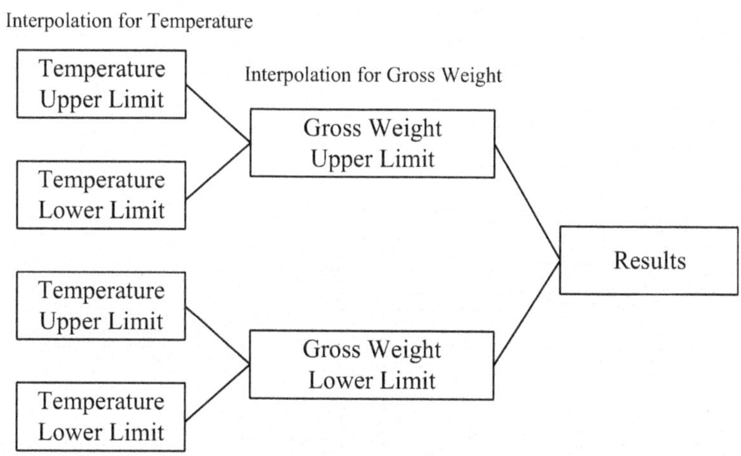

Fig. 2 Interpolation process. Temperature interpolation is executed first, followed by interpolations in the gross weight

As fuel is burned along the trajectory, the aircraft's mass diminishes over time. The fuel burned should thus be reduced from the total aircraft weight every time the aircraft flies a pre-defined distance. This distance is important, because the accuracy of the results depend upon how often the aircraft mass is updated: the more often it is updated, the more accurate the fuel consumption computation is. However, updating the aircraft mass too often will require the execution of too many interpolations that will result in increasing the computation time. As discussed in [39], the recommended pre-set distance to achieve a good compromise between accurate calculations and computation time should be between 20 and 35 nm. Finally, the fuel burn is the sum of the fuel that must be burned to fly every segment composing the trajectory, as shown in Fig. 3.

To validate this methodology, two different flights were simulated using the same parameters as those of the algorithm developed here, a FMS commercial simulator (PTT), and those of a specialized simulation software, FlightSIM® (developed by Presagis®). The PTT uses the same PDB as that of the algorithm, and FlightSIM® uses the complete aerodynamic and engines model of an aircraft. The results are shown in Table 1.

The results indicate that for the same aircraft compared across the three platforms, the FlightSIM® (FL) is considered to be the most exact method because it uses a flight dynamics precise model. As it can be seen in Table 1, the algorithm developed in this paper provides a difference as small as 1.09% (50 kg) in fuel burn, and as small as 0.52% (55.4) seconds in flight time compared to the

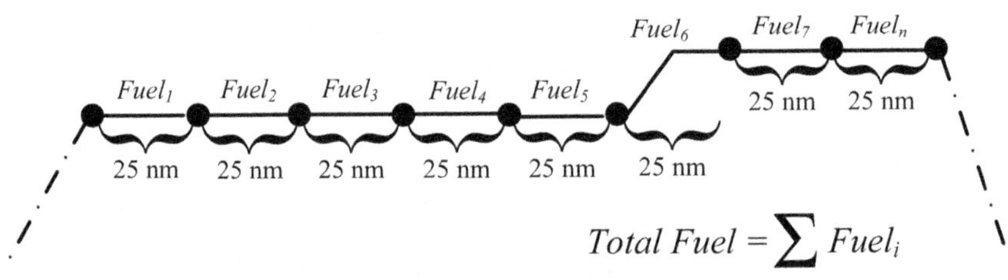

Fig. 3 Fuel burn is computed as the sum of the fuel burn required to fly each segment

Table 1 Fuel burn and flight time comparison between three different platforms

Flight	Parameter	FlightSIM	PTT	Algo	Difference FS* versus Algo	Difference Algo versus PTT
YUL–YVZ	Fuel (kg)	4518	4575	4568	50 (1.09%)	7 (0.15%)
	Time (s)	2491	2527	4455	35 (1.46%)	72.11 (2.85%)
LAX–MSP	Fuel (kg)	18,472	19,020	19,021	579 (2.89%)	1 (0.01%)
	Time (s)	10,619	10,479	10,564	55 (0.52%)	84 (0.81%)

*FS = FlightSIM

FlightSIM®'s results. Simulations using FlightSIM® take the real duration of flight, while the developed algorithm takes only seconds to compute the fuel burn.

Comparing the results of the algorithm to those of the PTT, the differences were as low as 0.01% (1 kg) in fuel burn, and as low as 0.81% (84.4) seconds for the flight time.

2.7 *The Weather Information and Its Effects on the Fuel Burn*

Weather has a direct impact on flight planning and routing processes. Storms can delay flights from their take-off and landing, while accumulated ice on the aircraft surfaces increases the aircraft weight and drag representing both cost and safety issues. Therefore, weather conditions can make an aircraft deviate from its trajectory.

Within the flight trajectory optimization framework, the most common uses of weather information is to detect storms so that they can be avoided in an efficient way [23], to take advantage of tailwinds and avoid headwinds to reduce flight costs [30, 40, 41], and to detect potential places where "contrails" (clouds generated by aircraft) can be formed in order to avoid them [42]. Open weather data sources include the National Oceanic and Atmospheric Administration (NOAA) in the United States and Weather Canada.

A detailed method to obtain the weather parameters can be found in [43]. This approach first identifies the four points of the grid surrounding the aircraft, which are further interpolated in altitude (pressure) and time. Finally, a bilinear interpolation can be executed to obtain the required weather parameter values (e.g. wind, temperature) at the aircraft position. This is explained with some detail in the following paragraphs.

The weather information for Weather Canada is provided in the form of a grid that contains the weather information for the world map for different altitudes at different hours. The grid is provided with a resolution of 0.6 km × 0.6 km at 3-h time blocs for different pressures. Pressure is used to measure altitude; if the barometer is set to a reference of 29.92 in hg, the measured altitude at a given altitude can be tracked. For example, for 35,000 ft the measured pressure would be 7.06 in Hg, which corresponds to 23.93 kPa under ISA. This type of altitude is called the altitude pressure or flight level and it is commonly used above 18,000 ft to safely separate aircraft.

Similar to the PDB explained above, the set of pressure altitudes provided by Weather Canada are not the exact ones that define the required altitude. Thus, once the four points enclosing the aircraft have been identified, the next thing to do is to identify the pressures that enclose the desired altitude, and the time blocks corresponding to the fly hours.

Once the pressure altitudes and the time blocks have been identified, interpolations in *altitude* (Altitude 1 and Altitude 2 in Fig. 4) for each time block are executed to determine the weather at the four points surrounding the aircraft. The next step is to interpolate in *time*, as shown in the Interpolation between Time block 1 and Time block 2 in Fig. 4.

At this stage, the weather information for the four points surrounding the aircraft contains the weather for the required altitude and time. The last step is to execute a bi-linear interpolation. In this type of interpolation, the weather for each point is weighted according to its distance from the aircraft, as indicated in Eq. (17).

$$P_x = \frac{1}{\sqrt{(\lambda_{airc} - \lambda_x)^2 + (\varphi_{airc} - \varphi_x)^2}} \tag{17}$$

where (λ_{airc}, φ_{airc}) represents the actual geographic aircraft position, and (λ_x, φ_x) represents the geographical position of a given point surrounding the aircraft. This weight is evaluated for each of the four points surrounding the aircraft. The bilinear interpolation in Eq (18) is then computed.

$$variable(lon, lat) = \frac{P_1 va_1 + P_2 va_2 + P_3 va_3 + P_4 va_4}{P_1 + P_2 + P_3 + P_4} \tag{18}$$

In other words, Eq. (18) means that the closer the waypoint is to the aircraft, the more influence it will have. The main values obtained from the weather are temperature and wind information (Fig. 4).

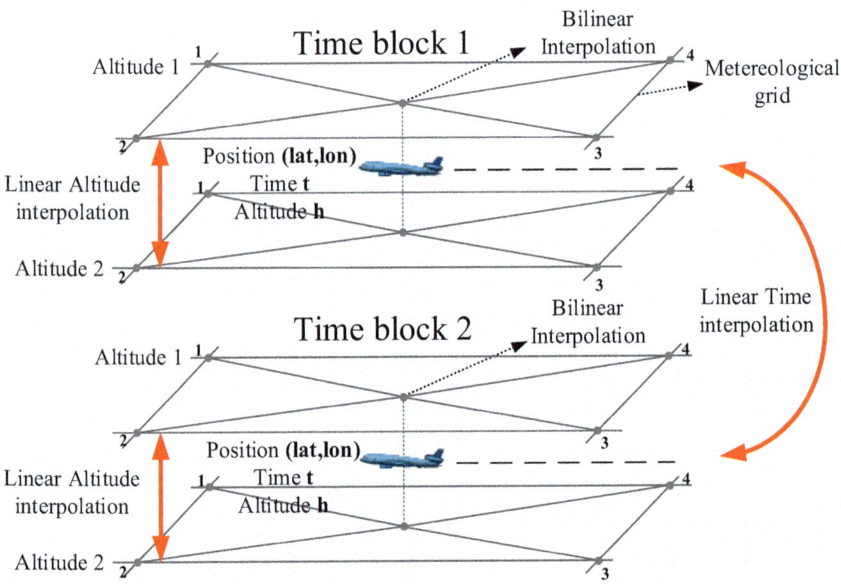

Fig. 4 Weather interpolation scheme for wind and temperature using open source data

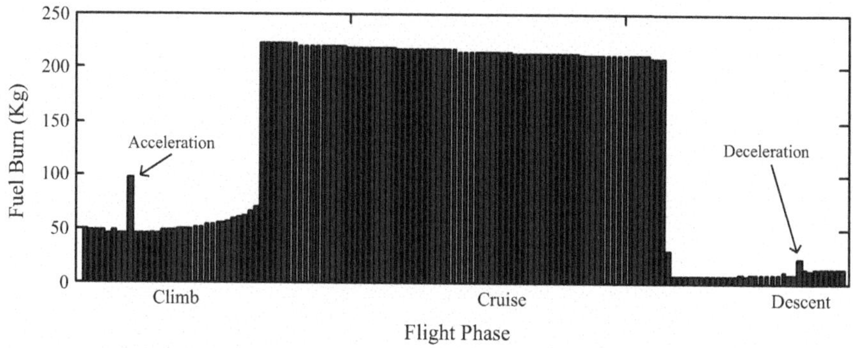

Fig. 5 Fuel consumption behaviour without the wind effect

Different flights were performed to observe the influence of the wind for a given trajectory. Figure 5 shows the fuel consumption during cruise for a given flight without taking weather information into account; the total fuel consumption for this flight was close to 17,969 kg. Figure 6 shows the fuel consumption for the same flight when taking into account the influence of weather. The fuel consumption for this flight was close to 17,185 kg. Thus, the flight consumption for the aircraft when weather is accounted for was around 780 kg less than not accounting for the weather. The weather thus plays an important role in the flight cost; in this example, there was a "tailwind" that actually incremented the aircraft ground speed.

These two figures show that, especially during the cruise phase, a linear reduction of fuel burn occurs in Fig. 5 as it takes the same time to fly each segment, while a non-linear behaviour occurs in Fig. 6 because the time required to fly each segment changes due to wind effects. Fuel burn depends on fuel flow, as revealed by this flight time behaviour.

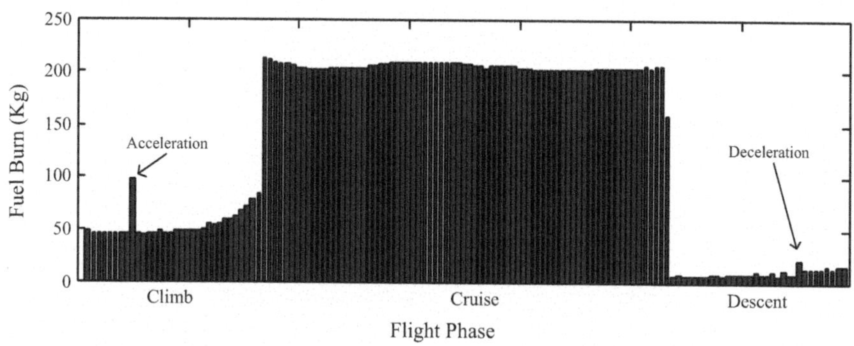

Fig. 6 Fuel consumption behaviour with the wind effect

3 Airspace Modelled as a Graph

Aircraft move in a 3D space, restricted to fly at altitudes that are multiples of 1000 ft and to respect no-fly zones, such as some city centres and restricted military areas. Attempting to evaluate every single part of this 3D space using trajectory models such as Dubin's particles would lead to an infinite number of combinations. To overcome this problem, the air space is modelled as a positive unidirectional weighted graph *G* (*E*, *V*) where *V* are the vertexes (waypoints, not necessarily published waypoints) and *E* are the paths connecting two vertexes. The graph is unidirectional as aircraft only move forward (it is impossible for an aircraft to move backwards). The weight for each edge in this case is the flight cost; this causes the graph to be positive, as negative weights mean that travelling between waypoints is producing fuel, an impossibility. This graph reduces the number of combinations to a finite number by modelling the airspace as a graph. The search space for the cruise phase, from the Top of Climb (ToC) to the Top of Descent (ToD), is shown in Fig. 7.

This type of graph is created to show where an aircraft can change altitudes in order to emulate step climbs and moderate its direction to take advantage of tailwinds or avoid headwinds. The waypoints in altitude are separated by 2000 ft, and the horizontal distance between two consecutive waypoints is 150 nautical miles. These distances are set to avoid abrupt changes of headings and altitudes.

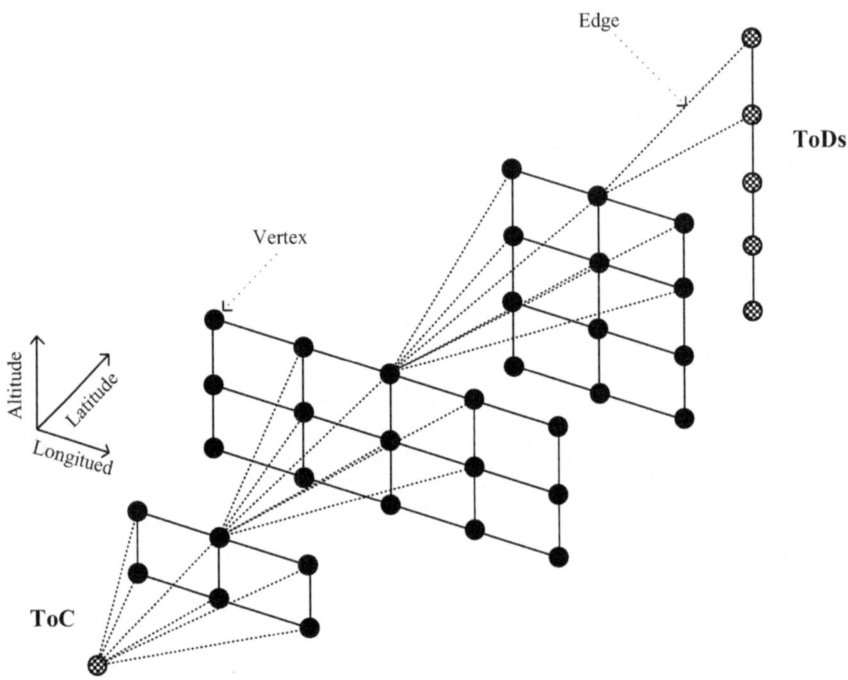

Fig. 7 The search space as a unidirectional graph

Using this search space graph, the trajectory optimization problem can be seen as a short path problem. The most economical combination of points can be defined as the optimal trajectory. These graphs offer a huge number of possible combinations. Felix and Botez in [44] computed that there could be a total of over 43 million combinations for a graph with only 12 reference waypoints with a maximal step climbs number of five. Brute force (exhaustive search) algorithms become impractical as they would require too much time to find a solution. Metaheuristic algorithms are therefore implemented to solve this trajectory optimization problem.

4 Using Metaheuristic Algorithms for Trajectory Optimization

Metaheuristic algorithms are normally used to find the maximal or minimal solution of an optimization problem, basically by trial and error, guiding the trial to solutions where the optimal solution might be found. Normally, but not necessarily, metaheuristic algorithms have a stochastic component in order to explore the search space and try to avoid results restricted to local minima. Metaheuristic algorithms normally try to focus in on a local optimal solution, anticipating that it is the optimal solution, while at a given point they use some sort of diversification to explore new parts of the search space. These algorithms might provide the optimal solution (or a very good sub-optimal), however, they are not guaranteed to find the optimal solution.

This Section explores four metaheuristic algorithms applied to four different commercial trajectory problems: Genetic Algorithms for vertical reference trajectory optimization; the Artificial Bee Colony algorithm for a lateral reference trajectory; the Particle Swarm Optimization for 3D reference trajectory optimization; and the Ants Colony Optimization algorithm to reach a given waypoint at a given time constraint.

4.1 Genetic Algorithms and the Vertical Reference Trajectory Optimization

Genetic algorithms have been used to find the best winds at a fixed altitude to optimize flight costs [41], and they have also been applied to couple the lateral reference and the vertical reference flight trajectory in 3D [45].

The vertical reference trajectory optimization can be computed by considering the whole flight, using various methods such as golden search Section, Branch and bound, genetic algorithms, and optimal control [24, 28, 40, 46]. These algorithms have the advantage that they can be useful for optimizing short flights of less than 500 nm of distance, in which the climb and the descent phases have a significant

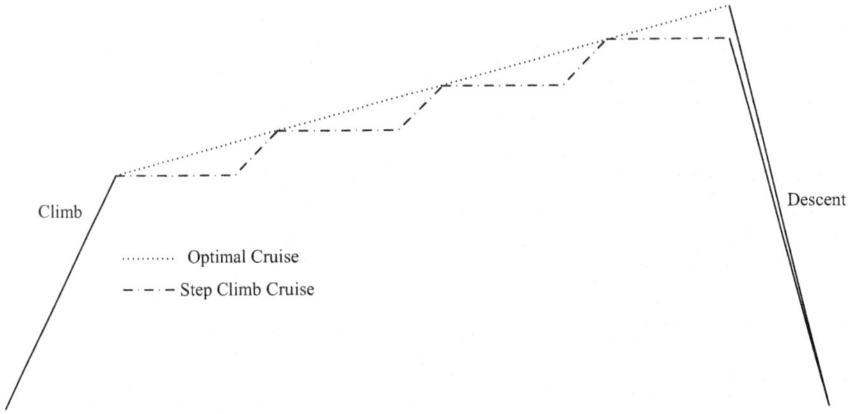

Fig. 8 Optimal cruise and the step climb cruise

impact on the flight cost. However, for long flights the best strategy during vertical trajectories is to perform changes in altitude during cruise, called step climbs, emulating what is called the constant climb cruise. This is explained in Fig. 8.

While the algorithms mentioned above allow for execution of the "step climb" function, they mostly rely on different strategies, such as distance [47], flight time [29], or a pre-computed weight-altitude relationship [46]. Another strategy is to evaluate each waypoint to determine if at that waypoint it is worthwhile to perform the step climb. Doing an exhaustive search (testing all of possible combinations) would take much too much time.

To find the best combination of altitudes, and to determine at which waypoints an aircraft should execute the step-climb procedure, a Genetic Algorithm (GA) was used.

The Genetic Algorithm mimics Darwin's theory of evolution where the strongest individual survives, so that through reproduction and genetic transmission, it is the stronger individuals who dominate the population.

To implement the Genetic Algorithm approach in this problem, a unidirectional weighted graph, such as the one in Fig. 9 is, required to allow the aircraft to change altitudes.

Before detailing the steps followed by the algorithm, some definitions are required. An "individual" is a set of waypoints showing the altitudes the aircraft should be flying at, as indicated in Eq. (19).

$$Individual = [Alt_1\, Alt_2\, Alt_3 \cdots Alt_n] \tag{19}$$

"Reproduction" takes place when two trajectories generate an "offspring". The offspring is generated by taking half of the data from one parent and half of the other parent's data. This is known as crossover and it is shown in Fig. 10.

The "tournament" is a selection process where two individuals face off against each other, and the most fit survives to the next round. To avoid having

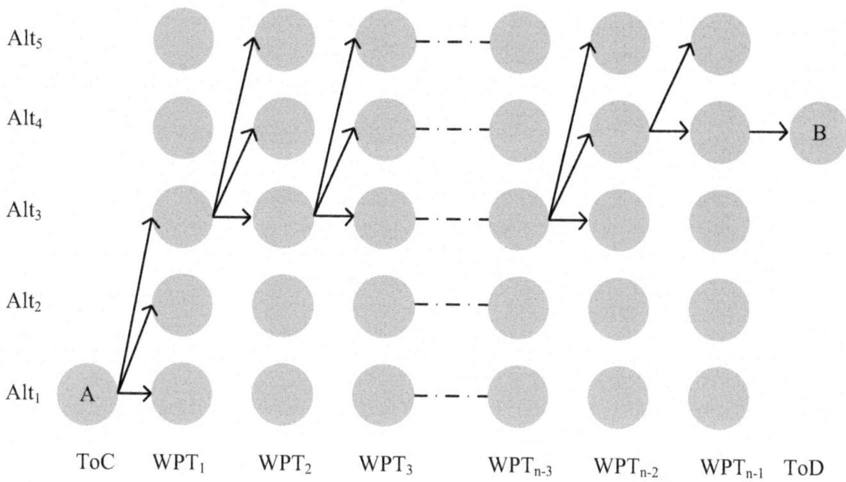

Fig. 9 Graph for altitude selection

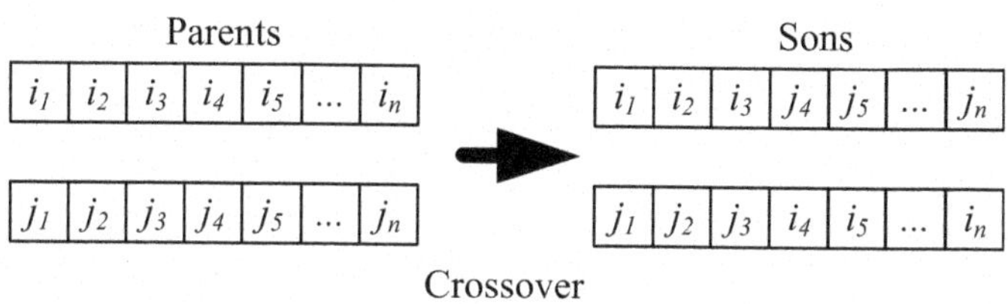

Fig. 10 Crossover to create the offspring

tournaments wherein the most-fit population always competes against the half less-fit individuals, the "brackets" are randomly defined. In this way, diversity is created to avoid determining a "local optimal".

With these two concepts, the algorithm is executed as follows.

Step 1: An initial population is generated with random trajectories. Each trajectory is evaluated to determine its fitness.
Step 2: The surviving individuals are selected via a tournament. The losing individuals are discarded.
Step 3: Surviving individuals reproduce to create a new generation. The new individuals are evaluated to determine their fitness.
Step 4: Population is sorted, and 20% of the individuals deemed the weakest are discarded.
Step 5: The discarded individuals are replaced by random trajectories to increase the diversity and thus to explore more of the search space.
Step 6: The optimal trajectory is determined.

Step 7: If the maximal number of iterations has been reached, the algorithm stops, otherwise the algorithms goes back to Step 2.

More details explaining this algorithm can be found in [48].

Results have shown that the selection of step climbs improves the trajectory cost compared to a flight plan computed by an airline. The airline flight plan required around 32.2 tons of fuel, and the optimized trajectory required around 30.7 tons of fuel, for a savings of around 4.6%. This reduction in the required amount of fuel occurred because the waypoints were selected at better locations than those computed by the airline flight plan [48].

4.2 Artificial Bees Colony and Lateral Reference Trajectory Optimization

Another aspect that should be studied when optimizing trajectories is to determine the lateral reference trajectory (also called ground track or route) where more favourable wind can be found. A favourable wind is one that moves in the same direction as the speed vector; these favourable trajectories avoid those winds that are in a different direction than the speed vector. Flying with tailwinds decreases the flight time for a given constant Mach number, thereby reducing the fuel burn. Different deterministic algorithms such as Dijksra's algorithm [49] and the Floyd-Warshall algorithm [50] have been also implemented for this problem.

The Artificial Bee Colony (ABC) is used here to find the combination of waypoints that minimize the flight time. ABC, one of the most recently-developed metaheuristic algorithms, is inspired by the behaviour of honey bees in their search for food sources. In nature, honey bees form a swarm able to successfully accomplish tasks through social cooperation. The ABC algorithm takes the behaviour of three types of bees: employed, onlooker, and scout bees. Employed bees gather food from different known food sources that they have in their memories. They share their known food sources' quality information with the onlooker bees. The onlooker bees tend to select the highest quality food sources from among those found by the employed bees. The scout bees are a few of the employed bees that abandon known food sources' vicinities in the search for new food sources. All bees slightly modify their trajectories every time they make a trip from the food source to the hive. It is this little modification which eventually helps bees to find the most efficient trajectory.

This behaviour can be adapted to the trajectory optimization utilizing the following steps:

Step 1—Inputs: Departure and arrival waypoints (the ToC and the ToD), weather information, cost index, number of employed bees (N_e), number of onlooker bees (N_o), number of iterations, maximal counter number.

Step 2— Initialization: A predefined number of trajectories are randomly created. Every trajectory is assigned to an employed bee. In other words, every single bee represents a different trajectory linking the ToC to the ToD. Every trajectory is evaluated using its positions and the weather at the aircraft's passage time.
Step 3—Employed bee: Every employed bee will create a random mutation on their assigned trajectory. The mutation is created by observing the behaviour of other trajectories. The more different trajectories there are, the more variations there will be in the mutations. The mutated trajectory cost is evaluated and compared to the original trajectory. If the mutated trajectory is more economical than the original trajectory, it becomes the new trajectory for that employed bee. Otherwise, the less economical mutated trajectory is discarded and a counter is incremented explained in Step 5, is incremented.
Step 4—Onlooker bee: Every trajectory explored by the employed bees is rated according to its cost. If the rating of a trajectory is high enough, it is likely to be selected by an onlooker bee. Low-rated trajectories have fewer chances to be selected by on-looker bees. Once a trajectory is selected by the onlooker bee, the same process followed by the employed bee is applied. Then, the best trajectory found by both the onlooker bees and an employed bee is memorized. Steps 3 and 4 are computed until the maximal number of iterations is reached.
Step 5—Scout bee phase: Every time a mutation fails to create a more economical solution, a counter is incremented. There is one counter per trajectory. If the counter reaches a pre-defined number, the associated trajectory is discarded, since it is assumed that the trajectory fell into a local optimal. If the discarded trajectory was the most economical overall, it is kept in memory. The discarded trajectory is replaced by a new random trajectory on the same model as in the initialization phase. This process is executed to explore the rest of the search space.

To evaluate the algorithm, a constant altitude flight at constant Mach number was evaluated. The reference trajectory was a Montreal—Paris flight with a distance

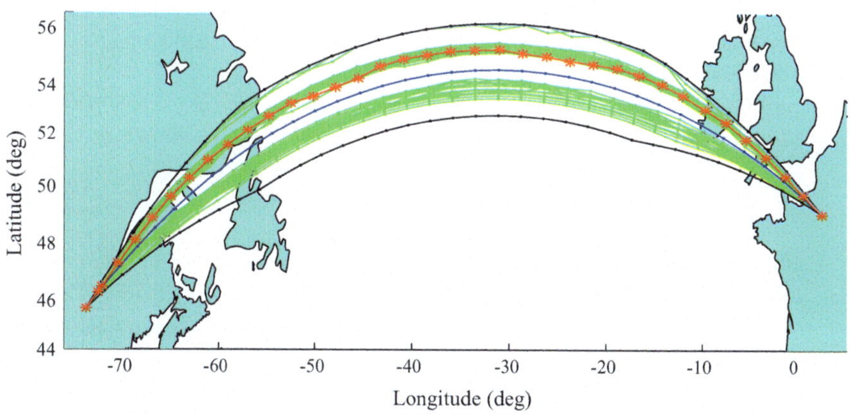

Fig. 11 The lateral reference trajectory for a flight between Montreal and Paris. The optimal trajectory is shown in red

of 2970 nm. 36 waypoints were considered taking into account weather information.

In Fig. 11, the black lines are the original flight plan in the centre and the search area boundaries. The optimal trajectory is represented in red. Other trajectories used in the algorithm that were less efficient than the optimal are shown in green. The comportment of the non-optimal trajectories is clearly visible.

This flight was simulated over 6000 runs under the same conditions. It was observed that the tests provided flight time savings ranging from 66 to 110 s. The average time saving was on the order of 102.5 s. 92% of the optimized flights provided trajectories with a flight time reduction of at least 100 s. Fuel savings were reported to be from 102 to 140 kg. The amount of savings was highly dependent on weather conditions, as seen in [51] and [41], where is some cases no optimization was reported as the weather was more favourable in the reference trajectory.

4.3 Particle Swarm Optimization and the 3D Reference Trajectory Optimization

The vertical and the lateral reference trajectories were optimized in the preceding two sub-sections. The optimized vertical reference trajectory offered more fuel savings than the optimized lateral reference trajectory. However, to achieve the maximal optimization requires optimizing both dimensions—3D reference trajectory optimization. The Particle Swarm Optimization (PSO) is implemented here to find the combination of waypoints that deliver the most economical 3D trajectory in terms of fuel burn. Different implementation of this algorithm to reference trajectory optimization can be found in [52].

The PSO is an iterative metaheuristic algorithm that mimics the behaviour of social organisms such as bird flocks. These organisms are able to organize high numbers of members without direct communication. Members are also called particles that continuously move through the search space following local and global leaders.

In utilizing the PSO for the trajectory optimization problem, each particle is defined as the set of waypoints linking the ToC to the ToD, as shown in Eq. (20)

$$WPT_i = \left[wpt_i^1, wpt_i^2, \ldots, wpt_i^n\right] \tag{20}$$

Each parameter of the WPT is composed by the waypoint position in terms of (lat., long.) and altitude, indicating the lateral and the vertical dimensions. Particles are grouped in clusters in which a local leader (the most economical trajectory in a cluster) is defined. The most economical leader is defined as the global leader. At each iteration, particles move through the search space influenced by the local and the global leaders' positions as well as by the displacement inertia. The equations defining each particle's movement are shown in Eqs. (21) and (22).

$$D_i\{j\}(k+1) = \omega R_\omega D_i\{j\}(k) + c_1 R_1 (B_l\{j\} - X_i\{j\}(k)) + c_2 R_2 \left(B_g\{j\} - X_i\{j\}(k)\right) \tag{21}$$

$$X_i\{j\}(k+1) = X_i\{j\}(k) + D_i\{j\}(k+1) \tag{22}$$

where i is the *current* trajectory and k is the current iteration. D represents the particle's displacement associate with the parameter j, which indicates the dimension (vertical or lateral—randomly selected). X represents the particle's location. B_g refers to the best global trajectory's position, and B_l refers to the best local trajectory's position. The parameter ω refers to the particle's inertia, the particle's tendency to keep the same direction. The role of the three random parameters R_ω, R_1 and R_2 is to determine the influence of the values of ω, B_l, and B_g, respectively, on the particle's movement. The parameters c_1 and c_2 are positive constants that influence the particle's movement in reference to the global and the local leaders, respectively.

The developed algorithm can be summarized in the following seven steps:

Step 1: A pre-defined number of random trajectories are generated and grouped in a pre-defined number of clusters. These trajectories can either be at fixed altitudes or contain multiple altitude changes. These trajectories are evaluated, thereby identifying the values of B_l and the B_g.
Step 2: At the beginning of each iteration, an initialization is performed, which consists of making a backup for the trajectory being worked on. If the modification to be done does not improve the trajectory cost, the original trajectory can be recovered. The random vectors R_w, R_1 and R_2 are generated here.
Step 3: The dimension to be mutated (j) for each trajectory is randomly selected.
Step 4: The displacements D (k + 1) for each trajectory in every cluster are computed.
Step 5: The position X (k + 1) is updated.
Step 6: The cost of the generated trajectories is computed. If they are more economical than the original reference trajectory, they are kept, otherwise they are rejected.
Step 7: The new local and the new global optimal are identified. If the current local best within a given cluster is more expensive than any of the new generated trajectories, then the most economical trajectory becomes the new local optimal. If any of the local best trajectories are more economical than the current global best, then the most economical trajectory becomes the global optimal. If the total number of iterations is reached, the algorithms ends, otherwise Step 2 is executed to begin another iteration.

The results were evaluated for two different long-haul flights: Montreal to Paris with a distance of 2918 nm, and Toronto to London with a distance of 3085 nm. The reference flight was the shortest distance between two airports flown at a constant altitude of 34,000 ft. It was observed that fuel savings for the

Montreal-Paris flight was close to 3.35%, and the savings for the London-Toronto flight was around 3.56%.

4.4 The Ant Colony Optimization Algorithm and Fullfiling the Required Time of Arrival Constraint

As explained in the Introduction, the future airspace will be "time-based". This means that a given waypoint must be reached at the "provided time", called the Required Time of Arrival (RTA).

This problem can easily be solved by knowing the distance to travel, the weather conditions and the RTA. A constant Mach number can be computed, based on the distance and the RTA. However, an adjustment of this computed Mach number will be required to account for the winds the aircraft will find along the trajectory, as they will modify the ground speed.

The RTA problem consists of finding the most economical combination of Mach numbers that fulfils the RTA constraint. An adaption of the Ant Colony Optimization (ACO) algorithm was implemented to find this combination. This algorithm mimics the behaviour of ants in their search for food. In nature, ants wonder around looking for food sources. When an ant finds a food source, it transports food back to the hive and comes back to the food source gather more. Ants are able to remember the path to/from the food source because they release pheromones along their route, a chemical substance that evaporates over time.

The stronger the pheromone trail is, the more that ants will be attracted to it, following the same trail and depositing more pheromone, thus incrementing the pheromone concentration. Other ants will find different paths towards the same food. Eventually the shortest path will have the most pheromone concentration, attracting ants from other paths and creating a unique path from the colony to the food source. Every time a given ant finds a pheromone trail, there is a probability for an ant to follow that path or not. It is this cooperative behaviour that it is simulated in the ABC algorithm.

To find the most economical combination of Mach numbers for a fixed altitude and a provided lateral reference trajectory, the space search consists of the possible Mach numbers the aircraft can flight at. These Mach number options can be modelled in the form of the grid given in Fig. 12, where the aircraft can change its Mach number at every waypoint.

Note that to create a Mach number path, the algorithm can select it from among three different pre-defined Mach numbers.

To obtain the probability of a given ant to select a given Mach number (or trail) that corresponds to the best compromise between fuel consumption and flight time, the relationship shown in Eq. (23) is used:

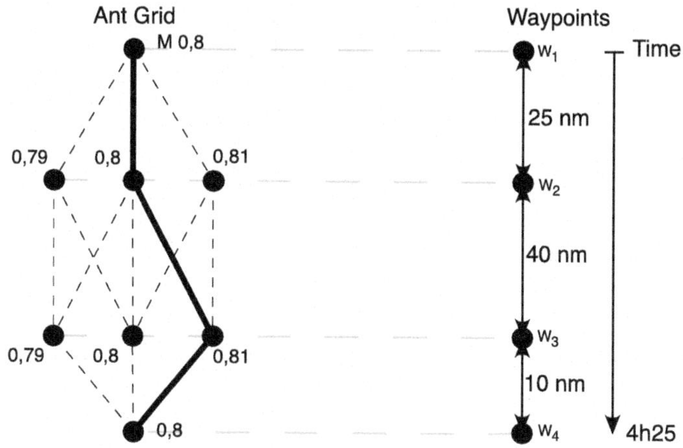

Fig. 12 Graph search for the Mach number selection

$$P_{ij} = \frac{F_{ij}^{\alpha}/\left(e_{ij} \cdot Cons_{ij}\right)^{\beta}}{\sum\left(F_{ik}^{\alpha}/\left(e_{ik} \cdot Cons_{ik}\right)^{\beta}\right)} \tag{23}$$

The pheromone concentration F_{ij} is updated at each iteration as follows:

$$F_{ij}(t+1) = F_{ij}(t)(1-\gamma) + N_{ij} \cdot P_{ij} \cdot C \tag{24}$$

where the pheromone evaporation rate is γ and the pheromone released by a given ant is C. β and α are parameters that influence the system convergence. The objective is to reduce fuel requirements, and so the fuel consumption is considered with $Cons_{ij}$, which is the amount of fuel required to fly from waypoint i to waypoint j. The sub-indices ik represent the total number of available Mach numbers. e_{ij} is the flight time difference between the required time to reach the RTA, and the arrival time that a given Mach number will require.

The algorithm can be summarized in the following four steps:

Step 1: The total number of ants, the LNAV reference flight plan, the initial Mach number, the RTA, the fixed altitude, and the weather information are given as the inputs.
Step 2: At the current waypoint, Eq. (23) is solved to select the current Mach number.
Step 3: Pheromone is deposited in the selected pathway between two points.
Step 4: At the new waypoint, Step 2 is executed. If the new waypoint is the ToC, the ant is discarded and the pheromone in the graph is updated using Eq. (24); a new ant is then placed at the ToC.

Table 2 Fuel burn at a given constant number

Mach number	Flight time (s)	Fuel burn (kg)
0.78	5h48m57s	24,112
0.80	5h40m14s	24,401
0.82	5h31m57s	15,260
0.84	5h25m03s	24,345

Table 3 Fuel burn and flight time difference with different RTAs

RTA	ETA	Difference (s)	Fuel burn (kg)
5 h30	5h29m59s	1	22,056
5 h35	5h34m59s	1	23,115
5 h50	5h50m00s	0	24,170

Stopping criteria: The algorithm stops when the last ant reaches the ToD. At this point, the path with the highest pheromone concentration is the "optimal" trajectory.

To validate this algorithm, a flight was performed at different constant Mach numbers to observe the algorithm behaviour. The flight costs (fuel burned) at a constant Mach number are presented in Table 2.

The same flight was then simulated imposing different RTA constraints, which produced the results presented in Table 3. In this table, it can be seen that the RTA constraint was fulfilled for all cases. It can be seen that in two of the three cases, the fuel flow was reduced compared to flights with similar flight times in Table 2.

Fuel reduction depends on the RTA constraint and on the weather. For example, the 5h30 RTA required the aircraft to fly at speeds that made it possible to consume less fuel. Flying at that speed also allowed the aircraft to take advantage of favourable winds.

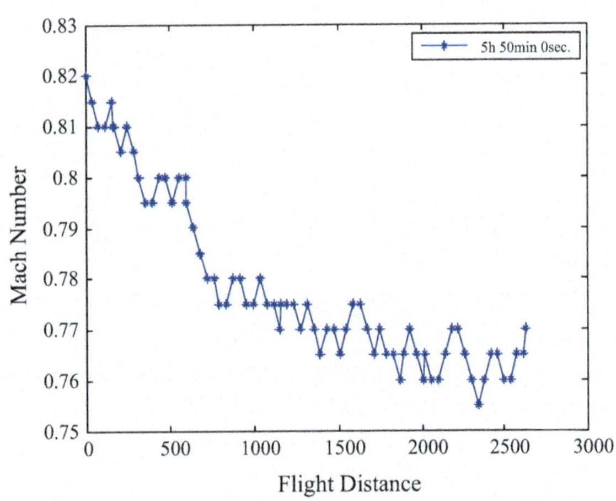

Fig. 13 Mach number selection to fulfil a given RTA

The Mach selection for the flight with a RTA constraint of 5h50 is illustrated in Fig. 13. It is clear that the aircraft reduced its Mach number as it advanced through the wind. This was possible because weather conditions were favourable to the flight, and the aircraft was thus to reduce its Mach number to fulfil the RTA constraint.

5 Conclusion

Aircraft cause pollution, leading to many negative impacts on human health as well as contributing to climate change. With air traffic only predicted to increase, flight reference trajectory optimization is seen as a means to reduce fuel consumption and thus to reduce the production of polluting emissions.

An aircraft model called a Performance Database was used here to optimize the flight trajectories. Considered proprietary information, performance databases are difficult to obtain from manufacturers. This work therefore introduced a methodology to create a database from a flight simulator, as well as an efficient methodology to compute fuel consumption from a database. A methodology to compute the fuel burn for a flight using the manufacturer's Performance Database was also explained.

The optimized navigational trajectories are the lateral, the vertical, and the combination of Mach numbers that fulfil the RTA constrain. Four optimization algorithms were explored: the artificial bee colony, ant colony optimization, genetic algorithms, and particle swarm optimization. It was observed that all algorithms were able to improve the reference trajectory. Weather conditions, which have a significant influence on fuel burn were obtained in the form of forecasts from Weather Canada. The interpolations required to compute the weather information were explained throughout this document.

Future work consists of implementing optimization algorithms to optimize flight trajectories in 4D. Ongoing work consists of implementing the Ant Colony Optimization and the Artificial Bee Colony algorithms to provide the optimal 4D trajectory.

Acknowledgements We would like to thank the team of the Business-led Network of Centres of Excellence Green Aviation Research & Development Network (GARDN), and in particular Mr. Sylvan Cofsky for the funds received for this project (GARDN II—Project: CMC-21). This research was conducted at the Research Laboratory in Active Controls, Avionics and Aeroservoelasticity (LARCASE) in the framework of the global project "Optimized Descent and Cruise". For more information please visit http://larcase.etsmtl.ca. We would like to thank Mr. Rex Haygate, Mr. Oussama Abdul-Baki, Mr. Reza Neshat and Mr. Yvan Blondeau from CMC-Electronics—Esterline. We would also like to thank Mr. Antoine Hamy, Mr. Audric Bunel, Mr. Charles Romain, Mr. Paul Mugnier, Mr. Jocelyn Gagné, Mr. Roberto Felix- Patron, Mr. Hugo Ruiz, Ms. Sonya Kessaci, and Mr. Oscar Carranza from LARCASE for their invaluable contributions. We also would like to thank the CONACYT (Mexico) and the FQRNT (Quebec, Canada) for their financial support.

References

1. ATAG (2016) Aviation benefits beyond borders. Air Transport Action Group, Geneva, Switzerland, July (2016). https://aviationbenefits.org/
2. Crutzen PJ (1970) The influence of nitrogen oxides on the atmospheric ozone content. QJR Meteorol Soc 96:320–325. http://dx.doi.org/10.1002/qj.49709640815
3. Robinson E (1978) Hydrocarbons in the atmosphere. Pure appl Geophys 116:372–384. https://doi.org/10.1007/bf01636892
4. Black DA, Black JA, Issarayangyun T, Samuels SE (2007) Aircraft noise exposure and resident's stress and hypertension: a public health perspective for airport environmental management. J Air Transp Manag 13:264–276. https://doi.org/10.1016/j.jairtraman.2007.04.003
5. Clarke JP, Brooks J, Nagle G, Scacchioli A, White W, Liu SR (2013) Optimized profile descent arrivals at los angeles international airport. J Aircr 50(2):360–369. http://dx.doi.org/10.2514/1.c031529
6. Kwok-On T, Anthony W, John B (2003) Continuous descent approach procedure development for noise abatement tests at louisville international airport, KY. In: AIAA's 3rd annual aviation technology, integration, and operations (ATIO) Forum
7. ICAO (2010) Aviation's contribution to climate change. International Civil Aviation Organization, Montreal
8. IATA (2009) Beginner's guide to aviation biofuels. International Air Transport Association, Geneva, Switzerland
9. McConnachie D, Wollersheim C, Hansman RJ (2013) The impact of fuel price on airline fuel efficiency and operations. In: Aviation Technology, Integration, and Operations Conference, Los Angeles, CA
10. Jensen L, Tran H, Hansman JR (2015) Cruise fuel reduction potential from altitude and speed optimization in global airline operations. Presented at the Eleventh USA/Europe air traffic management research and development seminar (ATM2015), Lisbon, Portugal,
11. Jensen L, Hansman JR, Venuti J, Reynolds T (2014) Commercial airline altitude optimization strategies for reduced cruise fuel consumption. In: 14th AIAA aviation technology, integration, and operations conference, Atlanta, GA. http://dx.doi.org/10.2514/6.2014-3006
12. Jensen L, Hansman JR, Venuti JC, Reynolds T (2013) Commercial airline speed optimization strategies for reduced cruise fuel consumption. Presented at the 2013 aviation technology, integration, and operations conference, Los Angeles, USA. http://dx.doi.org/10.2514/6.2013-4289
13. Turgut ET, Cavcar M, Usanmaz O, Canarslanlar AO, Dogeroglu T, Armutlu K, et al (2014) Fuel flow analysis for the cruise phase of commercial aircraft on domestic routes. Aerosp Sci Technol 37:1–9. http://dx.doi.org/10.1016/j.ast.2014.04.012
14. Murrieta-Mendoza A, Botez RM, Ford S (2016) New method to compute the missed approach fuel consumption and its emissions. Aeronaut J 120(1228):910–929. http://dx.doi.org/10.1017/aer.2016.37
15. Lovegren JA (2011) Quantification of fuel burn reduction in cruise via speed and altitude optimization strategies. Massachusetts Institute of Technology Cambridge, MA 02139 USA. https://dspace.mit.edu/handle/1721.1/97360
16. Bonami P, Olivares A, Soler M, Staffetti E (2014) Multiphase mixed-integer optimal control approach to aircraft trajectory optimization. J Guid Control Dyn 36(5):1267–1277. https://doi.org/10.2514/1.60492
17. Cobano JA, Alejo D, Heredia G, Ollero A (2013) 4D trajectory planning in ATM with an anytime stochastic approach. In: Proceedings of the 3rd international conference on application and theory of automation in command and control systems, Naples, Italy. 978-1-4503-2249-2, http://dx.doi.org/10.1145/2494493.2494494
18. Korn B, Helmke H, Kuenz A (2006) 4D trajectory management in the extended TMA: coupling AMAN And 4D FMS for optimized approach trajectoreis. In: 25th congress of international council of the aeronautical sciences, Hamburg, Germany. 0-9533991-7-6

19. Soler-Arnedo M, Hansen M, Zou B, Contrail sensitive 4D trajectory planning with flight level allocation using multiphase mixed-integer optimal control. In: AIAA guidance, navigation, and control (GNC) conference, 2013. Los Angeles, CA, http://dx.doi.org/10.2514/6.2013-5179
20. Tsotskas C, Kipouros T, Savill M, Biobjective optimisation of preliminary aircraft trajectories. In: Evolutionary multi-criterion optimization vol 7811. Springer Berlin Heidelberg, 2013 pp. 741–755
21. Valenzuela A, Rivas D (2014) Optimization of aircraft cruise procedures using discrete trajectory patterns. J Aircr 51:1632–1640. https://doi.org/10.2514/1.C032041
22. Rippel E, Bar-Gill A, Shimkin N (2005) Fast graph-search algorithms for general-aviation flight trajectory generation. J Guid Control Dyn 28:801–811. https://doi.org/10.2514/1.7370
23. Dicheva S, Bestaoui Y (2014) Three-Dimensional A* dynamic mission planning for an airborne launch vehicle. J Aerosp Inf Syst 11:98–106. http://dx.doi.org/10.2514/1.I010070
24. Felix Patron RS, Botez RM, Labour D (2013) New altitude optimisation algorithm for the flight management system CMA-9000 improvement on the A310 and L-1011 aircraft. Aeronaut J 117(117):787–805. https://doi.org/10.1017/S0001924000008459
25. Gagné J, Murrieta-Mendoza A, Botez R, Labour D (2013) New method for aircraft fuel saving using flight management system and its validation on the L-1011 aircraft. In: 2013 Aviation Technology, Integration, and Operations Conference, Los Angeles, CA. http://dx.doi.org/10.2514/6.2013-4290
26. Murrieta-Mendoza A, Botez RM (2014) Vertical navigation trajectory optimization algorithm for a commercial aircraft. AIAA/3AF Aircraft Noise and Emissions Reduction Symposium, June 16–20 http://dx.doi.org/10.2514/6.2014-3019
27. Dancila B, Botez RM, Labour D (2013) Fuel burn prediction algorithm for cruise, constant speed and level flight segments. Aeronaut J 117(1191):491–504
28. Murrieta-Mendoza A, Beuze B, Ternisien L, Botez R (2015) Branch & bound-based algorithm for aircraft VNAV profile reference trajectory optimization. In: 15th AIAA aviation technology, integration, and operations conference, aviation forum, Dallas, TX, USA. 978-1-5108-0818-8. http://dx.doi.org/10.2514/6.2015-2280
29. Murrieta-Mendoza A, Botez RM (2015) Aircraft vertical route optimization deterministic algorithm for a flight management system. Presented at the SAE AeroTech Congress & Exhibition, Seattle, USA. http://dx.doi.org/10.4271/2015-01-2541
30. Ng HK, Sridhar B, Grabbe S (2014) Optimizing aircraft trajectories with multiple cruise altitudes in the presence of winds. J Aerosp Inf Syst 11:35–47. https://doi.org/10.2514/1.I010084
31. Hagelauer P, Mora-Camino F (1998) A soft dynamic programming approach for on-line aircraft 4D-trajectory optimization. Eur J Oper Res 107:87–95. https://doi.org/10.1016/S0377-2217(97)00221-X
32. Miyazawa Y, Wickramasinghe NK, Harada A, Miyamoto Y (2013) Dynamic programming application to airliner four dimensional optimal flight trajectory. In: AIAA guidance, navigation, and control (GNC) conference, Boston, USA (2013). http://dx.doi.org/10.2514/6.2013-4969
33. Bronsvoort J (2014) Contributions to trajectory prediction theory and its application to arrival management for air traffic control. Ph. D. Escuela Tecnica Superior de Ingenieros de Telecomunicacion, Universidad Politecnica de Madrid
34. Murrieta-Mendoza A, Demange S, George F, Botez RM (2015) Performance database creation using a flight D simulator for cessna citation X aircraft in cruise regime. In: The 34th IASTED international conference on modelling, identification, and control (MIC2015), Innsbruck, Austria. http://dx.doi.org/10.2316/P.2015.826-028
35. Ghazi G, Botez RM, Tudor M (2015) Performance database creation for cessna citation X aircraft in climb regime using an aero-propulsive model developed from flight tests. Presented at the AHS sustainability, Montreal, Canada
36. Kimlberlin R (2003) Flight testing of fixed-wing aircraft, 1st edn. AIAA, Reston, Virginia

37. Ghazi G, Tudor M, Botez R (2015) Identification of a cessna citation X aero-propulsive model in climb regime from flight tests. Presented at the international conference on air transport INAIR, Amsterdam, the Netherlands
38. Murrieta-Mendoza A, Botez RM (2014) Method to calculate aircraft VNAV trajectory cost using a performance database. In: International mechanical engineering congress & exposition, montreal, Canada. 978-0-7918-4642-1, http://dx.doi.org/10.1115/IMECE2014-37568
39. Murrieta-Mendoza A, Botez RM (2015) Methodology for vertical-navigation flight-trajectory cost calculation using a performance database. J Aerosp Inf Syst 12:519–532. https://doi.org/10.2514/1.I010347
40. Félix Patrón RS, Schindler M, Botez RM (2015) Aircraft trajectories optimization by genetic algorithms to reduce flight cost using a dynamic weather model. In: 15th AIAA aviation technology, integration, and operations conference, Dallas, TX
41. Félix-Patrón RS, Kessaci A, Botez R (2014) Horizontal flight trajectories optimisation for commercial aircraft through a flight management system. Aeronaut J 118(1210):1499–1518. https://doi.org/10.1017/S0001924000010162
42. Sridhar B, Ng H, Chen N (2013) Aircraft trajectory optimization and contrails avoidance in the presence of winds. J Guid Control Dyn 34:1577–1584. https://doi.org/10.2514/1.53378
43. Murrieta-Mendoza A (2013) Vertical and lateral flight optimization algorithm and missed approach cost calculation. Master, École de Technologie Supérieure, Montreal
44. Félix-Patrón RS, Botez RM (2015) Flight trajectory optimization through genetic algorithms for lateral and vertical integrated navigation. J Aerosp Inf Syst 12:533–544. http://dx.doi.org/10.2514/1.I010348
45. Félix-Patrón RS, Berrou Y, Botez RM (2014) New methods of optimization of the flight profiles for performance database-modeled aircraft. Proceedings of the institution of mechanical engineers, Part G: J Aerosp Eng 229(10):1853–1867. http://dx.doi.org/10.1177/0954410014561772
46. Sidibe S, Botez RM (2013) Trajectory optimization of FMS-CMA 9000 by dynamic programming. Presented at the ASI AÉRO 2013 conference, 60th Aeronautics Conference and AGM, Toronto, Canada
47. Felix-Patron RS, Botez RM, Labour D (2012) Vertical profile optimization for the flight management system CMA-9000 using the golden section search method. Presented at the IECON 2012—38th Annual Conference on IEEE, Montreal, QC. https://doi.org/10.1109/iecon.2012.6389517
48. Murrieta-Mendoza A, Félix-Patrón RS, Botez RM (2015) Flight Altitude Optimization Using Genetic Algorithms Considering Climb and Descent Costs in Cruise with Flight Plan Information," presented at the SAE AeroTech Congress & Exhibition, Seattle, USA, 2015. http://dx.doi.org/10.4271/2015-01-2542
49. Murrieta-Mendoza A, Botez RM (2014) Lateral navigation optimization considering winds and temperatures for fixed altitude cruise using The Dijkstra's algorithm. In: International mechanical engineering congress & exposition, Montreal, Canada. http://dx.doi.org/10.1115/IMECE2014-37570
50. Murrieta-Mendoza A, Romain C, Botez RM (2016) Commercial aircraft lateral flight reference trajectory optimization. In: 20th IFAC symposium on automatic control in aerospace ACA 2016, 19(17):1–6. 2405–8963, Sherbrooke, QC, Canada. http://dx.doi.org/10.1016/j.ifacol.2016.09.001
51. Murrieta-Mendoza A, Hamy A, Botez RM (2016) Lateral reference trajectory algorithm using ant colony optimization. In: 16th AIAA aviation technology, integration, and operations conference, Washington, D.C. http://dx.doi.org/10.2514/6.2016-4209
52. Murrieta-Mendoza A, Ruiz S, Kessaci S, Botez RM (2017) Vertical reference flight trajectory optimization with the particle swarm optimisation. In The 36th IASTED international conference on modelling, identification and control (MIC 2017), Innsbruck, Austria

Uncertainty Quantification and Robust Optimization in Engineering

D. Kumar, S. B. Alam, Dean Vučinić and C. Lacor

Abstract The application and use of engineering components such as engines, wings, or complete airplanes are all subject to uncertainties, either of operational nature (variations in speed, angle of attack, pressure, etc) or of geometrical nature (manufacturing tolerances or uncertainties due to wearing). These uncertainties can have an important effect on the performance (output) of these components. The effect of these uncertain parameters should be quantified and included in the final solution to assure and improve the quality of the results. Polynomial chaos is a recent methodology to account for uncertainties that can be described by a distribution function. The method allows to obtain the distribution of the output for given input distributions. Over the last decade, with increasing computational resources and hardware power, design optimization is receiving more and more interest in aeronautical applications. Due to the uncertainties in a design process, the objective is also uncertain. Robust optimization is an extension of conventional optimization where uncertainties are also included in the design procedure. Using polynomial chaos expansion, the uncertain objective can be characterized by its mean and its variance. Therefore, it becomes a multi-objective problem and gradient based optimization requires the gradient of both quantities. These gradients can be obtained from the polynomial chaos expansion of the gradient of the objective. In this chapter, first, a brief introduction to polynomial chaos approach for uncertainty quantification is provided. Further its formulation with adjoint methods is described for gradient based robust optimization. The approach is applied to the optimal shape design of a transonic airfoil under uncertainties.

D. Kumar (✉)
Department of Physics and Astronomy, Uppsala University,
P.O. Box 516, 751 20 Uppsala, Sweden
e-mail: dinesh.kumar@physics.uu.se

D. Kumar · D. Vučinić · C. Lacor
Department of Mechanical Engineering,
Vrije Universiteit Brussel, Pleinlaan 2, 1050 Brussels, Belgium

S. B. Alam
Department of Engineering, University of Cambridge, Cambridge, UK

D. Vucinic et al. (eds.), *Advances in Visualization and Optimization Techniques for Multidisciplinary Research*, Lecture Notes in Mechanical Engineering, https://doi.org/10.1007/978-981-13-9806-3_3

Keywords CFD · Uncertainties · Non-intrusive · Polynomial chaos · Robust optimization

1 Introduction

Nowadays, mathematical modeling and computer simulations are used in almost every field of science to understand real world phenomena. Usually, the mathematical model of a real world problem is a set of equations based on some hypothesis. These governing equations are then solved by applying proper numerical schemes. In the last few decades, a major revolution is seen with the emergence of computer simulation. Due to the development of powerful computer hardware and efficient numerical algorithms, it became affordable to analyze realistic problems with complex mathematics more accurately. The fidelity of the mathematical models is significantly improved. Mathematical modeling is nowadays an unavoidable tool for scientists and engineers, and is fundamental in the designing of products such as airplanes, trains, cars, turbines, nuclear power plants, bridges, etc.

However, in spite of increasing computational power and accuracy of the mathematical results, models are still not able to predict the solution of a real world problem accurately. The results from simulations usually differ from experiments. The possible sources of these discrepancies may be due to modeling errors, numerical errors and/or uncertainties due to inadequate information of the parameters and geometric domain considered in the computational modeling. By definition, mathematical models are idealizations of reality, i.e. they always rely on simplified assumptions and neglect some aspects of the phenomena of interest. Owing to the numerical algorithms developed in the art of computing, it can be assumed that numerical schemes are sufficiently accurate to predict the underlying phenomena. Numerical errors can always be avoided or reduced by mesh refinement, thanks to developments in the field of computer hardware. In computational modeling, there are always numerous input parameters such as material properties, geometry, boundary conditions, initial conditions, etc. whose values are often inaccurate or uncertain. The discretization errors are usually small in comparison to uncertainties.

Due to the lack of information of several sources, i.e., uncertainties present in the mathematical modeling, the computational output is also uncertain. The computational domain (including geometrical uncertainties due to manufacturing variability and imprecise geometrical definitions) and model parameters are commonly encountered as uncertain by engineers when designing new products. Geometrical uncertainties usually arise from production tolerances, loading deformation or wear of the product. The effect of these uncertain parameters should be quantified and included in the final solution to assure and improve the quality of the results. In a model, some sources of uncertainties may have a minor influence, and some sources of uncertainties may have a major influence. In chaotic systems, an extreme case is known as the butterfly effect. According to this hypothesis, the flap of a butterfly's wings would be enough to alter the course of the weather forever.

For example, in weather forecasting, meteorologists obtain initial conditions from weather stations, and feed it to a mathematical model. Measurement errors introduce uncertainties in the initial conditions. Other parameters such as wind speed, cloud formation, precipitation, etc. also introduce uncertainties in the computer simulations. The effect of uncertainties in a real life problem and how they effect our daily life is visualized here with a situation that you may encounter regularly, when you use a weather forecast website to obtain information about weather for a nice vacation. On 25th August 2016 you extracted information about weather for 8th September 2016 in Brussels, Belgium using a popular website timeanddate.com. It predicts a minimum temperature of 10 degrees and a maximum temperature of 19 degrees (see Fig. 1). However, when the weather for the same day is checked after a few days i.e. on 31st August 2016, the temperature predicted is completely different from the previous prediction. The minimum temperature is now 17 degrees which is almost equal to the maximum temperature predicted a few days before i.e. on 25th September (see Fig. 2). The day seems to be relatively warm with a very clear sky. Here, the simulation results are not reliable, and completely depend on the initial conditions provided to the model. To obtain simulation results more clearly, i.e. with a variation in the output solution, one needs to provide the information of input uncertainties to the mathematical model.

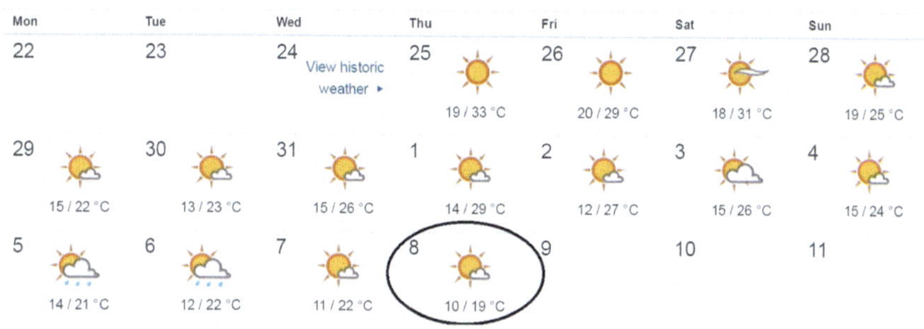

Fig. 1 Weather forecast on 25th September 2016 for 8th October 2016

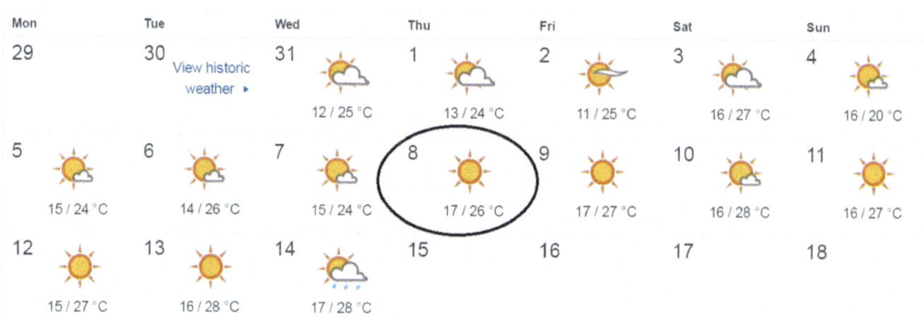

Fig. 2 Weather forecast on 31st September 2016 for 8th October 2016

The presence of uncertainties brings several difficulties to the optimization process. For a given criterion, selecting a best element from a set of available alternatives is optimization. Conventional optimization deals with the problem of finding numerically the minimum (or maximum) of an objective function. In aeronautics, an example of optimization can be considered as finding the optimal shape of an aircraft wing that minimizes the drag while maintaining the prescribed lift and geometry constraints. In case of a multi-objective optimization, one can either combine all objective functions into one single objective function by a weighted sum, or, alternatively, keep the different objectives which usually leads to a so-called Pareto front. These are different designs, where, if one compares design x on the front with design y on the front, x is not better than y in all objectives. With the increase in computational power and resources, optimization under uncertainties has become a very active area of research. For optimization, sampling based random search algorithms (gradient free, such as Genetic algorithm) [1, 2] and gradient based methods [3, 4] are common methodologies. Evolutionary algorithms are less dependent to the starting point, as the whole design space is explored. Therefore there is less risk of ending up in a local minimum. Gradient based methods show fast convergence to the optimum but strongly dependent on the starting point and hence higher risk of ending up in a local minimum. The derivatives needed in gradient based optimization can be computed efficiently by using the adjoint approach. Adjoint methods are nowadays widely used for optimization with a large number of design variables.

Introducing uncertainties in the design process, the objective becomes stochastic (or non-deterministic) and can be characterized by its mean, variance and its higher order moments. **Robust optimization** is an extension of conventional optimization where uncertainties are also included in the design procedure. Several methods for robust optimization have been proposed independently and applied in different scientific fields. Wang et al. [2], applied non-intrusive polynomial chaos with multi-objective genetic algorithm for the robust optimization of rotor blades. Palar et al. [5], recently applied robust optimization in the aerodynamic shape optimization of a transonic airfoil using non-intrusive polynomial chaos with gradient free evolutionary algorithm. Sriram et al., in 2011 [3] proposed a robust optimization framework to optimize the mean value of a given objective function using polynomial chaos and adjoint based gradient methods. Schillings et al., in 2014 [6] proposed a polynomial chaos and gradient based robust optimization framework to optimize the objective function defined as a combination of the mean and the variance. They applied it to aerodynamic shape optimization under uncertainties in 2D Euler flows. Recently, Maruyama et al. [7], proposed a robust optimization framework using surrogate models and applied it for robust design of a 2D airfoil under uncertainties.

In this chapter, first, some popular uncertainty quantification methods based on polynomial chaos approach are described. Further, for robust optimization, a non-intrusive polynomial chaos method is combined with adjoint methods. Having obtained the polynomial expansions of the objective functions, the gradient of the mean objective is combined with the gradient of its variance using weights. Gradient based optimization of the mean objective and of the variance of the objective therefore requires the gradient of both quantities. By changing the weights the Pareto front (if any, i.e. if the 2 objectives are conflicting) or at least part of it can be recovered. By varying the weights, a robust design is obtained leading to multiple designs that favor either the mean objective or the variance of the objective. The approach is applied to the optimal shape design of the transonic RAE2822 airfoil under operational uncertainties.

2 Uncertainty Quantification

Over the last decades, uncertainty quantification became an essential tool for the robustness of aeronautical applications. New methodologies are therefore required to quantify the effect of these uncertainties efficiently. Efficient uncertainty quantification of a model due to uncertain input parameters is the first step for the robust design of a component. The second one is to consider these uncertainties in the optimization process to have robust design of the component.

Uncertainties are associated to the selection of the models and to the input parameters (operational and geometrical) of the physical model considered for analysis. For example, in computational fluid modeling, the physical properties of the fluid or the boundary conditions may not be known exactly. Geometrical uncertainties mainly arise due to manufacturing tolerances and imprecise geometrical definitions in the model. Wearing and possible small deformations during operation can also be counted as geometrical uncertainties. However, large geometrical changes, e.g. aeroelastic wing deformations, can not be considered as random behavior. Aeroelastic effects can be treated with unsteady fluid-structure interaction models, where, if needed, at each time step UQ methods can be used to model wing manufacturing uncertainties.

The uncertainties associated to the system of consideration can have different origins and can be classified as **epistemic** and **aleatory** [8]. The epistemic uncertainties are related to the properties of the model considered. While, the aleatory uncertainties are related to the properties of the system being considered for the analysis.

Epistemic uncertainties are due to any lack of information in any phase of the modeling process. These uncertainties may arise from mathematical assumptions or simplifications used in the derivation process of the governing equations. The parameters with epistemic uncertainty are not associated to any probability density functions and are not well characterized by stochastic approaches. Typically they are specified as an interval. In uncertainty quantification of epistemic uncertainty, the output result is also specified as an interval. Epistemic uncertainties can be quantified using sampling based methods (suh as random sampling, Latin hypercube sampling [8, 9]), interval analysis, fuzzy set theory [10, 11], etc. Turbulence modeling assumption is a typical example of the source of epistemic uncertainty.

Aleatory uncertainties are associated with the physical system or the environment under consideration. Material properties, operating conditions, geometrical variability, initial conditions, boundary conditions etc. are some of the examples of aleatory uncertainties. Uncertain parameters of aleatory uncertainty are always associated to a particular probability distribution function. Using appropriate uncertainty quantification (UQ) methods, the probability distribution function of the output quantity can be determined. In UQ methods, the quantities of interest are probability distribution function (PDF), cumulative distribution function (CDF), the mean, variance and other higher order moments of the solution.

Computer simulations without the inclusion of uncertainties in the model are called deterministic simulations, and the output solution is called deterministic solution. However, when uncertainties are considered in the model, the simulations are called non-deterministic simulations, and the results are called non-deterministic or stochastic. In deterministic simulations, one solves a system of governing equations using proper numerical schemes. For example, in aeronautics and CFD the Navier-Stokes equations are solved. In Fig. 3, a typical computational framework for deterministic simulations is shown. The mathematical model consists of numerical parameters and a suitable discretization scheme to solve the governing system of equations. The model parameters are provided to the model to compute the output solution.

In non-deterministic simulations, uncertainties associated with model parameters and geometry are propagated to the output of the simulation. Introducing the probabilistic nature of the input uncertainties in the model, the deterministic process transforms into non-deterministic process governed by stochastic partial differential equations (SPDE).

A mathematical model can be written as:

$$K(u, a) = 0 \tag{1}$$

where u is the output solution and a is a set of input parameters $\{a_1, a_2 \dots a_n\}$. If a is uncertain and parameterized as a vector of independent random variables $\boldsymbol{\xi} = \{\xi_1, \xi_2 \dots \xi_n\}$, the output solution u also becomes stochastic i.e. $u(\boldsymbol{\xi})$. The stochastic output solution $u(\boldsymbol{\xi})$ can be computed using intrusive or non-intrusive stochastic methods by specifying the probability distribution of $\boldsymbol{\xi}$.

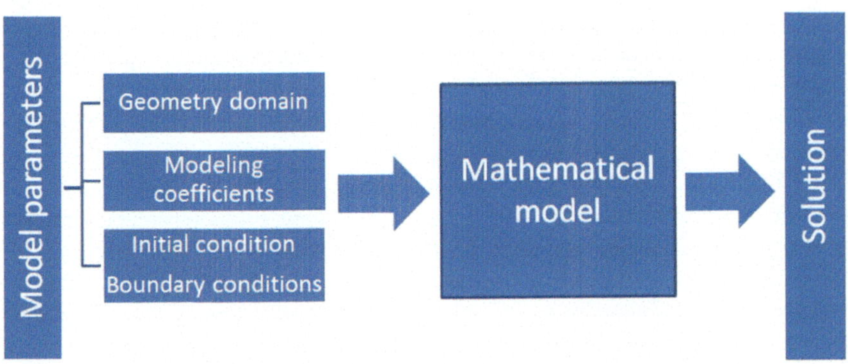

Fig. 3 Computational framework: deterministic

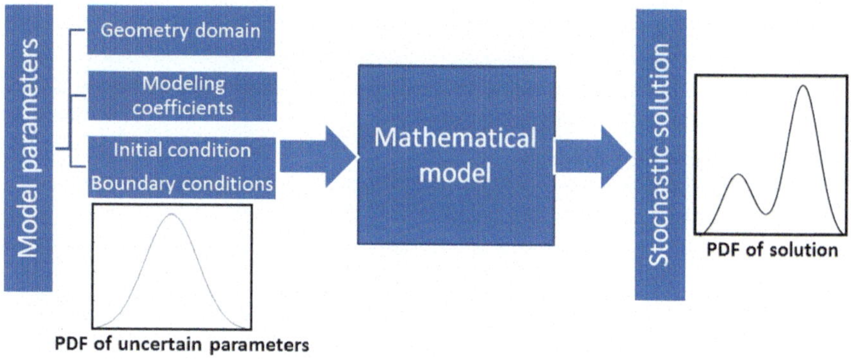

Fig. 4 Computational framework: non-deterministic

A deterministic approach leads to a single value of the output solution. When uncertainties in the input parameters are introduced by their probability density function (PDF), the results from non-deterministic simulations are interpreted in terms of their statistical moments (i.e. mean, variance, skewness, flatness) and probability distribution functions (PDF). In Fig. 4, a typical computational framework for non-deterministic simulations is shown. Using an appropriate uncertainty quantification method, a confidence level of the estimated quantities (or the solution of a model) can be calculated and the reliability of the results can be characterized in a stochastic or probabilistic sense.

A unique set of methods are required to quantify each form of uncertainty. Epistemic uncertainties are considered as reducible uncertainties. They can be reduced by research advancement. However, aleatory uncertainties are unavoidable. The work in this chapter is restricted to stochastic methods to quantify aleatory uncertainties only.

3 Methods for Uncertainty Quantification

For the reproducibility of experiments, typically the results from laboratory experiments are reported with error bars due to the uncertainties associated to the measurements. The main objective of the stochastic methods is to build a framework to predict similar error bars in the computational results due to uncertainties in a mathematical model.

Over the past decade, UQ methods have received a lot of interest and attention. Depending on the properties of the input uncertainties, UQ methods can be divided into probabilistic (for aleatory) and non-probabilistic (for epistemic) methods. The uncertainty quantification methods described in this article can be employed if the probability distribution functions of the described uncertainties is known or defined.

For uncertainty quantification, a range of **intrusive** and **non-intrusive** methods have been developed. The implementation of intrusive approaches requires considerable modification of the deterministic codes. The non-intrusive methods are sampling based methods where the original deterministic code does not need any modifications and it can be used as a black box.

The objective of all uncertainty quantification methods is to compute the statistical moments and/or PDF of the output quantity $u(\boldsymbol{\xi})$ due to the uncertainties in the input parameters $\boldsymbol{\xi}$. The mean value $< u >$ and the variance σ_u^2 of the output quantity u can be written as:

$$E[u] =< u >= \int_{-\infty}^{\infty} u f_u(u) du$$

$$\begin{aligned} E[(u- < u >)^2] = \sigma_u^2 &= \int_{-\infty}^{\infty} (u- < u >)^2 f_u(u) du \\ &= E[u^2] - (E[u])^2 \end{aligned} \tag{2}$$

where $f_u(u)$ is the probability density function of the output u. The above integrals for the statistical moments can be computed numerically by using UQ methods.

Usually one is interested mostly in the first two moments as they give the mean value and an idea of the spreading of the value. Nevertheless sometimes higher order moments are also needed such as the skewness (3rd moment), which is a measure of the lack of symmetry of the pdf, and the kurtosis (4th moment), which is a measure of the tails of the pdf i.e. the outliers.

A variety of different UQ methods such as Monte Carlo methods [12], polynomial chaos methods [13–17], perturbation method [18–21] and sensitivity methods [22–25] have been proposed for uncertainty quantification. More recently, the polynomial chaos based methods have been successfully applied by various researchers for a wide range of problems [26–30]. Here, some popular UQ methods using polynomial chaos approach are introduced in the following section.

4 Polynomial Chaos for UQ

The polynomial chaos method (PCM), is a very recent method for uncertainty quantification (UQ), which offers a large potential for non-deterministic simulations. In PCM, stochastic variables with different distributions can be handled both with intrusive and non-ntrusive approaches. The properties of the input random variables and the output stochastic solution can be described in terms of its statistical moments and probability density functions. The polynomial chaos methodology was originally formulated by Wiener [31] in an intrusive framework and was further expanded by Xiu and Karniadakis [13]. Several non-intrusive polynomial chaos methods (NIPCM) were also developed during recent years [15, 32, 33]. NIPCM are sampling based methods and can be implemented easily into any in-house and commercial CFD code.

In this section, the basic aspects of orthogonal polynomials are presented and several uncertainty qualification methods using polynomial chaos theory are discussed. The polynomial chaos expansion is the spectral representation of a random variable, which decomposes the random variable into separable deterministic and stochastic parts. Let u be the response of a stochastic system with n random variables. In polynomial chaos expansion, the uncertain variable $u(x; \boldsymbol{\xi})$ is decomposed in a basis of complete orthogonal polynomials as:

$$u(x; \boldsymbol{\xi}) = \sum_{i=0}^{P} u_i(x)\psi_i(\boldsymbol{\xi}) \tag{3}$$

where u_i are deterministic PC coefficients. ψ_i are the multi-dimensional orthogonal polynomials which are function of the random variables $\boldsymbol{\xi} = \{\xi_1, \xi_2 \ldots, \xi_n\}$ and are used to span the n-dimensional stochastic space. Here ξ_j is a random variable with a specific probability density function. In polynomial chaos, it is assumed that $\boldsymbol{\xi}$ is a set of n independent random variables. For independent random variables, the PDF of $\boldsymbol{\xi}$ is the joint PDF of all ξ's, i.e. the product of the the PDF's of each uncertain variable.

If p is the highest order of polynomial considered to approximate the stochastic response and n is the number of random variables, the total number of terms $P + 1$ in Eq. 3 depends on the order of polynomial (p) and on the number of random dimensions (n) as:

$$P + 1 = \frac{(n + p)!}{n!p!} \tag{4}$$

The polynomials ψ_i are orthogonal with respect to the probability distribution function $\boldsymbol{W}_{\xi}$ of the random variable $\boldsymbol{\xi}$. Orthogonality means that:

$$< \psi_i\psi_j >=< \psi_i^2 > \delta_{ij} \tag{5}$$

where δ_{ij} is the Kronecker delta and $< \psi_i\psi_j >$ is the inner product defined as:

$$< \psi_i\psi_j >= \int_{\xi} \psi_i(\boldsymbol{\xi})\psi_j(\boldsymbol{\xi})\boldsymbol{W}_{\xi}(\boldsymbol{\xi})d(\boldsymbol{\xi}) \tag{6}$$

The expectation or the mean of u can be written as:

$$\begin{aligned} E[u] &= \int_{\xi} u(x; \boldsymbol{\xi})\boldsymbol{W}_{\xi}(\boldsymbol{\xi})d(\boldsymbol{\xi}) \\ &= u_0 \end{aligned} \tag{7}$$

Due to the orthogonality of the polynomials, the coefficient u_0 is the mean of the uncertain parameter $u(x; \boldsymbol{\xi})$. Further, the variance of the uncertain parameter can be written as:

$$\begin{aligned} \sigma_u^2 &= E[(u - E[u])^2] \\ &= \int_{\boldsymbol{\xi}} \{\sum_{i=1}^{P} u_i(x)\psi_i(\boldsymbol{\xi})\}^2 W_{\boldsymbol{\xi}}(\boldsymbol{\xi}) d(\boldsymbol{\xi}) \\ &= \sum_{i=1}^{P} u_i^2(x) < \psi_i^2 > \end{aligned} \tag{8}$$

Similarly, the higher order moments of $u(x; \boldsymbol{\xi})$ can also be derived in terms of the PC coefficients. However, the expressions are not as simple as they are for the mean and the variance.

To study the influence of each random variable of the model onto the final solution we need to construct multi-dimensional polynomials starting from the one dimensional polynomials. The multi-dimensional polynomial chaos of order p in terms of 1D polynomials can be written as explained below.

If $\{\Psi_0, \Psi_1, \Psi_2, \Psi_3\}$ is the set of 1D orthogonal polynomials of PC order 3, then a 2D stochastic quantity $u(\xi_1, \xi_2)$ can be decomposed in terms of PC expansion of order 3 as:

$$\begin{aligned} u(\xi_1, \xi_2) = &\ u_{00}\Psi_0 \\ &+ u_{10}\Psi_1(\xi_1) + u_{01}\Psi_1(\xi_2) \\ &+ u_{20}\Psi_2(\xi_1) + u_{11}\Psi_1(\xi_1)\Psi_1(\xi_2) + u_{02}\Psi_2(\xi_2) \\ &+ u_{30}\Psi_3(\xi_1) + u_{21}\Psi_2(\xi_1)\Psi_1(\xi_2) + u_{12}\Psi_1(\xi_1)\Psi_2(\xi_2) + u_{03}\Psi_3(\xi_2) \end{aligned} \tag{9}$$

or

$$u(\xi_1, \xi_2) = u_0\psi_0 + u_1\psi_1(\xi_1, \xi_2) + u_2\psi_2(\xi_1, \xi_2) + \cdots + u_9\psi_9(\xi_1, \xi_2) \tag{10}$$

where $\{\psi_0, \psi_1, \psi_2, \ldots, \psi_9\}$ is the set of two-dimensional polynomials of PC order 3.

For example, the set of 2D Hermite basis of PC order 3 is shown in Table 1. The shape of the first six 2D Hermite polynomials is also shown in Fig. 5.

For uncertainty quantification, the generalized polynomials from the Wiener-Askey scheme are considered. Hermite, Legendre or Jacobi polynomials can be used to model the uncertainties described by normal, uniform or beta probability distributions, respectively. The support of these polynomials correspond to their probability distributions. For example, the support range of the Hermite polynomials $(-\infty, \infty)$ corresponds to the normal distribution. In Table 2, some orthogonal polynomials of the Wiener-Askey scheme and corresponding probability distributions are pre-

Table 1 2D Legendre polynomials of PC order 0, 1, 2, 3

j	Order of PC	jth polynomial
0	p = 0	1
1	p = 1	ξ_1
2		ξ_2
3	p = 2	$\frac{1}{2}(3\xi_1^2 - 1)$
4		$\xi_1\xi_2$
5		$\frac{1}{2}(3\xi_2^2 - 1)$
6	p = 3	$\frac{1}{2}(5\xi_1^3 - 3\xi_1)$
7		$\frac{1}{2}\xi_2(3\xi_1^2 - 1)$
8		$\frac{1}{2}\xi_1(3\xi_2^2 - 1)$
9		$\frac{1}{2}(5\xi_2^3 - 3\xi_2)$

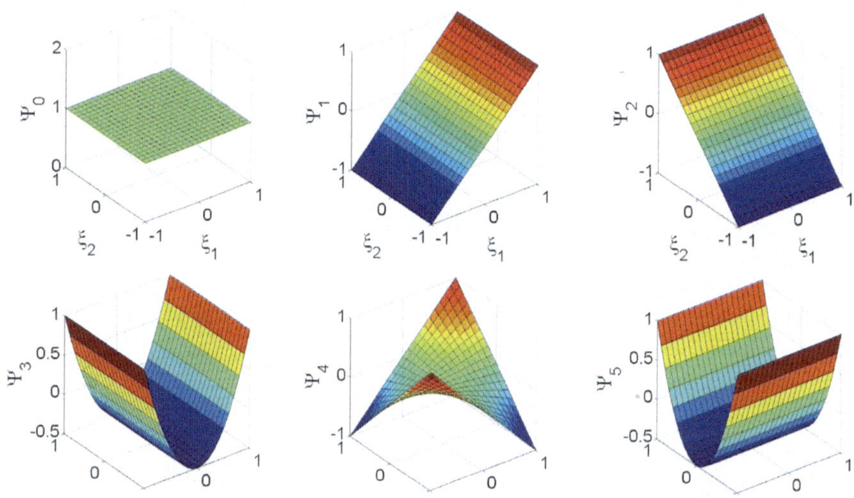

Fig. 5 The first six 2D Legendre polynomials

Table 2 Probability density functions and corresponding orthogonal polynomials from Wiener-Askey scheme

Distribution	Polynomial	Support
Normal	Hermite	$(-\infty, \infty)$
Uniform	Legendre	$[-1, 1]$
Beta	Jacobi	$[-1, 1]$
Gamma	Laguerre	$[0, \infty)$

sented. These polynomials are optimal for their corresponding distributions, i.e. for uniformly distributed random variables, the Legendre polynomials are optimal for efficient uncertainty quantification.

If $\boldsymbol{\xi} = (\xi_1, \xi_2 \ldots, \xi_n)$ is a set of n-dimensional independent random variables, the probability density function (PDF) of $\boldsymbol{\xi}$ can be written as:

$$\boldsymbol{W}(\boldsymbol{\xi}) = \prod_{i=1}^{n} W_i(\xi_i) \tag{11}$$

where $W_i(\xi_i)$ is the individual PDF of the random variable ξ_i.

The idea of a PC based UQ method is to compute the PC coefficients of the quantity of interest and estimate the statistics of the quantity (or statistical moments) in terms of its PC coefficients. The PC expansion coefficients can be computed using sampling based non-intrusive methods such as regression, projection or collocation methods.

4.1 Stochastic Galerkin Method

The original polynomial chaos approach (Ghanem and Spanos [14]; Xiu and Karniadakis [13]) or Galerkin polynomial chaos is an intrusive technique for uncertainty quantification where the polynomial chaos expansion of the uncertain variable and the solution are substituted into the governing equations. If we consider the following stochastic model:

$$\mathbb{A}(u(x,t,\boldsymbol{\xi}); b(x,t,\boldsymbol{\xi}), x, t, \boldsymbol{\xi}) = \mathbb{B}(x,t) \tag{12}$$

where $\mathbb{A}$ is a differential operator, $\mathbb{B}(x,t)$ is a source term, $b(x,t,\boldsymbol{\xi})$ is an uncertain input parameter and $u(x,t,\boldsymbol{\xi})$ is the stochastic solution of the governing equation.

Introducing the PC expansion of the uncertain parameter and the response into the governing equation:

$$\mathbb{A}(\sum_{i=0}^{P} u_i(x,t)\psi_i(\boldsymbol{\xi}); \sum_{i=0}^{P} b_i(x,t)\psi_i(\boldsymbol{\xi}), x, t, \boldsymbol{\xi}) = \mathbb{B}(x,t) \tag{13}$$

A Galerkin projection (or inner product) with ψ_k provides a system of P+1 deterministic equations as:

$$\left\langle \mathbb{A}\left(\sum_{i=0}^{P} u_i(x,t)\psi_i(\boldsymbol{\xi}); \sum_{i=0}^{P} b_i(x,t)\psi_i(\boldsymbol{\xi}), x, t, \boldsymbol{\xi}\right).\psi_k(\boldsymbol{\xi})\right\rangle = \langle \mathbb{B}(x,t).\psi_k(\boldsymbol{\xi})\rangle, \tag{14}$$

where $k = 0, 1, 2 \ldots P$ and $< a(\boldsymbol{\xi}).b(\boldsymbol{\xi}) >= \int_{\boldsymbol{\xi}} a(\boldsymbol{\xi})b(\boldsymbol{\xi})W_{\boldsymbol{\xi}}(\boldsymbol{\xi})d(\boldsymbol{\xi})$.

The above set of P + 1 deterministic equations is solved simultaneously for the PC coefficients $u_i(x, t)$ by using a conventional numerical discretization method. Once the PC coefficients are determined, the statistical quantities of interest can be computed as explained in the previous sections.

4.1.1 Example: Stochastic ODE

The ODE is given by

$$\frac{du(x)}{dx} = -ku \;\; ; \;\; u(0) = u^0 \;\; ; \;\; x \in [0, 1] \tag{15}$$

The exact solution of the above ODE is:

$$u(x) = u^0 e^{-kx} \tag{16}$$

If the coefficient k in the above ODE is a Gaussian distributed uncertain parameter with a mean k_0 and a standard deviation k_1, the above ODE becomes a stochastic ODE. Using Hermite polynomials, the PC expansion of the uncertain parameter k and the output response u can be written as:

$$k(\xi) = \sum_i k_i \psi_i(\xi(\theta)) = k_0 + k_1 \xi \tag{17}$$

$$u(x, \xi) = \sum_i u_i(x) \psi_i(\xi(\theta)) \tag{18}$$

Substituting the PC expansion of k and u in the original ODE:

$$\sum_i \frac{du_i}{dx} \psi_i = -\sum_j \sum_l u_j k_l \psi_j \psi_l \tag{19}$$

Multiplying with ψ_k and performing the scalar products, one obtains

$$\frac{du_k}{dx} = -\frac{1}{< \psi_k^2 >} \sum_j \sum_l < \psi_j \psi_l \psi_k > u_j k_l \tag{20}$$

or

$$\frac{du_k}{dx} = -\frac{1}{< \psi_k^2 >} \sum_j \left(< \psi_j \psi_0 \psi_k > u_j k_0 + < \psi_j \psi_1 \psi_k > u_j k_1 \right) \tag{21}$$

If the mean and standard deviation of the input uncertainty are $k_0 = 0$ and $k_1 = 1$, by considering the 4th order Hermite chaos, we get a system of deterministic ODE as:

$$\frac{du_k}{dx} = -\frac{1}{< \psi_k^2 >} \sum_{j=0}^{4} < \psi_j \psi_1 \psi_k > u_j \tag{22}$$

where $k = 0, 1, 2, 3, 4$

For 1D-Hermite polynomials, the double products $< \psi_k^2 >$ and the triple products $< \psi_i \psi_j \psi_k >$ can be computed analytically as:

$$< \psi_k^2 >= k! \tag{23}$$

$$\begin{aligned} < \psi_i \psi_j \psi_k > &= 0 \text{ if } i + j + k \text{ is odd or } \max(i, j, k) > s \\ &= \frac{i!j!k!}{(s-i)!(s-j)!(s-k)!} \text{ otherwise} \end{aligned} \tag{24}$$

where $s = (i + j + k)/2$.

Substituting these values in Eq. 22, the system of deterministic ODE can be written as:

$$\begin{aligned} \frac{du_0}{dx} &= -u_1; \\ \frac{du_1}{dx} &= -u_0 - 2u_2; \\ \frac{du_2}{dx} &= -u_1 - 3u_3; \\ \frac{du_3}{dx} &= -u_2 - 4u_4; \\ \frac{du_4}{dx} &= -u_3 \end{aligned} \tag{25}$$

With the given initial condition at $x = 0 : u_0 = 1, u_1 = u_2 = u_3 = u_4 = 0$; the above system of ODE can be solved by using an appropriate numerical technique. Numerical results and detailed analysis for this test case can be found in [13].

4.2 Projection Method

The spectral projection method for uncertainty quantification is in the context of non-intrusive polynomial chaos. The orthogonal projection of the system output is computed by using numerical integration in terms of deterministic sampling in the stochastic space. If $u(x;\boldsymbol{\xi})$ is the stochastic solution and is decomposed in a basis of complete orthogonal polynomials as:

$$u(x;\boldsymbol{\xi}) = \sum_{i=0}^{P} u_i(x)\psi_i(\boldsymbol{\xi}) \tag{26}$$

An orthogonal projection or the inner product with respect to ψ_j is expressed as:

$$\begin{aligned}\int_{\xi} u(\boldsymbol{\xi})\psi_j(\boldsymbol{\xi})\boldsymbol{W}(\boldsymbol{\xi})d\boldsymbol{\xi} &= \int_{\xi}\sum_{i=0}^{P}\{u_i(x)\psi_i(\boldsymbol{\xi})\}\psi_j(\boldsymbol{\xi})\boldsymbol{W}_{\xi}(\boldsymbol{\xi})d\boldsymbol{\xi} \\ &= \sum_{i=0}^{P} u_i(x)\int_{\xi}\psi_i(\boldsymbol{\xi})\psi_j(\boldsymbol{\xi})\boldsymbol{W}_{\xi}(\boldsymbol{\xi})d\boldsymbol{\xi} \\ &= u_j < \psi_j^2 > \end{aligned} \tag{27}$$

In the above equation, $\boldsymbol{W}_{\xi}(\boldsymbol{\xi})$ is the joint PDF of the random variables. The above equation can be considered as a problem of computing the unknown coefficients u_j. The integral term $\int u(\boldsymbol{\xi})\psi_i(\boldsymbol{\xi})\boldsymbol{W}_{\xi}(\boldsymbol{\xi})d\boldsymbol{\xi}$ can be evaluated with standard numerical quadrature schemes using Monte Carlo simulation. Efficient sampling techniques such as Latin Hypercube, Sobol sampling, multi-level Monte Carlo, quasi Monte Carlo can be used for faster convergence.

The integral $\int u(\boldsymbol{\xi})\psi_i(\boldsymbol{\xi})\boldsymbol{W}_{\xi}(\boldsymbol{\xi})d\boldsymbol{\xi}$, can be approximated in terms of a weighted sum of the function values at the quadrature points as:

$$\int u(\boldsymbol{\xi})\psi_j(\boldsymbol{\xi})\boldsymbol{W}_{\xi}(\boldsymbol{\xi})d\boldsymbol{\xi} = \sum_{i=1}^{N} u(\boldsymbol{\xi}^i)\psi_j(\boldsymbol{\xi}^i)\boldsymbol{w}^i \tag{28}$$

where $\boldsymbol{\xi}^i$ is the quadrature point (or sampling point) and $\boldsymbol{w}^i$ is the corresponding weight. $u(\boldsymbol{\xi}^i)$ is the function evaluation and $\psi_j(\boldsymbol{\xi}^i)$ is the value of the polynomial ψ_j at the quadrature point $\boldsymbol{\xi}^i$.

Gauss quadrature based numerical integration is a very common method in solving numerical integration due to its accuracy. For each distribution i.e. $\boldsymbol{W}_{\xi}(\boldsymbol{\xi})$, there is an adapted Gaussian quadrature method. E.g., for uniform it is Gauss-Legendre as the Legendre polynomials are orthogonal with respect to the pdf of a uniform distribution. The sampling points correspond to the roots of the Legendre polynomial. Similarly, for Gauss distribution it is Gauss-Hermite with sampling points the roots of the Hermite polynomials.

By using $p + 1$ Gauss quadrature points, the Gauss quadrature method integrates a polynomial function of maximum order $2p + 1$ exactly in 1D. So if $u(\boldsymbol{\xi})$ is described by a polynomial of maximum order $p + 1$, the integral in 28, evaluated with $p + 1$ quadrature points will be exact. Therefore, for uncertainty quantification of a stochastic system of dimension n, the total number of samples $N = (p + 1)^n$ will be needed, where p is the order of PC approximation. For high PC order or for large number of uncertainties, Gauss quadrature becomes very expensive. Sparse grid sampling [6, 34] can be used as quadrature points, to avoid the full tensorial sampling.

Combining Eqs. 27 and 28, the PC coefficients can be evaluated as:

$$\begin{pmatrix} <\psi_0^2> u_0 \\ <\psi_1^2> u_1 \\ <\psi_2^2> u_2 \\ . \\ . \\ <\psi_P^2> u_P \end{pmatrix} = \begin{pmatrix} \psi_0(\boldsymbol{\xi}^1)w^1 & \psi_0(\boldsymbol{\xi}^2)w^2 \dots & \psi_0(\boldsymbol{\xi}^N)w^N \\ \psi_1(\boldsymbol{\xi}^1)w^1 & \psi_1(\boldsymbol{\xi}^2)w^2 \dots & \psi_1(\boldsymbol{\xi}^N)w^N \\ \psi_2(\boldsymbol{\xi}^1)w^1 & \psi_2(\boldsymbol{\xi}^2)w^2 \dots & \psi_2(\boldsymbol{\xi}^N)w^N \\ . & \dots & . \\ . & \dots & . \\ \psi_P(\boldsymbol{\xi}^1)w^1 & \psi_P(\boldsymbol{\xi}^2)w^2 \dots & \psi_P(\boldsymbol{\xi}^N)w^N \end{pmatrix} \begin{pmatrix} u(\boldsymbol{\xi}^1) \\ u(\boldsymbol{\xi}^2) \\ u(\boldsymbol{\xi}^3) \\ . \\ . \\ u(\boldsymbol{\xi}^N) \end{pmatrix} \tag{29}$$

or

$$A = BC \tag{30}$$

The inner product $<\psi_i^2>$ in the above equation, depends on the polynomials only. It can be calculated and stored upfront as it does not depend on the solution. It can be computed analytically or using numerical quadrature method. Similarly, the coefficient matrix B, is also independent of the solution and can be computed using polynomial values at sample points and the weights. The terms $u(\boldsymbol{\xi}^i)$ in the column C are the function evaluations at the sample points. Once PC coefficients, $u_0, u_1, u_2 \dots u_P$ are computed from Eq. 29, the mean and the variance of the solution u can be computed easily as explained in the previous section.

4.3 Regression Method

The regression method is a statistical method used to estimate the relationship between dependent and independent variables and to approximate the solution of an over-determined system in linear algebra. It is also known as the least squares method. The regression based non-intrusive polynomial chaos method was first introduced by Walters [35] to compute the polynomial coefficients. In the sampling based regression method, the uncertain variables are written in terms of their polynomial expansions. The stochastic quantity of interest $u(x; \boldsymbol{\xi})$ can be approximated in terms of PC expression as:

$$u(x; \boldsymbol{\xi}) = \sum_{i=0}^{P} u_i(x)\psi_i(\boldsymbol{\xi}) \tag{31}$$

If one chooses a set of m observations ($\boldsymbol{\xi}^j = \{\xi_1, \xi_2 \ldots \xi_{n_s}\}^j; j = 1, 2, 3 \ldots m$) in the stochastic space and evaluates the deterministic output $u(x; \boldsymbol{\xi}^j)$ at these sampling points, the PC coefficients can be determined by solving a system of equations as:

$$\begin{pmatrix} \psi_0(\boldsymbol{\xi}^1) & \psi_1(\boldsymbol{\xi}^1) & \psi_2(\boldsymbol{\xi}^1) & .. & \psi_P(\boldsymbol{\xi}^1) \\ \psi_0(\boldsymbol{\xi}^2) & \psi_1(\boldsymbol{\xi}^2) & \psi_2(\boldsymbol{\xi}^2) & .. & \psi_P(\boldsymbol{\xi}^2) \\ \psi_0(\boldsymbol{\xi}^3) & \psi_1(\boldsymbol{\xi}^3) & \psi_3(\boldsymbol{\xi}^3) & .. & \psi_P(\boldsymbol{\xi}^3) \\ . & . & . & .. & . \\ . & . & . & .. & . \\ \psi_0(\boldsymbol{\xi}^m) & \psi_1(\boldsymbol{\xi}^m) & \psi_2(\boldsymbol{\xi}^m) & .. & \psi_P(\boldsymbol{\xi}^m) \end{pmatrix} \begin{pmatrix} u_0(x) \\ u_1(x) \\ u_2(x) \\ . \\ . \\ u_P(x) \end{pmatrix} \tag{32}$$

$$= \begin{pmatrix} u(x; \boldsymbol{\xi}^1) \\ u(x; \boldsymbol{\xi}^2) \\ u(x; \boldsymbol{\xi}^3) \\ . \\ . \\ u(x; \boldsymbol{\xi}^m) \end{pmatrix} \tag{33}$$

or

$$Au = b \tag{34}$$

The least squares solution of the system of Eq. 34 is given by

$$u = (A^T A)^{-1} A^T b \tag{35}$$

The coefficient matrix A is very crucial in the least squares method. One can solve the over-determined system of equations by choosing more than $P + 1$ samples as its necessary that the problem is well posed i.e. the matrix $(A^T A)$ has full rank. Different sampling approaches such as random sampling, Latin Hypercube, sampling from quadrature points, Sobol sampling etc., can be used in the regression-based PC. Consistent with the literature [33], it was found that over-sampling with $2(P + 1)$ simulations is sufficient to achieve results accurately. It was also observed that considering more sampling points does not improve the accuracy of the statistical moments.

The issue of the sampling scheme and the number of samples have already been investigated in detail in the past by Hosder et al. [36] who compared random, LHS and Hammersley sampling schemes. They find substantial differences between the 3 methods in terms of accuracy and computational efficiency, but the last 2 methods exhibit a much smoother convergence with increasing PC order. Concerning the number of points, 2 times the number of unknowns appeared to be enough in their studies.

5 Gradient Based Optimization

In literature, several methods (such as gradient methods, genetic algorithm and response surface) are proposed for shape optimization. The adjoint methods have long been considered as a preferable choice for gradient based optimization. In gradient based optimization, the required gradient of the objective function can be computed using adjoint information as described in the subsection below.

5.1 The Adjoint Method

In a design optimization problem, an objective function $J(U, \alpha)$ is a function of the state variables U and the design variables α. Based on the derivation of the adjoint equations, adjoint solvers can be classified as discrete and continuous. In a continuous adjoint solver, the adjoint equations are derived from the governing equation and then they are discretized. However, in a discrete adjoint solver, the governing equations are first discretized and then the adjoint equations are obtained. Both approaches have their positive and negative aspects. However, the discrete adjoint formulations is explained below. The adjoint method aims at computing the gradient of the objective function with respect to the design variables α. The total derivative of the objective function, $G = \frac{dJ}{d\alpha}$ can be written as:

$$G = \frac{dJ}{d\alpha} = \frac{\partial J}{\partial \alpha} + \frac{\partial J}{\partial U}\frac{dU}{d\alpha} \tag{36}$$

The partial derivatives in the above equation can be evaluated directly by varying the design variables and re-evaluating the objective function in the numerator. However, evaluation of the term $\frac{dU}{d\alpha}$ requires the solution of the governing equations. If $R(U, \alpha) = 0$ represents the residuals of the flow equations of an aerodynamic problem, the total derivative of the flow equations with respect to the design variables α can be expressed as:

$$\frac{dR}{d\alpha} = \frac{\partial R}{\partial \alpha} + \frac{\partial R}{\partial U}\frac{dU}{d\alpha} = 0 \tag{37}$$

or

$$\frac{dU}{d\alpha} = -\left(\frac{\partial R}{\partial U}\right)^{-1}\frac{\partial R}{\partial \alpha} \tag{38}$$

which gives the change in the state variables U with respect to the design variables α. By combining Eqs. 36 and 38, we obtain the following expression for the total gradient:

$$G = \frac{dJ}{d\alpha} = \frac{\partial J}{\partial \alpha} - \frac{\partial J}{\partial U}\left(\frac{\partial R}{\partial U}\right)^{-1}\frac{\partial R}{\partial \alpha} \tag{39}$$

Now let assume λ satisfies the following linear equation:

$$\left(\frac{\partial R}{\partial U}\right)^T \lambda = \left(\frac{\partial J}{\partial U}\right)^T \tag{40}$$

which is also known as the adjoint equation and λ is the vector of the adjoint variables or the adjoint solution. Combining Eqs. 39 and 40, one finally obtains the following expression for the gradient G:

$$G = \frac{dJ}{d\alpha} = \frac{\partial J}{\partial \alpha} - \lambda^T \frac{\partial R}{\partial \alpha} \tag{41}$$

6 Optimization Under Uncertainties

When introducing uncertainties in a design process, the objective function is no longer deterministic and can be characterized by its mean and variance, i.e. in a robust design the optimization becomes multi-objective. Gradient based optimization of the mean objective and of the variance of the objective therefore requires the gradient of both quantities. The gradient of the mean objective is combined with the gradient of its variance using weights.

In this work, the non-intrusive polynomial chaos uncertainty quantification method is combined with gradient based methods for optimization under uncertainties. Having obtained the polynomial expansions of the objective functions, the gradient of the mean objective can be combined with the gradient of its variance using weights. Gradient based optimization of the mean objective and of the variance of the objective therefore requires the gradient of both quantities.

6.1 Stochastic Objective and Its Gradient

In stochastic applications, the objective function (J) will also be stochastic and can be written in terms of polynomial chaos expansion as:

$$J = \sum_{i=0}^{P} J_i \psi_i(\boldsymbol{\xi}) \tag{42}$$

where the mean value of J is $\bar{J} = J_0$ and the variance of J is $\sigma_J^2 = \sum_{i=1}^{P} J_i^2 < \psi_i^2 >$.

The gradient of the objective function J with respect to the design variable can be expressed as:

$$G = \nabla J = \sum_{i=0}^{P} \nabla J_i \psi_i(\boldsymbol{\xi}) = \sum_{i=0}^{P} \frac{dJ_i}{d\alpha} \psi_i(\boldsymbol{\xi}) = \sum_{i=0}^{P} G_i \psi_i(\boldsymbol{\xi}) \tag{43}$$

In stochastic applications, the stochastic objective function is usually written as the weighted sum of its statistical moments. Considering a multi-objective optimization, a new objective function can be defined as a linear combination of the mean and the standard deviation of the original objective function as:

$$I = K_1 \bar{J} + K_2 \sigma_J \tag{44}$$

In order to optimize (or minimize) the objective function I using gradient based methods, one needs to compute the gradient of the objective function I with respect to the design variable as:

$$\nabla I = K_1 \nabla \bar{J} + K_2 \nabla \sigma_J \tag{45}$$

From Eq. 43

$$\nabla \bar{J} = \nabla J_0 = G_0 \tag{46}$$

where $\nabla \bar{J}$ is the gradient of the mean and G_0 is the mean of the objective gradient. The gradient of the variance can be expressed as:

$$\nabla \sigma_J^2 = \sum_{i=1}^{P} \nabla J_i^2 < \psi_i^2 > \tag{47}$$

or

$$2\sigma_J \nabla \sigma_J = 2 \sum_{i=1}^{P} J_i G_i < \psi_i^2 > \tag{48}$$

The above equation can further simplified as:

$$\nabla \sigma_J = \frac{1}{\sigma_J} \sum_{i=1}^{P} J_i G_i < \psi_i^2 > \tag{49}$$

where J_i and G_i are the PC expansion coefficients of the objective and its gradient respectively.

7 Aerodynamic Shape Optimization of the RAE2822 Airfoil Under Uncertainties

To validate the developed approach, here the robust optimization approach is applied to a 2D transonic airfoil, the RAE2822 under operational uncertainties. The adjoint solver and the flow solver of SU2 (an open source CFD solver) are coupled with the polynomial chaos method for the optimal shape design under uncertainties. The optimization procedure is performed using a Python based optimizer SciPy. The SciPy is a gradient based optimizer for solving nonlinear optimization problems with constraints using the sequential least square programming (SLSQP) algorithm [37].

7.1 Test Case Description

In this section, the robust design methodology is applied to the RAE2822 airfoil in transonic viscous flow. The airfoil geometry is described in Cook et al. [38]. The nominal flow conditions correspond to Mach number $M_\infty = 0.729$, angle of attack $\alpha = 2.31^o$ and Reynolds number $Re_\infty = 6.5 \times 10^6$. The operating flow conditions for the test case correspond to case 6 (AGARD report, [38]) with corrected wall interference (see Table 3). The Mach number and the angle of attack are considered as uncertain parameters with standard deviations of $\sigma_M = 0.01$ and $\sigma_\alpha = 0.4^o$ respectively. Both uncertain inputs are considered as uniformly distributed parameters. The stochastic information of the input data is tabulated in Table 4.

For optimization, the objectives considered are the mean drag coefficient and its standard deviation with a constraint in the mean lift. For uncertainty quantification, the quadrature based second order polynomial method is used. The location of the quadrature points for PC order 2 are the roots of the third order Legendre polynomials (for uniformly distributed uncertainties). Two uncertainties and second order

Table 3 Flow conditions for case 6: wind tunnel data used in AGARD report and corrected values for wall interference

	Mach number	Angle of attack	Reynolds number
Wind tunnel test	0.725	2.92°	6.5×10^6
Corrected values	0.729	2.31°	6.5×10^6

Table 4 UQ data for RAE2822 airfoil

Variables	Mean	Std	Distribution
Mach number	0.729	0.01	Uniform
Angle of attack	2.31°	0.4°	Uniform

polynomial requires $(p+1)^n = 9$ individual runs of the flow and the adjoint solver in each iteration of the optimization process. The optimization problem becomes as:

$$\begin{aligned} min\ K_1\bar{C_d} + K_2\mu_{C_d} \\ subjected\ to: \\ \bar{C_l} \geq C_o \end{aligned} \tag{50}$$

where $\bar{C_d}$ is the mean drag coefficient and μ_{C_d} is the standard deviation of the drag coefficient. $\bar{C_l}$, the mean lift coefficient is an inequality which is maintained greater than a constant C_o during the optimization process.

7.2 Deterministic Solution

For the RAE2822 airfoil, the unstructured hybrid mesh with 13,937 cells, used for the CFD simulation is shown in Fig. 6. The computational domain and grid cells near the airfoil surface are depicted in the zoomed area (see Fig. 6). A verified and validated solver SU2 is used for deterministic CFD solutions. The Spalart-Allmaras (SA) model is used for modeling of turbulence. In addition to solving the RANS equations, the adjoint equations can be also solved in SU2 to calculate the surface sensitivities for the optimization purposes. In Fig. 7, the pressure coefficient for the test case is depicted. The pressure coefficient distribution obtained with the SU2 solver is also compared with experimental results (see Fig. 7). It can be observed that the numerical predictions of the C_p using the SA turbulence model are in close agreement with the measurements.

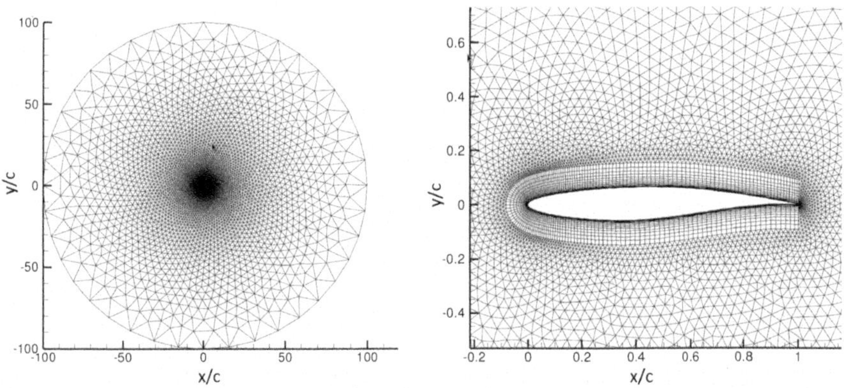

Fig. 6 RAE2822: Computational domain (left) and zoomed mesh near to airfoil (right)

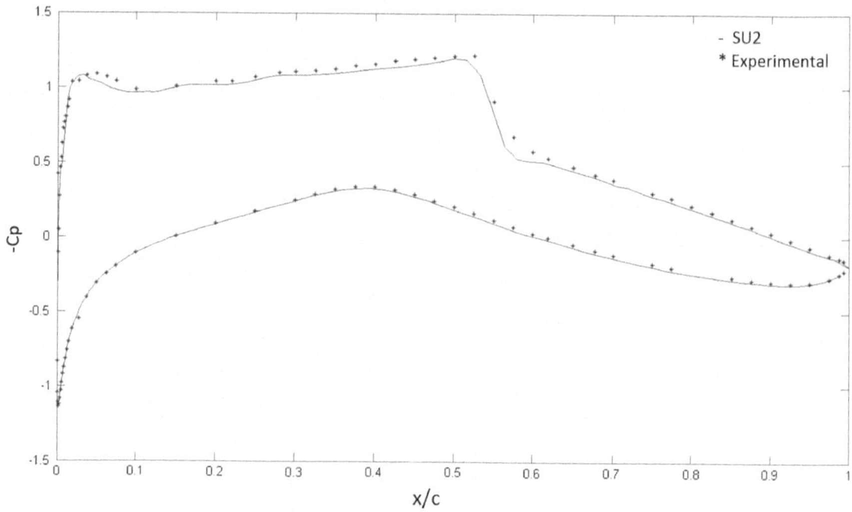

Fig. 7 RAE2822: Pressure coefficient comparison (SU2 and experimental results)

In a gradient based aerodynamic shape optimization model, the chosen objective function is optimized by computing its gradient with respect to the shape or design parameters. The gradient of the objective function with respect to the deformation in the geometry is computed using the adjoint approach. The adjoint solver of SU2 computes the surface sensitivities at each node of the airfoil surface.

7.3 *Shape Parameterization*

For aerodynamic shape optimization problems, design variables are related to the surface boundaries. The design variables are important aspect of an optimization process. A simple way to construct the design variables is to consider surface mesh points. During the optimization process, the independent movement of these surface points may lead to a non-smooth configuration. The shape parameterization ensures smooth geometry deformation. Therefore, in an optimal shape design process, it is essential to parameterize the geometry using an appropriate parameterization technique. In this way, the design parameters can be defined and geometry changes with respect to the design parameters can be provided to the optimizer. Shape parameterization is needed for evolutionary and genetic algorithms as well as for gradient based optimization.

For robust optimization, the shape of the airfoil is parameterized and modified using a set of Hicks-Henne bump functions. Hicks-Henne bump functions recently received popularity in modeling small perturbations in many shape optimization problems [39].

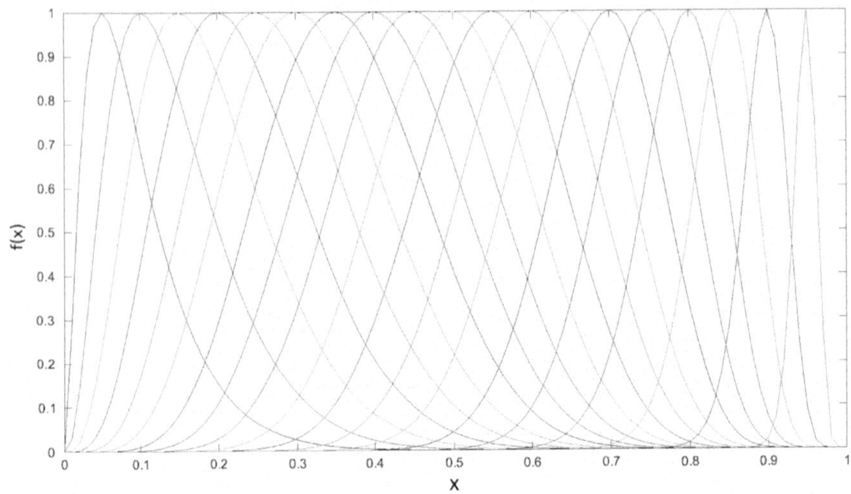

Fig. 8 Hicks-Henne bump functions for $t_2 = 10$ and $t_1 = 0.05, 0.10, 0.15 \ldots 0.95$

The Hicks-Henne functions are defined as:

$$f(x) = [sin(\pi x^{\frac{\log 0.5}{\log t_1}})]^{t_2}; 0 \leq x \leq 1 \tag{51}$$

where, t_1 is the x location of the maximum and t_2 is the width of the bump. In Fig. 8, the Hicks-Henne bump functions for $t_2 = 10$ and $t_1 = 0.05, 0.10, 0.15....0.95$ are shown. The bump functions reach maximum at the given values of $t_1 = 0.05, 0.10, 0.15....0.95$. In Fig. 9, the effect of t_2 parameter on the width of the bump is shown. Increasing the value of t_2, decreases the width of the Hick-Henne function.

The modified airfoil shape can be expressed as a weighted sum of sin functions (Hicks-Henne):

$$y = y_{base} + \sum_{i=1}^{N} a_i f_i(x) \tag{52}$$

where $(a_1, a_2 \ldots a_N)$ are N design variables which are used to control the magnitude of the shape functions. In Fig. 10, the modified shapes of the baseline function $y = x$ are shown where all 19 coefficients a_i (located at $x = 0.05, 0.10, 0.15 \ldots 0.95$) of the Hicks-Henne functions are seeded with some particular values.

The bump functions are applied to the baseline airfoil to modify the shape of the airfoil. A total of 38 Hicks-Henne bump functions are applied to the upper and the lower surface of the airfoil at 5, 10 ..., 95% of the chord length. In this way, 38 coefficients of the bump functions are considered as the design variables. The leading and trailing edges are kept fixed in the design process by not adding any shape function to those points.

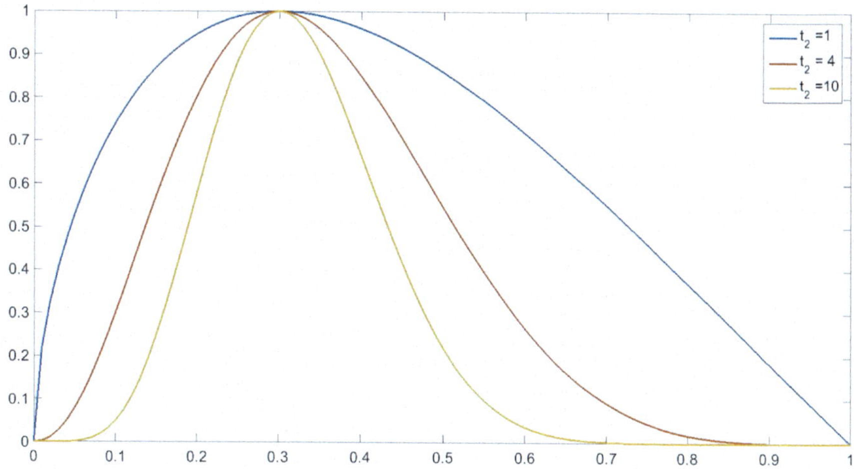

Fig. 9 Hicks-Henne bump functions for $t_2 = 1, 4, 10$ and $t_1 = 0.30$

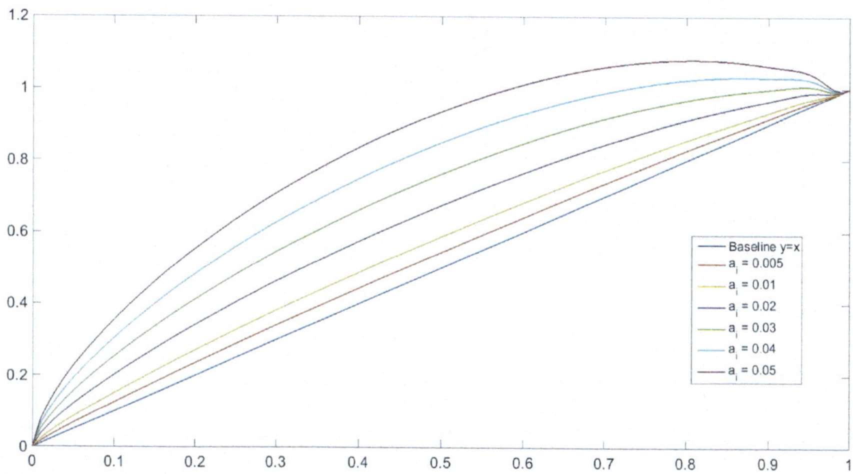

Fig. 10 Shape modification of line y = x with Hicks-Henne coefficients

7.4 Robust Optimization

In this section, the initial geometry of the RAE2822 airfoil is taken for the shape optimization in order to minimize the drag as the objective function. A built-in gradient based optimizer using SLSQP algorithm in SU2 framework combines the flow solver, the adjoint solver and the mesh deformation tools to minimize the objective function. By using a simple chain rule, the surface sensitivities (computed with adjoint solver) are projected into the design space. For uncertainty quantification at each iteration of the optimization process, the input values of the Mach number and angle of attack for all CFD runs are shown in Table 5.

Table 5 Mach number and AoA samples for RAE2822 airfoil

	S_1	S_2	S_3	S_4	S_5	S_6	S_7	S_8	S_9
M	0.715	0.729	0.742	0.715	0.729	0.742	0.715	0.729	0.742
AoA	1.773°	1.773°	1.773°	2.31°	2.31°	2.31°	2.846°	2.846°	2.846°

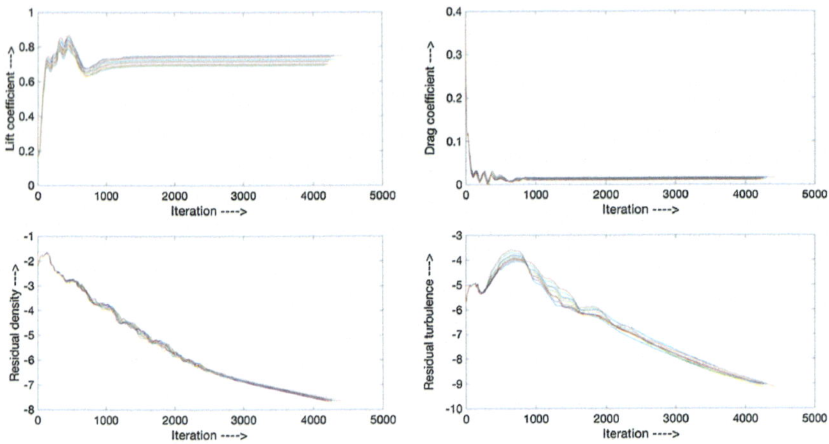

Fig. 11 CFD solution and residuals for all CFD runs of RAE2822 airfoil

In Fig. 11, the convergence history and CFD solutions are plotted for all CFD runs. As it can be observed, the CFD solutions and residuals for all the samples are converged.

In the robust optimization process, the objective function (a weighted sum of the mean and its standard deviation) and the gradient of the objective function are provided to the optimizer. By changing the weights, the results for three different test cases are obtained. The first case is where a higher weight is given to the mean (i.e. $K_1 = 0.9$ and $K_2 = 0.1$), for the second test case, more weight is given to the standard deviation (i.e. $K_1 = 0.1$ and $K_2 = 0.9$) and the last case is where equal weights are given to the mean and the standard deviation.

In Fig. 12, the optimization history of the mean and the standard deviation for the three test cases is shown. From the optimization history, one can see that for all the test cases, the mean and the standard deviation of the drag coefficients are optimized after a few iterations only.

In Table 6, the mean and the standard deviation of the drag coefficient are shown for the baseline RAE2822, deterministic optimized airfoil and for the robust design of the three test cases. From Table 6, it can be seen that for the case 1, where more weight is given to the mean, the optimized mean drag is lower than in all other designs. Similarly for case 2, where more weight is given to the standard deviation, the standard deviation is lower than in all other designs. Case 3, where equal weights are given to the mean and the standard deviation, optimizes both values i.e. the mean

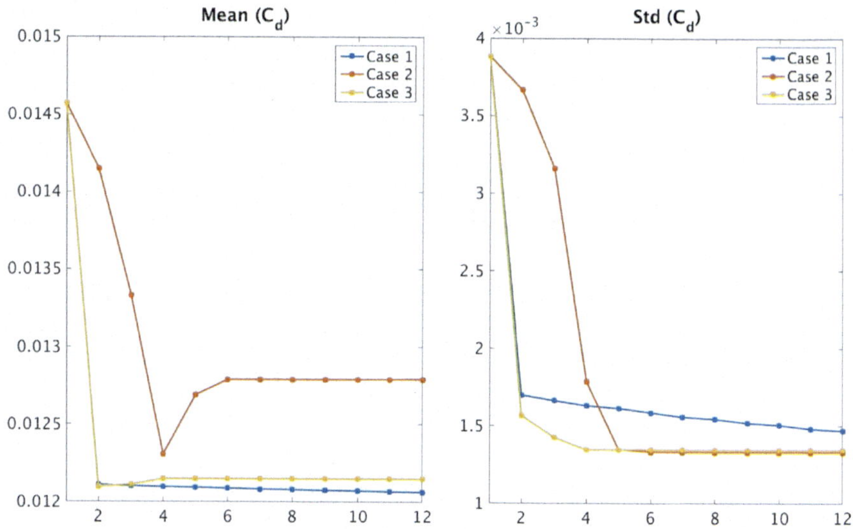

Fig. 12 Optimization history for the mean and the standard deviation of the 3 cases

Table 6 Mean and std of the drag coefficient for baseline RAE, optimized and robust airfoils

Case	Mean C_d	Std C_d	CoV
RAE2822 baseline	1.46e−02	3.89e−03	26.66% mean
Det. optimized	1.25e−02	2.52e−03	20.18% mean
Case 1, K=(0.9,0.1)	1.21e−02	1.47e−03	12.20% mean
Case 2, K=(0.1,0.9)	1.28e−02	1.34e−03	10.44% mean
Case 3, K=(1,1)	1.22e−02	1.35e−03	11.11% mean

and the standard deviation simultaneously. For case 3, the optimized mean value is very close to the mean value of case 1 and the optimized standard deviation is very close to the standard deviation of case 2.

When the uncertainties are applied to the airfoil obtained with a deterministic optimization, the mean drag coefficient is reduced by approximately 15% and the coefficients of variance by approximately 25% compared to the original design (see Table 6). When the uncertainties are included in the design process (case 1 and case 2), the coefficients of variance are reduced by approximately 50%, making the final designs robust with respect to the uncertainties. In Fig. 13, the airfoil designs for all cases, i.e. the baseline RAE2822, optimal shape (without uncertainties) and robust shape for both the test cases (under uncertainties) are shown. It can be seen that during the optimization process, mostly the area near the shock region is deformed. Three different robust airfoil designs (for the cases under uncertainties) are obtained. The shapes of these airfoils are quite different from the one without uncertainties.

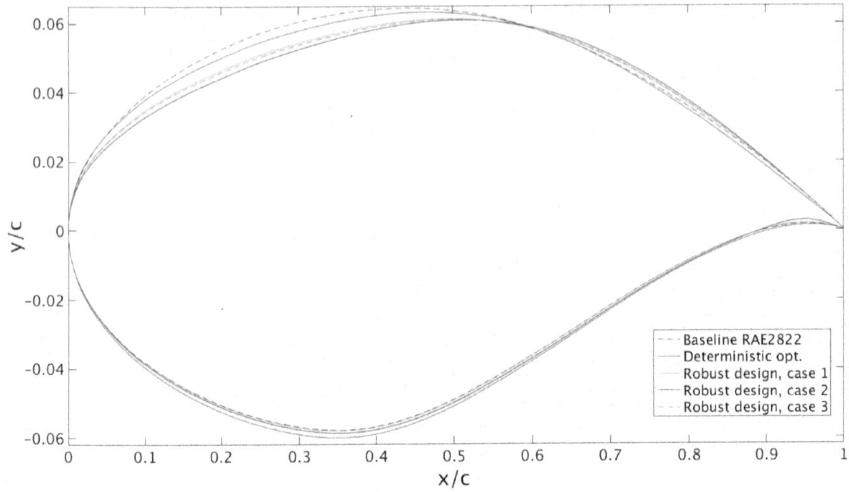

Fig. 13 Airfoil shape: RAE2822 baseline and optimized

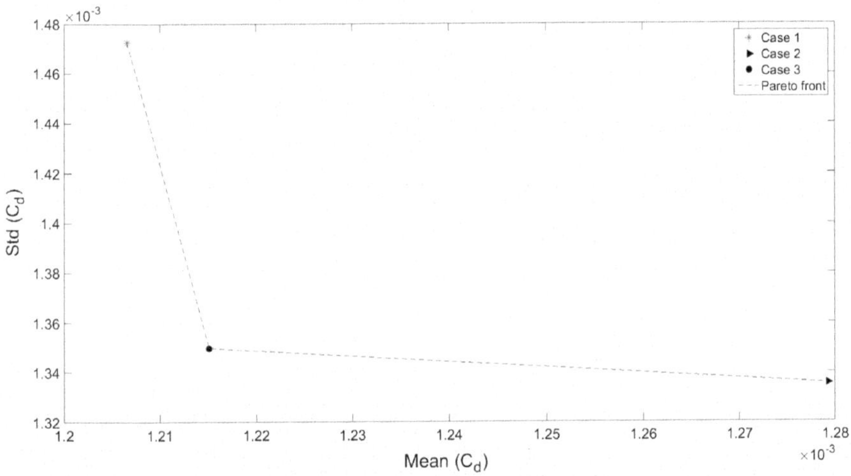

Fig. 14 Three points on the Pareto front

In Fig. 14, the standard deviation of C_d is plotted against the mean C_d for these three cases of robust designs making a Pareto front.

It is clear that this is just a part of the full Pareto front. This is related to the use of a gradient based method where only part of the design space is explored in the neighborhood of the original airfoil. Putting it in another way: a gradient based method will allow shape optimizations of e.g. a VW car (for drag reduction) but it will not change it into a Ferrari (with even lower drag). For this an evolutionary method would be needed exploring the design space to a large extent.

8 Conclusion

In this chapter, some popular non-intrusive polynomial chaos methods for uncertainty quantification are discussed and further the quadrature based non-intrusive polynomial chaos method is combined with the adjoint based gradient methods for optimization under uncertainties. The developed robust optimization method is applied to optimal shape design of the RAE2822 airfoil using the flow solver and the adjoint solver of SU2 (an opensource CFD solver). The surface of the RAE2822 airfoil is parameterized using the Hicks-Henne bump functions and the optimization procedure is performed using a Python based optimizer SciPy. For the RAE2822 airfoil, three different optimal designs are obtained by considering three different points on the Pareto front. Currently, the quadrature based polynomial chaos is used for uncertainty quantification as only two operation uncertainties are considered in the robust optimization process. However, for a large number of uncertainties (e.g., geometrical uncertainties), an efficient non-intrusive UQ method such as sparse quadrature, reduced basis, etc. can be employed easily to the robust optimization approach.

References

1. Poloni C, Mosetti G (1996) Aerodynamic shape optimisation by means of hybrid ge-netic algorithm. Z Angew Math Mech 76:247–250
2. Wang X, Hirsch C, Liu Z, Kang S, Lacor C (2013) Uncertainty-based robust aerodynamic optimization of rotor blades. Int J Numer Meth Engng 94:111–127. https://doi.org/10.1002/nme.4438
3. Sriram A (2011) Jameson, robust optimal control using polynomial chaos and adjoints for systems with uncertain inputs. In: 20th AIAA computational fluid dynamics conference, Honolulu, Hawaii, 27–30 June 2011
4. Congedo PM, Corre C, Martinez J-M (2011) Shape optimization of an airfoil in a BZT flow with multiple-source uncertainties. Comput Methods Appl Mech Engg 200, 216–232
5. Palar PS, Tsuchiya T, Parks G (2015) Decomposition-based evolutionary aerodynamic robust optimization with multi-fidelity point collocation non-intrusive polynomial chaos. In: 17th AIAA non-deterministic approaches conference, AIAA SciTech, AIAA paper 2015–1377
6. Schillings C, Schulz V (2015) On the influence of robustness measures on shape optimization with stochastic uncertainties. Optim Eng 16:347
7. Maruyama D, Liu D, Gortz S (2016) An efficient aerodynamics shape optimization framework for robust design of airfoils using surrogate models. ECCOMAS congress, Crete Island
8. Helton JC, Johnson JD, Oberkampf WL, Sallaberry CJ (2008) Representation of analysis results involving aleatory and epistemic uncertainty. Sandia National Laboratories, Tech. Rep. SAND 2008-4379
9. Yager RR, Liu L (2008) Classic works of the dempster-shafer theory of belief functions. Studies in Fuzziness and Soft Computing Series. Springer, Berlin vol 219
10. Corveleyn S, Rosseel E, Vandewalle S (2013) Iterative solvers for a spectral Galerkin approach to elliptic partial differential equations with fuzzy coefficients. SIAM J Sci Comp 35(5):S420–S444
11. Zimmermann H-J (2000) An application-oriented view of modelling uncertainty. Eur J Oper Res 122:190–199
12. Fishman GS (1996) Monte Carlo: concepts, algorithms, and applications. Springer-Verlag, New York

13. Xiu D, Karniadakis GE (2002) The wiener-askey polynomial chaos for stochastic differential equations. SIAM J Sci Comput 24(2):619–644
14. Ghanem R, Spanos P (1991) Stochastic finite elements: a spectral approach. Springer-Verlag, New York
15. Babuska I, Nobile F, Tempone R (2007) A stochastic collocation method for elliptic partial differential equations with random input data. SIAM J Numer Anal 45(3):1005–1034
16. Mathelin L, Hussaini MY, Zang TA (2007) Stochastic approaches to uncertainty quantification in CFD simulations. Numer Algorithms 38:209–236
17. Wan X, Karniadakis G (2005) An adaptive multi-element generalized polynomial chaos method for stochastic differential equations. J Comput Phys 209:617–642
18. Nayfeh AH (1973) Perturbation methods. Wiley, New York
19. Chow P-L (1975) Perturbation methods in stochastic wave propagation. SIAM Review 17(1):57–81
20. Taylor AC, Green LL, Newman PA, Putko MM (2001) Some advanced concepts in discrete aerodynamic sensitivity analysis, vol 2001–2529. AIAA
21. Gaspar J, Judd KL (1997) Solving large scale rational expectations models. NBER Working Paper No. t0207
22. Atherton RW, Schainker RB, Ducot ER (1975) On the statistical sensitivity analysis of models for chemical kinetics. AIChE 21:441–448
23. Cacuci DG, Weber CE, Oblow EM, Marable JH (1980) Sensitivity theory for general systems of nonlinear equations. Nuc Sci Eng 75:88–110
24. Crick MJ, Hill MD, Charles D (1987) The role of sensitivity analysis in assessing uncertainty. In: Proceedings of an NEA workshop on uncertainty analysis for performance assessments of radioactive waste disposal systems, Paris, OECD, 1–258
25. Zirnmerman DA, Hanson RT, Davis EA (1991) A comparison of parameter estimation and sensitivity analysis techniques and their impact on the uncertainty in ground water flow model predictions. Albuquerque. Sandia National Laboratory, NM. Report No. NUREG/CR-5522
26. Kumar D, Lacor C (2012) Heat conduction in a 2D domain with geometrical uncertainty using intrusive polynomial chaos method. In: Proceedings of the 9th national congress on theoretical and applied mechanics, Brussels (Belgium) - ISBN 978-2-8052-0151-6, May 9–10
27. Le Maitre O, Reagan KM, Najm H, Ghanem R (2001) A stochastic projection method for fluid flow. I: basic formulation, J Comput Phys 173:481–511
28. Lacor C, Smirnov S (2008) Non-deterministic compressible navier - stokes simulations using polynomial chaos. Mini-symposium Uncertainty quantification methods in CFD and FSI problems, ECCOMAS Conference. Venice, 30/6–4/7/2008
29. Dinescu C, Smirnov S, Hirsch Ch, Lacor C (2010) Assessment of intrusive and non-intrusive non-deterministic CFD methodologies based on polynomial chaos expansions. Int J Eng Systems Modeling and Simulation 2:87–98
30. Wang X, Hirsch Ch, Liu Z, Kang S, Lacor C (2013) Uncertainty-based robust aerodynamic optimization of rotor blades. Int J Numer Meth Eng 94:111–127
31. Wiener N (1938) The homogeneous chaos. Am J Math 60(4):897–936
32. Eldred MS, Webster CG, Constantine P (2008) Evaluation of non-intrusive approaches for wiener-askey generalized polynomial chaos. In: Proceedings of the 10th AIAA nondeterministic approaches conference, No. AIAA-2008-1892, Schaumburg, IL, April 7–10, 2008
33. Hosder S, Walters RW, Perez R (2006) A non-intrusive polynomial chaos method for uncertainty propagation in CFD simulations. In: 44th AIAA aerospace sciences meeting and exhibit, Reno, Nevada, January 9–12, 2006
34. Smolyak S (1963) Quadrature and Interpolation formulas for tensor products of certain classes of functions, Dokl Adad Nauk USSR B, 240–243
35. Walters RW, Huyse L (2002) Uncertainty analysis for fluid mechanics with applications. NASA/CR 2002-211449
36. Hosder S, Walters RW, Balch M (2007) Efficient sampling for non-intrusive polynomial chaos applications with multiple input uncertain variables. In: 9th AIAA non-deterministic approaches conference, Honolulu, April 2007

37. Boggs Paul T, Tolleb Jon W (2000) Sequential quadratic programming for large-scale nonlinear optimization. J Comp and Appl Math
38. Cook PH, McDonald MA, Firmin MCP (1979) Aerofoil RAE 2822 - Pressure distributions, and boundary layer and wake measurements. Experimental Data Base for Computer Program Assessment, AGARD Report AR 138
39. Masters DA, Taylor NJ, Rendall T, Allen CB, Poole DJ (2015) Review of aerofoil parameterisation methods for aerodynamic shape optimisation. In: 53rd AIAA aerospace sciences meeting, AIAA SciTech (AIAA 2015-0761)

Determination of Linear and Non-linear Hydrodynamic Derivatives of a Surface Ship in Manoeuvring Using CFD Method

Sheeja Janardhanan and Parameswaran Krishnankutty

Abstract Ensuring navigational safety of a vessel entails the determination of the manoeuvring coefficients or the hydrodynamic derivatives and the subsequent simulation of its trajectory well in advance of constructing the ship and this task is indeed a very challenging one. The number of hydrodynamic derivatives to be determined is based on the mathematical model used for the representation of hydrodynamic forces and moments. This chapter presents the mathematical formulation of the problem and the numerical approach used for obtaining the hydrodynamic derivatives. For this, an attempt has been made to numerically simulate the conventional horizontal planar motion mechanism (HPMM) towing tank experiment using the RANSE CFD model. A container ship model has been modelled for performing simulation andanalysis. The numerical tank size has been set following the ITTC guidelines, applying the grid size of 600,000 hexahedral cells. The free-surface effects have been taken into account. The prescribed body motions have been modelled by using the mesh morphing ANSYS CFX technique with 3-modes motion viz. pure sway, pure yaw and combined sway and yaw motions, referred as combined motions in this chapter. The time histories of forces and moments have been approximated with the Fourier series in order to enable the comparison with the corresponding equations for forces and moments, as represented by the developed mathematical model defined with linear and non-linear hydrodynamic derivatives. These computed derivatives have been compared with the experimental results, showing reasonably good agreement.

Keywords Ship manoeuvring · Hydrodynamic derivatives · Planar motion mechanism · CFD simulation

---

S. Janardhanan (✉) · P. Krishnankutty
Mechanical Engineering, SCMS School of Engineering and Technology, Karukutty, Ernakulam, India
e-mail: sheejajanardhanan@scmsgroup.org

D. Vucinic et al. (eds.), *Advances in Visualization and Optimization Techniques for Multidisciplinary Research*, Lecture Notes in Mechanical Engineering, https://doi.org/10.1007/978-981-13-9806-3_4

Nomenclature

a_{dMm}	Fourier constants associated with cosine terms
A	Derivatives evaluated and deemed very important
b_{dMm}	Fourier constants associated with sine terms
B	Beam of the vessel
C	Derivatives evaluated and deemed of minor importance
C_B	Block coefficient of the vessel
D	Derivatives evaluated and deemed negligible
D_t	Depth of the vessel
L	Length between perpendiculars of the vessel
m	Integer for determining harmonics of Fourier series
N	Hydrodynamic reaction moment in yaw
N_{HN}	Hydrodynamic reaction moment in yaw in pure yaw mode
N_{HY}	Hydrodynamic reaction moment in yaw in pure sway mode
N_{HYN}	Hydrodynamic reaction moment in yaw in combined motion
r	Yaw rate
r_a	Amplitude of yaw rate
r'_a	Non-dimensional amplitude of yaw rate
$\dot{y}'_{0a}$	Non-dimensional amplitude of yaw rate
t	Instantaneous time
T	Time period of oscillation
T_a	Draft at aft of the vessel
T_f	Draft at fore of the vessel
T_m	Mean draft of the vessel
u	Forward velocity in ship-fixed co-ordinate system
u_0	Forward velocity in early-fixed co-ordinate system
u^+	Non-dimensional velocity
u_{aN}	Amplitude of surge velocity in pure yaw mode
u_{aYN}	Amplitude of surge velocity in pure yaw mode
u_{cN}	Constant in surge velocity in pure yaw mode (Eq. 26)
u_{cYN}	Constant in surge velocity in combined motion (Eq. 44)
v	Sway velocity in ship-fixed co-ordinate system
v_0	Sway velocity in earth-fixed co-ordinate system
v_{aYN}	Amplitude of sway velocity in combined motion
V	Resultant velocity
x	Position in longitudinal direction in ship-fixed co-ordinate system
x_0	Position in longitudinal direction in earth-fixed co-ordinate system
X	Non-dimensional hydrodynamic reaction force in surge
X_{HN}	Hydrodynamic reaction force in surge in pure yaw mode
X_{HY}	Hydrodynamic reaction force in surge in pure sway mode
X_{HYN}	Hydrodynamic reaction force in surge in combined motion
y	Position in transverse direction in ship-fixed co-ordinate system
y^+	Non-dimensional wall distance
y_0	Position in transverse direction in earth-fixed co-ordinate system

y_{0a} Amplitude of sway motion
Y Non-dimensional hydrodynamic reaction force in sway
Y_{HN} Hydrodynamic reaction force in sway in pure yaw mode
Y_{HY} Hydrodynamic reaction force in sway in pure sway mode
Y_{HYN} Hydrodynamic reaction force in pure sway in combined motion

Greek Symbols

β Drift angle
β_a Drift angle amplitude
γ Arbitrary value of phase which depends on the start value of oscillation
∇ Volumetric displacement of the vessel
ω Circular frequency of oscillation
ω' Non-dimensional circular frequency of oscillation
ψ Heading/yaw angle
ψ_a Amplitude of yaw angular motion

1 Introduction

Manoeuvring or steering of ships is generally concerned about their motions produced as a result of excitation forces applied through the deflection of control surfaces in the absence of excitation from the sea (calm water). The motion of the surface ship can be described using the 2 dimensional reference frame. The fixed earth reference frame defines the inertial frame of reference, while the ship reference frame is fixed to the ship gravity centre and moves together with the ship. Figure 1 shows the definition of the ship co-ordinate systems. In a pure mathematical sense, the ship manoeuvrability depends on the hydrodynamic coefficients or derivatives, which are the acceleration coefficients, defined as the velocity terms in the manoeuvring equations of motion.

The most challenging part of the ship manoeuvring simulation is the estimation of these derivatives. These derivatives need to be estimated and the manoeuvrability of the ship must be assessed well in advance to meet the standards of safe navigation. Experimental, theoretical and numerical methods are used for the hydrodynamic derivatives estimation. The experimental methods employing captive model tests have been used for the calculation of hydrodynamic derivatives from the measured forces and moments. The free-running model tests have yet another experimental method, which is more straightforward for predicting the trajectory and controllability assessment, while the captive model tests are more versatile, as the stability indices and forces, contributing to the manoeuvring motion, can be quantified with these tests. The hydrodynamic derivatives, which are determined through the captive model tests, can also be used in designing the autopilot.

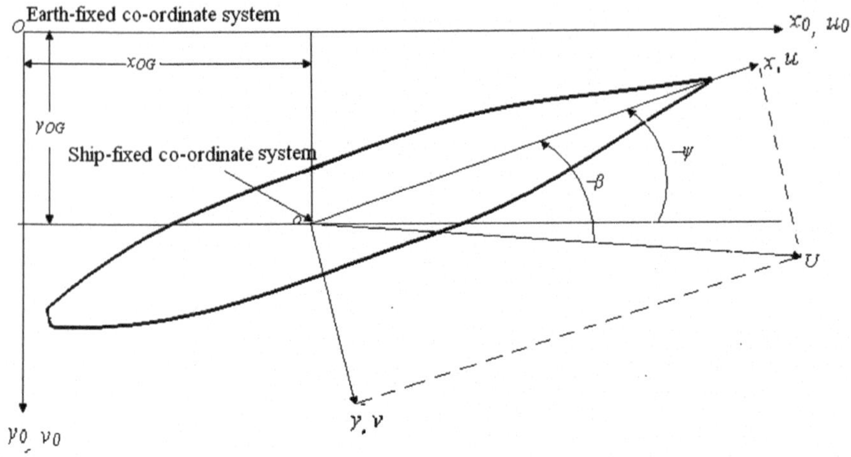

Fig. 1 Manoeuvring reference frames

The drawback of the experimental methods is the huge cost and involved experimental resources. The theoretical methods, based on a number of assumptions, on the other side have their own limitations as far as the accuracy is concerned. These difficulties have paved a glorious path in applying the numerical methods. The numerical methods are quite efficient in accomplishing simulations (duplicating experiments) and with related flow visualizations, they reduced the tedious model-testing task to become the computer modelling and simulation task. As indirectly mentioned with the "reducing" term, these techniques are not full-fledged, and will never replace the experiments, but will surely reduce their number to minimum. Nevertheless, they definitely seem to be very promising to be researched and developed in the years to come.

The past decade has brought with it, a host of numerical methods, which find easy implementation, with the tremendous advance in digital computing. Initially, numerical methods have been based on the potential flow theory, and have been used for the determination of the first order hydrodynamic derivatives. However, these methods have failed in predicting these derivatives successfully as they lacked adequate flow physics by not taking into account the viscous effects. With the emergence of the CFD based Reynolds Averaged Navier-Stokes Equations (RANSE), a powerful tool in flow prediction and associated phenomenon, the research focus has been set to explore its efficacies in the field of ship hydrodynamics, such as resistance and propulsion, sea keeping, manoeuvring etc. Especially, manoeuvring estimations demand attention, as safety issues in navigation have become prime concern as per IMO regulations. This factor has further strengthened the need for using CFD methods in this domain, and thus, brings to light the present research scenario in this field. For instance, Toxopeus [1] developed a mathematical model based on the bare hull forces and moments in steady drift, steady yaw and combined drift and yaw motions using the viscous flow calculations. Carrica and Stern [2] used DES simulations to assess the manoeuvrability of a tanker, by bringing about 35-degree rudder turn and 20/20 zig-zag manoeuvre.

Janardhanan and Krishnankutty [3] determined the sway velocity dependency on the linear and non-linear hydrodynamic derivatives of a container ship, by simulating straight line tests using the RANSE equations for wider range of drift angles. Nonaka et al [4] estimated the hydrodynamic forces acting on the ship during manoeuvring motion using CFD. Sulficker [5] has explained the simulation of HPMM for the evaluation of linear derivatives, where the derivatives have been determined by using the forces and moments obtained at the maximum velocities and acceleration points, without creating an extensive formulation problem to carry out such simulations.

Cura-Hochbaum [6] determined the linear hydrodynamic derivatives of a passenger ship model by simulating the viscous flow around the hull, undergoing forced motions using a RANSE solver. The free-surface effects have not been considered in this work. This method was extended to the tanker models by simulating the combined sway and yaw by Cura-Hochbaum et al. [7]. The trajectories of the vessels were also simulated using the estimated derivatives. Xing-Kaeding and Jensen [8] presented the CFD modelling of the hull, rudder and propeller, together with the estimation of forces and moments during a turning circle and zig-zag manoeuvres. Tyagi and Sen [9] have shown the procedure for estimating the transverse velocity, applying the linear and non-linear AUV derivatives. Ohmori [10] numerically simulated the straight-line test, circular motion test and planar motion mechanism tests, to find the transverse force acting on the hull based on the velocity dependent linear derivatives. All these remarkable contributions indicate the important impact of the RANSE based CFD calculations to the manoeuvring analysis.

This book chapter focuses on the method development for determining the hydrodynamic derivatives in conjunction with a numerical horizontal planar motion mechanism test based on the CFD simulations. The performed RANSE simulations employ the finite volume technique for solving the viscous fluid flow equations. A container ship model (S175) [11] has been used for the presented analysis. Figure 2 shows the container ship model geometry, as generated by using the ANSYS ICEM CFD software package, and Table 1 shows the model particulars.

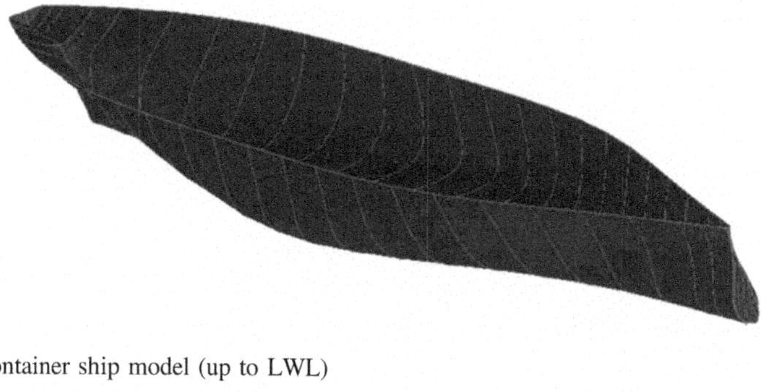

Fig. 2 Container ship model (up to LWL)

Table 1 Particulars of the container ship

Particular	Ship	Model (scale 1:36)
L (m)	175	4.86
B (m)	25.4	0.705
T_f (m)	8.0	0.22
T_a (m)	9.0	0.25
T_m (m)	8.5	0.236
D_t (m)	11	0.305
∇ (m^3)	21,222	0.4548
C_B	0.559	0.559

2 Mathematical Model

A mathematical model representing surface ship manoeuvring comprises the equations of motion in surge, sway and yaw modes, where the influence of other modes on manoeuvring are generally assumed as being insignificant. For ships, which take strong manoeuvres, the roll mode of the motion and its influence on the above-mentioned modes becomes considerable. A mathematical model is obtained by expanding the hydrodynamic reaction force or moment for each mode applying the Taylor series [12]. The retention of the higher order terms results in obtaining the non-linear models.

Any manoeuvring problem involving the estimation of hydrodynamic derivatives starts with the selection of an appropriate mathematical model. In this work the model proposed by Son and Nomoto [11] has been used after neglecting the roll effects where the speed and steering are given by simplifying Eqs. (1)–(3).

$$X = X(u) + (1-t)T + X_{vr}vr + X_{vv}v^2 + X_{rr}r^2 + X_\delta \sin\delta + X_{ext} \quad (1)$$

$$Y = Y_v v + Y_r r + Y_{vvv}v^3 + Y_{rrr}r^3 + Y_{vvr}v^2 r + Y_{vrr}vr^2 + Y_\delta \cos\delta + Y_{ext} \quad (2)$$

$$N = N_v v + N_r r + N_{vvv}v^3 + N_{rrr}r^3 + N_{vvr}v^2 r + N_{vrr}vr^2 + N_\delta \cos\delta + N_{ext} \quad (3)$$

The hydrodynamic terms can be separated out and the equations can be rearranged as:

$$\begin{aligned} X_H = {} & X_{\dot{u}}\dot{u} + Y_{\dot{v}}vr + X_{uu}u|u| + (1-t)T + X_{vr}vr \\ & + X_{vv}v^2 + X_{rr}r^2 + X_\delta \sin\,\delta \end{aligned} \quad (4)$$

$$\begin{aligned} Y_H = {} & X_{\dot{u}}ur + Y_{\dot{v}}\dot{v} + Y_{\dot{r}}\dot{r} + Y_v v + Y_r r + Y_{vvv}v^3 \\ & + Y_{rrr}r^3 + Y_{vvr}v^2 r + Y_{vrr}vr^2 + Y_\delta \cos\,\delta \end{aligned} \quad (5)$$

$$\begin{aligned} N_H = {} & N_{\dot{v}}\dot{v} + N_{\dot{r}}\dot{r} + N_v v + N_r r + N_{vvv}v^3 \\ & + N_{rrr}r^3 + N_{vvr}v^2 r + N_{vrr}vr^2 + + N_\delta \cos\,\delta \end{aligned} \tag{6}$$

where $X(u)$ is for instance, the velocity dependent damping function. This model does not involve modelling the rudder and propeller. In this study the hull derivatives are of prime interest and the effect of rudder and propeller can be brought in by some empirical relationships when required (trajectory simulation). Bilge keels have not been considered in this work.

3 Planar Motion Mechanism Tests

The Horizontal Planar Motion Mechanism (HPMM) is a device developed to determine the surface ships velocity and acceleration dependent derivatives for the sway and yaw modes. This results to be the most promising method among all the existing captive model test methods for the determination of hydrodynamic derivatives, as it contains all the derivatives defined in the mathematical model [Eqs. (4)–(6)] and contributing to the equations of motion [13, 14].

3.1 General Experimental Procedure

The HPMM consists of 2 oscillators, one of which produces a transverse oscillation at the bow and the other at the stern, while the model moves down the towing tank with a constant towing carriage velocity in the current procedure. The in-phase oscillations of the oscillators in transverse direction to the towing tank longitudinal axis produces pure sway motion of the model, while the out of phase oscillation brings in the pure yaw motion. The model pure yaw motion, initially subjected to a certain drift angle, brings together the drift and yaw motion. The model velocity is always tangential to the oscillating path. In each mode, the forces and moments acting on the model are measured and resolved using the special transducers and instrumentation. Hydrodynamic derivatives can then be determined from the measured forces and moments by using either a higher order curve-fit method or Fourier series expansion method.

3.2 Problem Formulation

The problem formulation has been tedious, as did involve a laborious procedure of many trials in order to arrive at the expressions for the body oscillations model, so that they in turn yield the most simplified expressions for the calculation of hydrodynamic derivatives. The Fourier series expansion method has been used for the determination of the derivatives and according to it the hydrodynamic forces and moments acting on the ship hull can be expressed as defined in Eq. (7a):

$$\text{Hydrodynamic force/moment} = \sum_{m=0}^{\infty} (a_m \cos m\omega t + b_m \sin m\omega t) \tag{7a}$$

where the Fourier coefficients are given by Eq. (7b):

$$a_0 = \frac{1}{T}\int_0^T f(t)dt \quad a_m = \frac{2}{T}\int_0^T f(t)\cos\ m\omega t dt \quad b_m = \frac{2}{T}\int_0^T f(t)\sin\ m\omega t dt \tag{7b}$$

in which, a truncated series up to 3rd harmonics is chosen, because the mathematical model considers only up to the 3rd order terms. Considering an arbitrary phase angle (γ) at the start value of oscillation, the above trigonometric function can be written as set in Eq. (7c)

$$\begin{aligned} \cos\ m\omega t &= \cos m(\omega t + \gamma)\cos m\gamma + \sin m(\omega t + \gamma)\sin m\gamma \\ \sin\ m\omega t &= \sin m(\omega t + \gamma)\cos m\gamma - \cos m(\omega t + \gamma)\sin m\gamma \end{aligned} \tag{7c}$$

Hence, the hydrodynamic force/moment can now be expresses as follows in the Eq. (7d)

$$\begin{aligned} H_M = \sum_{m=0}^{\infty} \{ & a_{dMm}(\cos\ m(\omega t + \gamma)\cos\ m\gamma + \sin\ m(\omega t + \gamma)\sin\ m\gamma) \\ & + b_{dMm}(\sin\ m(\omega t + \gamma)\cos\ m\gamma - \cos\ m(\omega t + \gamma)\sin\ m\gamma)\} \end{aligned} \tag{7d}$$

where H = X, Y, N; $d = x, y, n$; $M = Y, N, YN$; $m = 1, 2, 3$.

x represents the direction of surge force, y represents the direction of sway force, *n* represents yaw moment about z axis, *Y* represents pure sway mode, *N* represents pure yaw mode, *YN* represents combined mode and *m* represents the harmonics of Fourier series.

3.2.1 Pure Sway Mode

This part of HPMM comprises oscillating the model in transverse direction in which the oscillators fixed on the model oscillate in phase and the model kinematics parameters are represented as:

$$\text{Transverse displacement,} \qquad y_0 = -y_{0a} \sin \omega t = y \tag{8}$$

$$\text{Transverse velocity,} \qquad \dot{y}_0 = -y_{0a}\omega \cos \omega t = v \tag{9}$$

$$\text{Transverse acceleration,} \qquad \ddot{y}_0 = y_{0a}\omega^2 \sin \omega t = \dot{v} \tag{10}$$

$$\text{Forward velocity,} \qquad u_0 = V = u \tag{11}$$

$$\text{Forward acceleration,} \qquad \dot{u} = 0 \tag{12}$$

Yaw angular displacement, velocity and acceleration,

$$\psi = r = \dot{r} = 0 \tag{13}$$

In pure sway mode oscillations, the mathematical model given by Eqs. (4)–(6) will have only sway motion dependent terms and thus they reduce to:

$$X_{HY} = X_{u|u|}u|u| + X_{vv}v^2 \tag{14}$$

$$Y_{HY} = Y_{\dot{v}}\dot{v} + Y_v v + Y_{vvv}v^3 \tag{15}$$

$$N_{HY} = N_{\dot{v}}\dot{v} + N_v v + N_{vvv}v^3 \tag{16}$$

Expanding the hydrodynamic reaction forces and moment with applying Fourier series and substituting Eqs. (8)–(11) in Eqs. (14)–(16), and comparing the like terms in the corresponding equations by putting $r = 0$, we obtain the derivatives for the pure sway defined in Eqs. (17)–(24):

$$X_{vv} = \frac{2}{y_{0a}^2\omega^2} a_{xY2} \tag{17}$$

$$X_{u|u|} = \frac{1}{V^2}\left[a_{xY0} - a_{xY2}\right] \tag{18}$$

$$Y_{\dot{v}} = \frac{1}{y_{0a}\omega^2} b_{yY1} \tag{19}$$

$$Y_{vvv} = \frac{-4}{y_{0a}^3\omega^3} a_{yY3} \tag{20}$$

$$Y_v = \frac{1}{y_{0a}\omega}\left[3a_{yY3} - a_{yY1}\right] \tag{21}$$

$$N_{\dot{v}} = \frac{1}{y_{0a}\omega^2} b_{nY1} \tag{22}$$

$$N_{vvv} = \frac{-4}{y_{0a}^3 \omega^3} a_{nY3} \tag{23}$$

$$N_v = \frac{1}{y_{0a}\omega}[3a_{nY3} - a_{nY1}] \tag{24}$$

3.2.2 Pure Yaw Mode

In HPMM experiment, the oscillators are made to oscillate with a predetermined phase difference in order to take into account a rotation in the horizontal plane. In the present numerical model, the pure yaw mode is achieved by oscillating the vessel about z-axis, and at the desired frequency. The yaw acceleration causes a fluctuation in surge velocity. At any instant, the velocity will be tangential to the oscillating path. In model tests this condition is used for deriving the expression for the phase angle, which is not required for the numerical simulation The model kinematic parameters are represented as

$$\text{Yaw angle,} \quad \psi = -\psi_a \cos \omega t \tag{25}$$

$$\text{Yaw rate,} \quad \dot{\psi} = r = r_a \sin \omega t \tag{26}$$

$$\text{Yaw acceleration,} \quad \ddot{\psi} = \dot{r} = r_a \omega \cos \omega t \tag{27}$$

$$\text{Forward velocity,} \quad u = \sqrt{(V^2 + \dot{y}_0^2)} = u_{cN} + u_{aN} \cos 2\omega t \tag{28}$$

$$\text{Forward acceleration,} \quad \dot{u} = -2u_{aN}\omega \sin 2\omega t \tag{29}$$

$$\text{Sway displacement velocity and acceleration,} \quad y = v = \dot{v} = 0 \tag{30}$$

where

$$u_{aN} = \frac{y_{0a}^2 \omega^2}{4V} \quad \text{and} \quad u_{cN} = V + u_{aN}$$

In pure yaw mode oscillations, the mathematical model given by Eqs. (4)–(6) have only yaw motion dependent terms. Hence these equations are simplified to:

$$X_{HN} = X_{\dot{u}}\dot{u} + X_{uu}u|u| + X_{rr}r^2 \tag{31}$$

$$Y_{HN} = Y_{\dot{r}}\dot{r} + Y_r r + Y_{rrr}r^3 \tag{32}$$

$$N_{HN} = N_{\dot{r}}\dot{r} + N_r r + N_{rrr} r^3 \tag{33}$$

By expanding the hydrodynamic reaction forces and moment in Fourier series, and substituting Eqs. (25)–(29) in Eqs. (31)–(33) and following up by comparing like terms in the corresponding equations by putting $v = 0$, we get the derivatives of the pure yaw defined in Eqs. (34)–(41):

$$X_{\dot{u}} = \frac{-1}{2u_{aN}} b_{xN2} \tag{34}$$

$$X_{rr} = \frac{2}{r_a^2}\left[\frac{2u_{cN}u_{aN}}{V^2}(a_{xY0} - a_{xY2}) - a_{xN2}\right] \tag{35}$$

$$Y_{\dot{r}} = \frac{1}{r_a \omega} a_{yN1} \tag{36}$$

$$Y_{rrr} = \frac{-4}{r_a^3}\left[\frac{b_{xN2} r_a}{4\omega} + b_{yN3}\right] \tag{37}$$

$$Y_r = \frac{1}{r_a}\left[b_{yN1} + \frac{b_{xN2} r_a}{2u_{aN}\omega}\left(u_{cN} - \frac{u_{aN}}{2}\right) + 3\left(\frac{b_{xN2} r_a}{4\omega} + b_{yN3}\right)\right] \tag{38}$$

$$N_{\dot{r}} = \frac{1}{r_a \omega} a_{nN1} \tag{39}$$

$$N_{rrr} = \frac{-4}{r_a^3} b_{nN3} \tag{40}$$

$$N_r = \frac{1}{r_a}\left[b_{nN1} + 3b_{nN3}\right] \tag{41}$$

3.2.3 Combined Motion

In this mode the combined sway and yaw motions are induced upon the model by producing yaw motion on a drifted model and repeating the procedure for various drift angles in the experiment. In a numerical simulation both sway and yaw oscillations can be simultaneously imposed on the model. The kinematic parameters of the model in earth-fixed co-ordinate system are transformed to ship fixed co-ordinate system using ψ:

$$u = V\cos\psi + \dot{y}_0 \sin\psi \tag{42}$$

$$v = -V\sin\psi + \dot{y}_0 \cos\psi \tag{43}$$

we consider,

$$\psi \leq 12^\circ;\text{then } \cos\psi = 1; \sin\psi = \psi \tag{44}$$

$$u = u_{cYN} + u_{aYN}\cos 2\omega t \tag{45}$$

$$\dot{u} = -2u_{aYN}\omega \sin 2\omega t \tag{46}$$

$$v = v_{aYN}\cos\omega t \tag{47}$$

$$\dot{v} = -v_{aYN}\omega \sin\omega t \tag{48}$$

where $v_{aYN} = V\psi_a - y_{0a}\omega$, $u_{aYN} = \frac{y_{0a}\omega\psi_a}{2}$ and $u_{cYN} = V + u_{aYN}$.

The yaw motion equations remain same, due to the pure yaw mode as defined in Eqs. (25)–(27), for the both earth-fixed and ship-fixed co-ordinate systems. As the model undergoes combined sway and yaw mode oscillations, the mathematical model defined by Eqs. (4)–(6) have both sway and yaw motion dependent terms. Hence,

$$X_{HYN} = X_{\dot{u}}\dot{u} + Y_{\dot{v}}vr + X_u(u) + X_{vr}vr + X_{vv}v^2 + X_{rr}r^2 \tag{49}$$

$$Y_{HYN} = X_{\dot{u}}ur + Y_{\dot{v}}\dot{v} + Y_{\dot{r}}\dot{r} + Y_v v + Y_r r + Y_{vvv}v^3 + Y_{rrr}r^3 + Y_{vvr}v^2 r + Y_{vrr}vr^2 \tag{50}$$

$$N_{HYN} = N_{\dot{v}}\dot{v} + N_{\dot{r}}\dot{r} + N_v v + N_r r + N_{vvv}v^3 + N_{rrr}r^3 + N_{vvr}v^2 r + N_{vrr}vr^2 \tag{51}$$

By expanding the hydrodynamic forces and moments in Fourier series and substituting Eqs. (25)–(27) and (45)–(48) in Eqs. (49)–(51) and comparing the like terms in the corresponding equations putting $\gamma = 0$, we get the derivatives in combined motion given by the Eqs. (52)–(56):

$$X_{vr} = \frac{2}{v_{aYN}r_a}b_{xYN2} - \frac{b_{yY1}}{y_{0a}\omega^2} - b_{xN2}\frac{u_{aYN}}{u_{aN}} \tag{52}$$

$$Y_{vvr} = \frac{4}{v_{aYN}^2 r_a}\left[b_{yYN3} - \frac{b_{xN2}r_0}{4\omega}\left(\frac{u_{aYN}}{u_{aN}} + 1\right) - b_{yN3}\right] \tag{53}$$

$$Y_{vrr} = \frac{-4}{v_{aYN}r_a^2}\left[a_{yYN3} + a_{yY3}\left(\frac{v_{aYN}}{v_a}\right)^3\right] \tag{54}$$

$$N_{vvr} = \frac{4}{v_{aYN}^2 r_a}[b_{nYN3} - b_{nN3}] \tag{55}$$

$$N_{vrr} = \frac{-4}{v_{aYN} r_a^2}\left[a_{nYN3} + a_{nY3}\left(\frac{v_{aYN}}{v_a}\right)^3\right] \tag{56}$$

A detailed experimental procedure for carrying out model tests is explained in [15].

3.3 HPMM Numerical Simulations

Numerical methods are more flexible and thus numerous design versions can be analysed. They are efficient in performing simulations, analysing results through flow visualizations and creating related animations. Every small detail of flow patterns can be easily obtained spotting it any location around the ship hull, which may not be possible while conducting experiments. In model tests, such as the free-running tests are, it might be very easy to study the influence of appendages on the manoeuvrability of the vessel while in the numerical modelling the ship with appendages is a tedious task to create it. Once this is overcome, the rest of the processes can result simpler than experiments.

The HPMM simulation requires a numerical computational domain and a solver capable of solving the fluid flow equations, while the mesh in the fluid domain is re-oriented due to the hull motion. ANSYS ICEM CFD has been used for the mesh generation and the dynamic mesh motion option of the commercial CFD package ANSYS CFX has made possible to bring in the prescribed ship body motion.

3.3.1 Grid System

In the present work a structured grid system with hexahedral cells has been used with near wall refinement. The meshing has been carried out in ANSYS ICEM CFD. In order to avoid the near-wall deformation of the elements, the interface meshing strategy has been adopted, where the ship hull moves within an inner computational domain. CFX uses a 'mesh morphing' model that calculates the new node positions at each time step, while maintaining the basic mesh topology. However, the nodal displacements are calculated by using a spring analogy method. The outer and inner computational domains have a common merged interface and fluxes are transferred across it, keeping intact the near-wall mesh around the ship model. In order to avoid the complexity of sharp corners, a cylindrical inner domain with circular faces has been applied. This type of meshing strategy also helps in capturing the free-surface effects. The outer domain extends 1.5 L from the aft, 0.8 L from the bow, 1 L from each of port and star board sides and the bottom

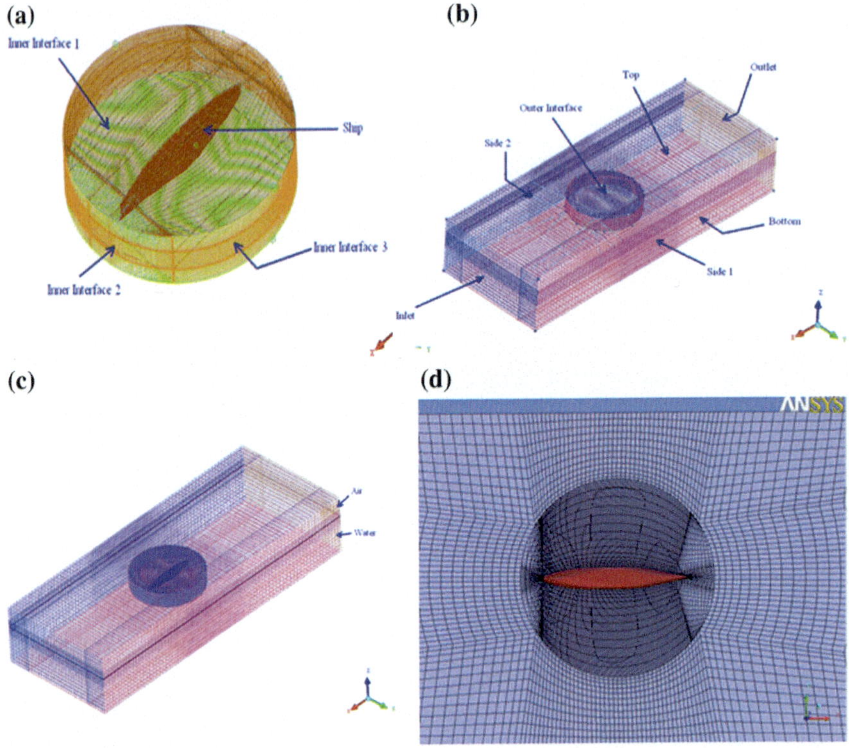

Fig. 3 3-dimensional structured grid system. **a** Inner domain, **b** outer domain, **c** combined domain and **d** mesh cut plane from the top

extends 0.6 L from the keel indicating the deep water conditions. The generated mesh has 0.6 million hexahedral cells. A more refined mesh couldn't be used because of both, computational time and resource constraints. The above described mesh is shown in Fig. 3.

3.3.2 Turbulence Model, Boundary Conditions and Solver Settings

The 2-equations k-ω shear stress transport (SST) model has been used for the calculations, by making use of wall and boundary distance formulations [16]. As a result, the near wall regions are well resolved ($y^+ < 2$).

No-slip wall boundary condition has been imposed on the ship hull, side and bottom boundaries [17]. Since the boundaries have been chosen sufficiently far away from the hull, it was assumed that there are no interactions between these boundaries due to viscous flow past them. The free-stream conditions prevail on the top. An inlet flow velocity of 1 m/s has been imposed at the inlet boundary for both air and water and the flow leaves the outlet with outflow conditions. The outer and inner interfaces have been merged to a single one and interface boundary condition has been assigned to it.

The space and time discretization use the high resolution and second order backward Euler schemes, respectively. A homogeneous coupled Volume Of Fluid (VOF) method has been adopted for capturing the free-surface effects. For convergence limiter, the RMS values have been set to a target value of 10^{-5}.

3.3.3 Simulations

Grid independency has not been carried out, but it has been well ensured that the grid is capable enough of producing reliable results. Steady state simulations have been initially carried out with objective to obtain converged results, and to initialize the mesh motions. The drag force on the ship hull has been found to match well with experimental and theoretical values. In order to carry out the HPMM tests in a numerical towing tank, ship body along with the inner domain has been subjected to pure sway, pure yaw and combined sway and yaw sinusoidal motions. The air and water flow velocity are taken as 1.0 m/s. Table 2 shows the values of various simulation constants. Motions have been brought in by using the transformations (translation and rotation) of co-ordinate axes, and the simulations time step has been calculated as 0.01 s. Since the solution follows an implicit scheme, the time step is not of major concern. The Courant number has been fixed to 0.5 and a smallest cell size to be 0.02 m in the flow direction. At each time step the mesh is updated with respect to its position obtained at the previous time step.

3.3.4 Pure Sway

The transverse motion of the ship along with the inner domain has been accomplished using the sinusoidal function given in Eq. (8). The mesh motions have been accomplished using the following translation of co-ordinate axes in CFX.

Table 2 Simulation constants

Constants	Values
T	13.33 s
ω	0.4714 rad/s
ω'	2.28
V	1 m/s
y_{0a}	0.3 m
y'_{0a}	0.0617
$\psi_a = \psi'_a$	0.19 rad
Re	4.86×10^6
Fr	0.145
β_a	12°

$$[x' \quad y' \quad z']^T = [x \quad (y + y_0) \quad z]^T \tag{59}$$

3.3.5 Pure Yaw

In order to rotate the ship model and the inner domain about z-axis, the sinusoidal function, see Eq. (25), has been used. The mesh motions have been accomplished by using the following rotation of co-ordinate axes in CFX.

$$\begin{Bmatrix} x' \\ y' \\ z' \end{Bmatrix} = \begin{Bmatrix} x \cos \psi + y \sin \psi \\ -x \sin \psi + y \cos \psi \\ z \end{Bmatrix} \tag{60}$$

3.3.6 Combined Sway and Yaw

The mesh motions in this case are the superimposition of the above 2 cases and the resulting transformations in CFX are shown in Eq. (61):

$$\begin{Bmatrix} x' \\ y' \\ z' \end{Bmatrix} = \begin{Bmatrix} x \cos \psi + y \sin \psi \\ -x \sin \psi + y \cos \psi + y_0 \\ z \end{Bmatrix} \tag{61}$$

4 Results and Discussions

The solution process has been carried out for 3 cycles in and for all 3 modes- pure sway, pure yaw and combined motions. The mesh displacements and pressure distributions around the hull for extreme motions have been shown in Figs. 4, 5 and 6. The first cycle has shown higher values with lots of fluctuations due to transience. The 2nd and 3rd cycles have shown constant trends with same maximum values. This is demonstrated as shown in Fig. 7. Hence the 3rd cycle was chosen for the analysis, as CFX uses a global reference frame for the calculations that does not move with the body. The conversion of the applied axes system to ship-fixed counterpart is almost impossible with CFX. Since the formulations had been with respect to Ship-Fixed Co-ordinate System (SFCS), the corresponding transformations have been done through a MATLAB code, where the time history of a force/ moment has been calculated in both the co-ordinate systems taken from experiments for setting the hydrodynamic derivatives and assumed body motions. The difference between the values in these 2 co-ordinate systems have been finally subtracted from the corresponding CFD results. These kind of transformations are

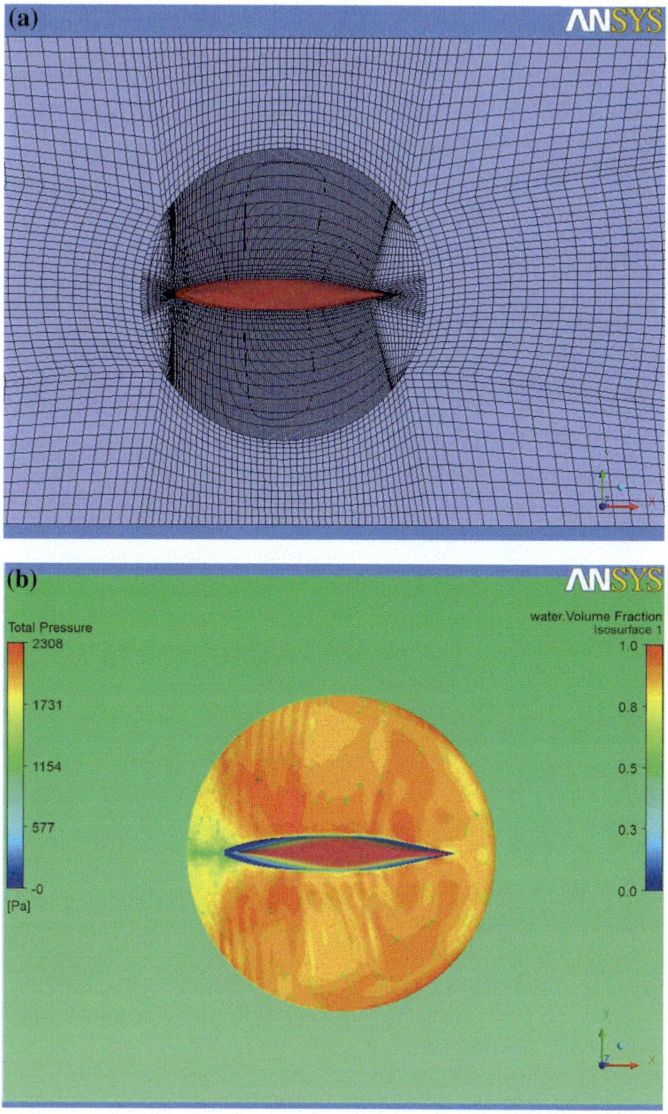

Fig. 4 Extreme sway motion. **a** Mesh displacement and **b** total pressure distribution around the hull

required in cases where the rotations of the co-ordinate system is involved and thus, no transformations were required for pure-sway case.

Section 3.2 shows the complete formulation of the problem and the derivation of the expressions for the calculation of the hydrodynamic derivatives. All the expressions for hydrodynamic derivatives have been defined in Eqs. (17)–(24),

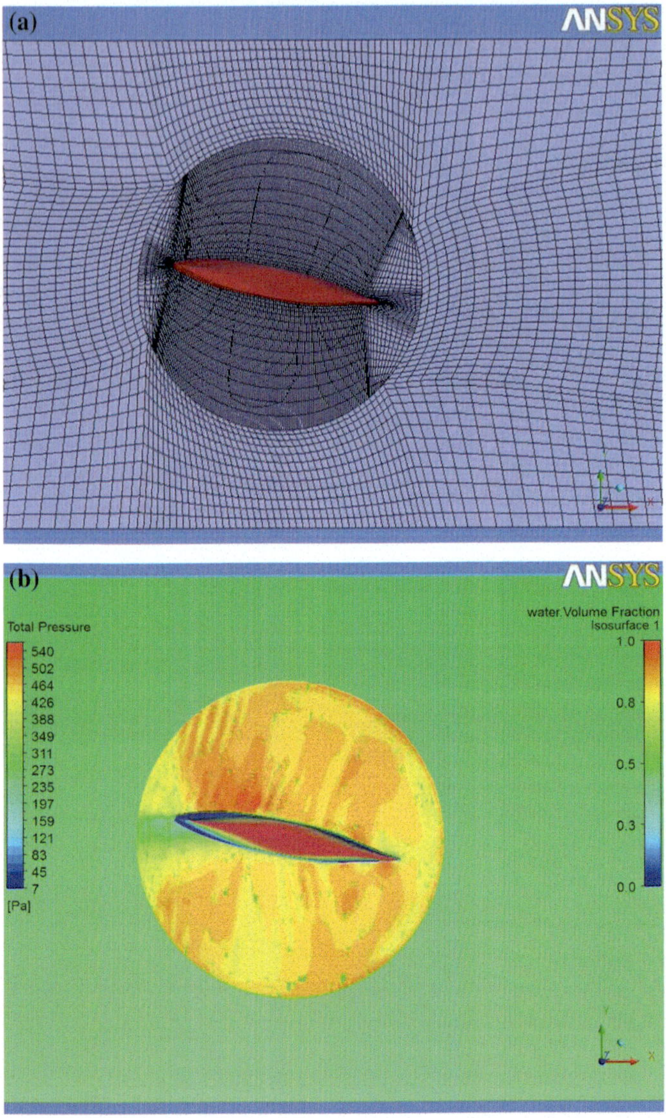

Fig. 5 Extreme yaw motion. **a** Mesh displacement and **b** total pressure distribution around the hull

Eqs. (34)–(41) and Eqs. (52)–(56) and have been expressed in terms of Fourier constants and hence, their evaluation required as influence a lot the simulations results. It is important to mention, that they have been obtained by the numerical integration of the time histories of forces and moments over a cycle using Simpson's rule or by obtaining the definite integral of the equation of the fitted

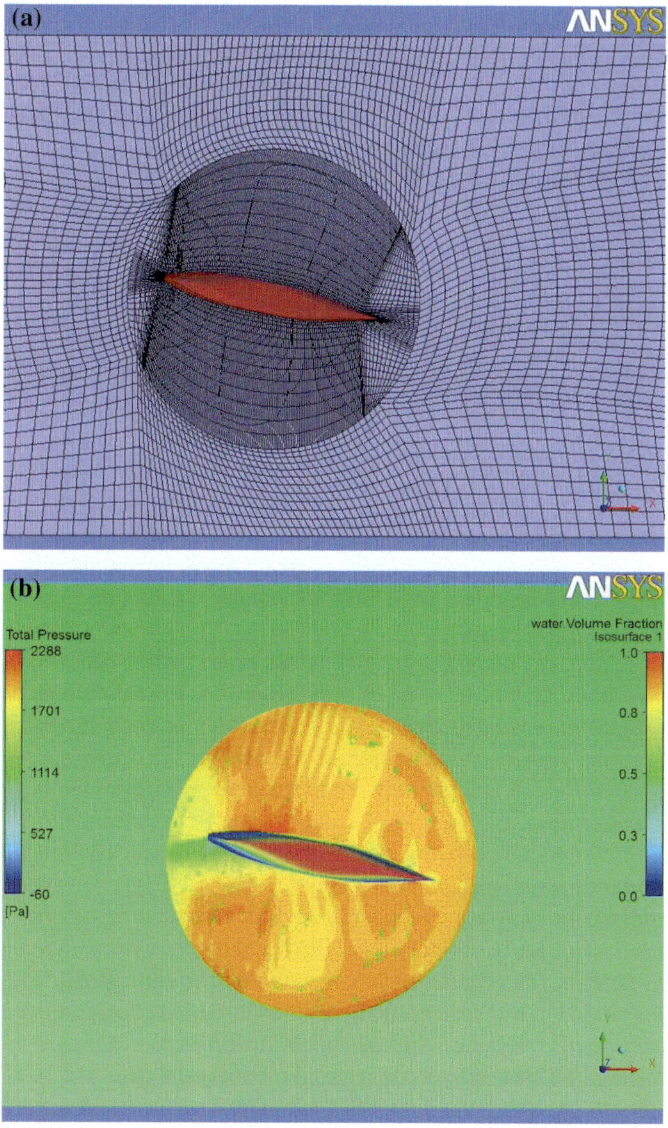

Fig. 6 Extreme combined motion. **a** Mesh displacement and **b** total pressure distribution around the hull

curve, as a function of time. The derivatives for the model have been obtained in the dimensional form initially, and then made non-dimensional to ease their representation. The non-dimensional factors for forces and moment are $\frac{1}{2}\rho L^2 V^2$ and $\frac{1}{2}\rho L^3 V^2$, respectively. Figures 8, 9 and 10 show the force and moment time histories in each mode of mesh motion. These figures show that the obtained numerical

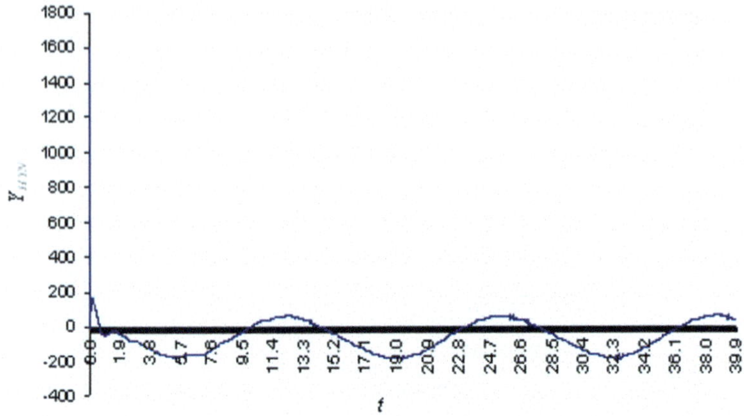

Fig. 7 Three cycles of combined motion simulation

simulations results do not represent a smooth curve, at each time step. Thus, it is appropriate to fit a smooth curve in order to obtain the Fourier constants accurately. As a relatively coarse mesh has been used, the values of forces and moments do not represent a very smooth curve at various time steps. The spikes must be eliminated and need to be replaced with interpolated values by using the values from previous and later time steps. However, some of them contribute towards higher order terms too. These values have been retained as they are. The Fourier coefficients have been obtained from the numerical integration of the curves. The experience has shown that Fourier coefficients are accurately predicted when smoother curves are used for integration.

Table 3 shows the values of hydrodynamic derivatives calculated through the presented method, as adopted in this chapter and their comparison with published experimental values [11] are given. The derivatives have an assigned symbols depending on their relative importance [18] for the trajectory prediction and stability characteristics. The forces and moments obtained through CFD simulations are smoothened and have been transformed into ship-fixed coordinate system (sfcs) and they have been compared with theoretical time histories and corresponding to the experimental values for hydrodynamic derivatives in sfcs. The deviations are found not very high. Since the forces and moments have been predicted well, it is implied that the derivatives have been also well predicted. The numerical simulations of HPMM using RANSE based CFD methods has been found to work well at least for moderate motion amplitudes. The Fourier series expansion method or the harmonic analysis, used in this work, for defining derivatives is indeed a promising, and has been successful enough to define all the 21 derivatives-coupled, uncoupled and cross-coupled, as required by the mathematical model. In the present case, the deviations with respect to the experimental results are found to follow a particular pattern. The uncoupled derivatives have been observed to be well predicted (0.8–6%). The coupled ones show more deviation (5–11%) and the cross-coupled

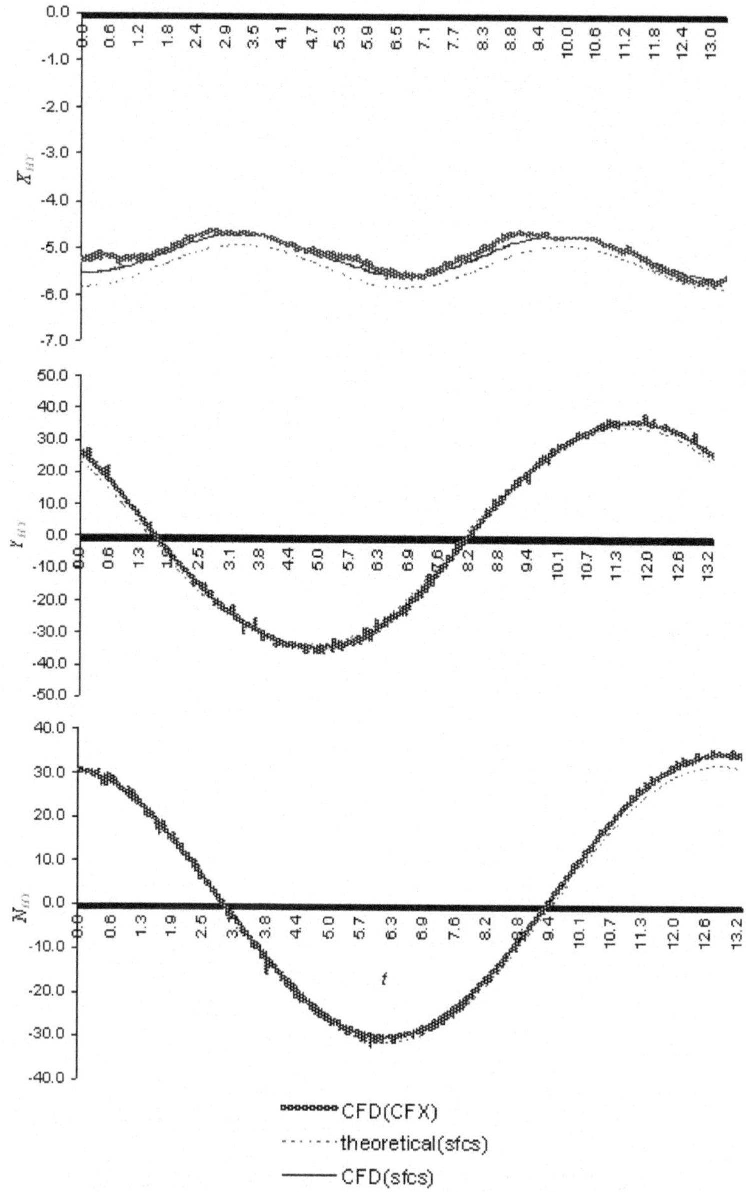

Fig. 8 Time histories of forces and moment in pure sway

show furthermore deviation (10–12%). The reason might be due to the relative coarseness of the mesh, for which the coupling effect of body motions and fluid flow has been not very well captured. However, the improvements might be possible applying better refinement techniques for meshing.

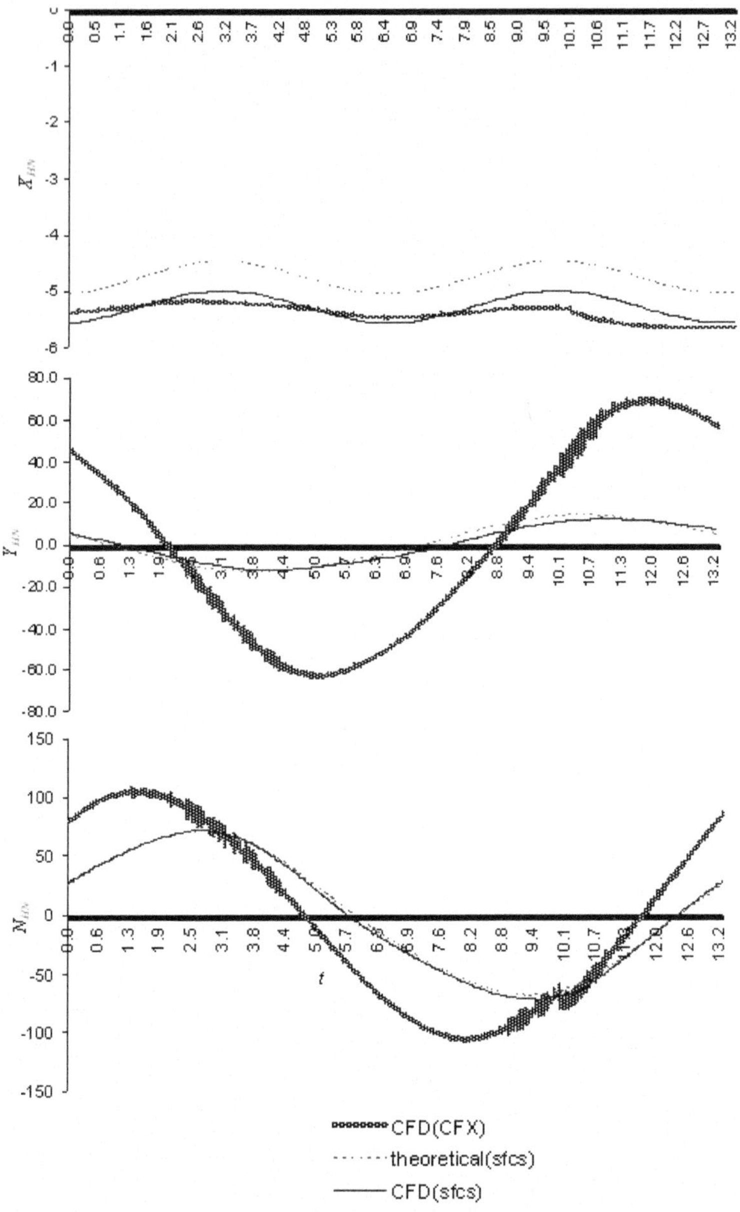

Fig. 9 Time histories of forces and moment in pure yaw

The variables assigned to the derivatives in assessing their relative importance in Table 3 indicates that the most of the derivatives with higher relative importance have been well predicted. As manoeuvring is concerned with low frequency,

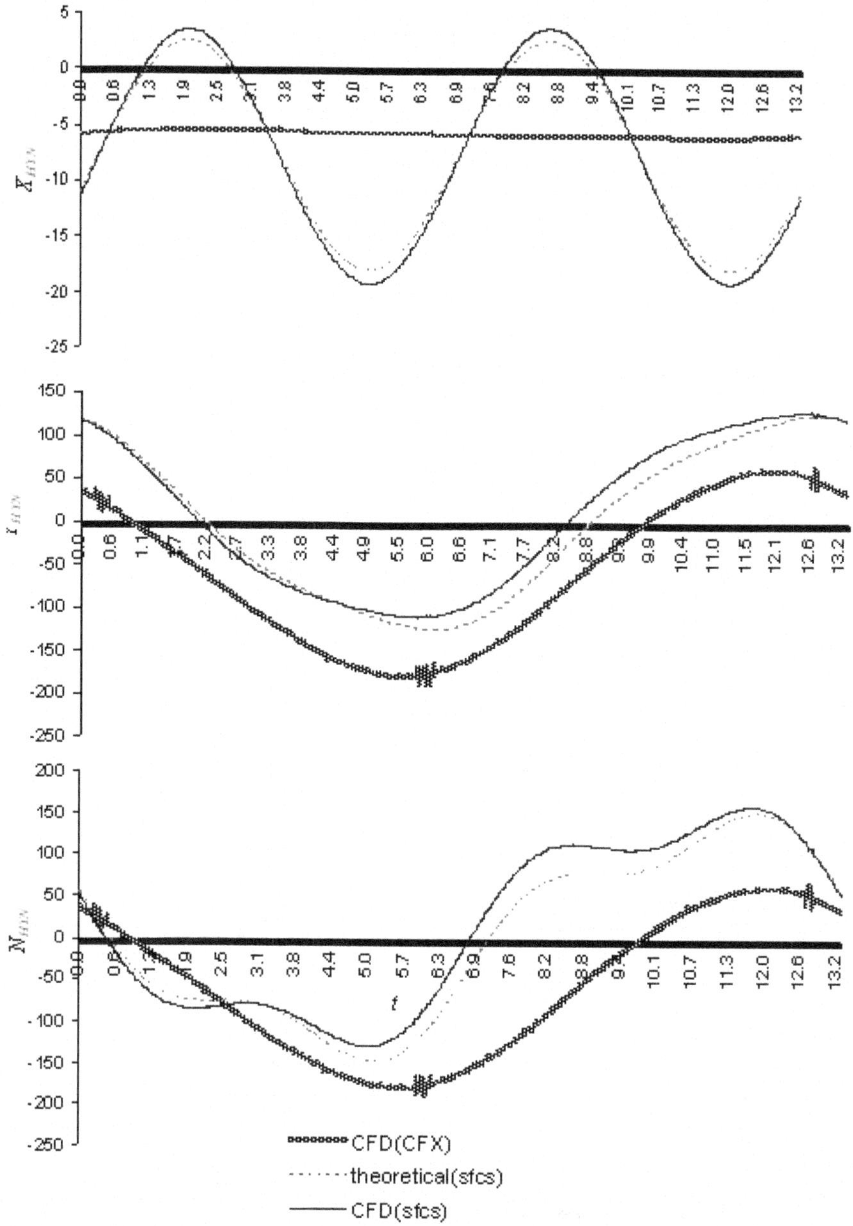

Fig. 10 Time histories of forces and moment in combined motion

the frequency of oscillation has been restricted to be at lowest possible level. Further frequency lowering has been found to result in unwanted oscillations in the time traces.

Table 3 Hydrodynamic coefficients in manoeuvring

Hyd. deri.	Non-dim. factor	Type	Present CFD	Expt. [11]	Deviation (%)	Category [18]
$X'_{\dot{u}}$	$0.5\rho L^3$	Uncoupled	−0.00025	−0.00024	4.16	*A*
$X'_{u\|u\|}$	$0.5\rho L^2$	,,	−0.00044	−0.00042	4.76	*A*
X'_{vv}	$0.5\rho L^2$	Coupled	−0.00365	−0.00386	−5.44	*C*
X'_{rr}	$0.5\rho L^4$	,,	0.00021	0.0002	5.01	*C*
X'_{vr}	$0.5\rho L^3$	Cross-coupled	−0.00345	−0.00311	10.93	*D*
$Y'_{\dot{v}}$	$0.5\rho L^3$	Uncoupled	−0.00699	−0.00705	−0.8	*A*
Y'_{v}	$0.5\rho L^2 V$	,,	−0.0118	−0.0116	1.7	*A*
Y'_{vvv}	$0.5\rho L^2 V$	,,	−0.111	−0.109	1.8	*C*
$Y'_{\dot{r}}$	$0.5\rho L^4$	Coupled	−0.000372	−0.00035	6.57	*A*
Y'_{r}	$0.5\rho L^3 V$	Coupled	0.00264	0.00242	9.01	*A*
Y'_{rrr}	$0.5\rho L^5/V$	Coupled	0.00192	0.00177	8.47	*D*
Y'_{vvr}	$0.5\rho L^3/V$	Cross-coupled	0.0239	0.0214	11.68	*A*
Y'_{vrr}	$0.5\rho L^4/V$	,,	−0.0507	−0.0405	10.09	*D*
$N'_{\dot{v}}$	$0.5\rho L^4$	Coupled	−0.000384	−0.00035	9.71	*A*
N'_{v}	$0.5\rho L^3 V$	,,	−0.00421	−0.00385	9.35	*A*
N'_{vvv}	$0.5\rho L^3/V$	,,	0.00166	0.001492	11.41	*C*
$N'_{\dot{r}}$	$0.5\rho L^5$	Uncoupled	−0.000408	−0.00042	−2.85	*A*
N'_{r}	$0.5\rho L^4 V$	,,	−0.00234	−0.00222	5.41	*A*
N'_{rrr}	$0.5\rho L^6/V$	,,	−0.00241	−0.00229	5.24	*D*
N'_{vvr}	$0.5\rho L^4/V$	Cross-coupled	−0.0471	−0.0424	11.08	*A*
N'_{vrr}	$0.5\rho L^5/V$	Cross-coupled	0.00175	0.00156	12.17	*D*

A—Derivatives deemed of major importance
C—Derivatives deemed of minor importance
D—Derivatives deemed negligible

Since most of the important derivatives have been predicted well, the trajectories do not deviate much from the actual trajectory in definitive manoeuvres. This fact has been illustrated using Fig. 11. A program has been developed in MATLAB for the numerical integration (4th order Runge-Kutta method) of the equations of motions represented by the mathematical model of Son and Nomoto [11]. Two manoeuvres have been simulated (1) turning circle and (2) zig-zag manoeuvres. From Fig. 11a, b, it is evident that the trajectories have been predicted quite well by using the presented CFD simulation approach.

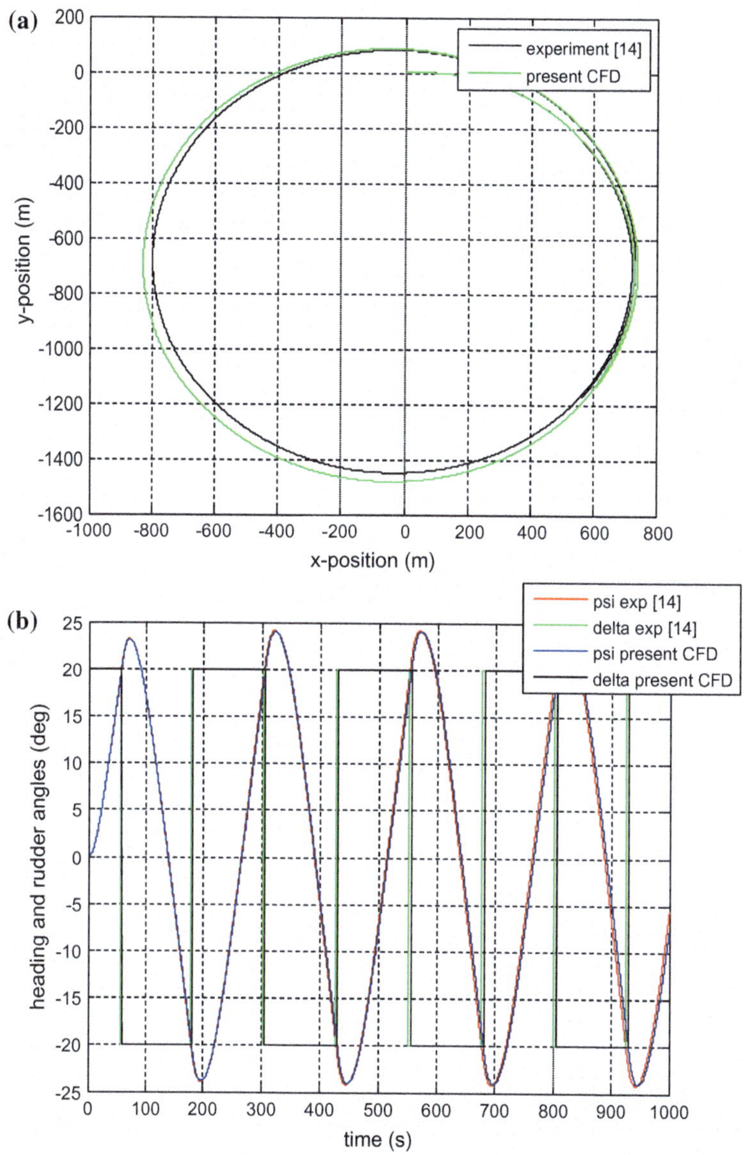

Fig. 11 Simulated trajectory of the vessel. **a** Turning circle manoeuvre for 35° rudder angle and **b** 20/20 zig-zag manoeuvre

5 Conclusions

The Fourier series expressions formulated here for the calculation of hydrodynamic derivatives can be used for any ship numerical simulations, provided that the time histories of forces and moments are obtained through simulations or model testing. The present work finds RANSE based CFD, as a promising alternative in the determination of all the hydrodynamic derivatives required in a mathematical model and hence can be considered as a reliable tool in the manoeuvrability assessment of vessels, especially useful for the early ship design phase. On the whole, this work finds its success in exploring and exploiting various features of a general-purpose fluid flow solver to solve a non-linear dynamic ship manoeuvring problem.

References

1. Toxopeus SL (2009) Deriving mathematical maneuvering models for bare ship hulls using viscous flow calculations. J Mar Sci Technol 14:30–38
2. Carrica PM, Stern F (2008) DES simulations of KVLCC1 in turn and zig-zag maneuvers with moving propeller and rudder. In: SIMMAN workshop on verification and validation of ship maneuvering simulation methods, pp F11–F16
3. Janardhanan S, Krishnankutty P (2009) Numerical estimation sway-dependent non-linear hydrodynamic derivatives in surface-ship maneuvering. In: Proceedings of the third international conference in ocean engineering, Chennai, India
4. Nonaka K, Miyazaki H, Nimura T, Ueno M, Hino T, Kodama Y (2007), Calculation of hydrodynamic forces acting on a ship in maneuvering motion. In: Research paper. Ship Research Institute, Japan, pp 307–317
5. Sulficker N (2007) RANSE based estimation of hydrodynamic forces acting on ships in dynamic conditions. M.S. thesis, Department of Ocean Engineering, IIT Madras, pp 50–82
6. Cura-Hochbaum A (2006) Prediction of hydrodynamic coefficients for a passenger ship model. In: International conference in marine hydrodynamics, pp 933–942
7. Cura-Hochbaum A, Vogt M, Gatchell S (2008) Maneuvering prediction for two tankers based on RANS simulations. In: SIMMAN workshop on verification and validation of ship maneuvering simulation methods, pp F23–F28
8. Xing-Kaeding Y, Jensen AG (2006) Simulation of ship motions during maneuvers. Ship Technol Res Schiffstechnick 53(4):14–28
9. Tyagi A, Sen D (2006) Calculation of transverse hydrodynamic coefficients using computational fluid dynamic approach. Ocean Eng 33:798–809
10. Ohmori T (1998) Finite-volume simulation of flows about a ship in maneuvering motion. J Mar Sci Technol 3:82–93
11. Son KH, Nomoto K (1981) On the coupled motion of steering and rolling of a high speed container ship. J Soc Naval Archit Jpn 150:73–83
12. Abkowitz MA (1964) Testing techniques used for the measurement of hydrodynamic derivatives. In: Lectures on ship hydrodynamics—steering and manoeuvrability, pp 53–62
13. Bishop RED, Parkinson AG (1970) On the planar motion mechanism used in ship model testing. Philos Trans Royal Soc Lond Ser A Math Phys Sci 266(1171):35–61
14. Crane CL, Eda H, Landsburg A (1989) Controllability, principles of naval architecture, vol 3, pp 200–258
15. ITTC (2002) ITTC recommended procedures manoeuvrability—captive model test procedure. In: Proceedings of 23rd ITTC

16. ANSYS CFX-solver theory guide, release 15 (2013)
17. Bertram V (2000) Practical ship hydrodynamics. Butterworth Publications, Oxford
18. Strom-Tejsen J (1965) A digital computer technique for the prediction of standard manoeuvres of surface ships. DTRC report 2130

Application of Pattern Recognition Method for Estimating Wind Loads on Ships and Marine Objects

Marko Valčić, Jasna Prpić-Oršić and Dean Vučinić

Abstract Wind loads on ships and marine objects are a complicated phenomenon because of the complex configuration of the above-water part of the structure. This study presents an extension of application capabilities of elliptic Fourier descriptors (EFDs) from the usual pattern recognition and classification problems to problems of very complex nonlinear multivariable approximations of multi-input and multi-output (MIMO) functions, where EFDs are used for ship frontal and lateral closed contour representation. This approach takes into account all aspects of the variability of the above-water frontal and lateral ship profile. It is very suitable for assessing wind loads on marine structures wherever we have a wind load database for a group of similar vessels. In this way the cheaper and faster calculation can bridge the gap between ship shapes for which calculations or experiments have already been made. The Generalized Regression Neural Network (GRNN) is trained by EFDs of closed contours as inputs and wind load data derived from wind tunnel tests for a group of ships are used as targets. The trained neural network is used for the estimation of non-dimensional wind load coefficients and results for a group of offshore supply vessels, car carriers and container ships are presented and compared with the experimental data. Finally, sensitivity analysis is performed with respect to the variability of lateral container vessel contours in order to investigate how small changes in the contour geometry affect the overall estimation of wind load coefficients.

Keywords Ships · Wind loads · Elliptic Fourier descriptors · Generalized regression neural network · Sensitivity analysis

M. Valčić (✉) · J. Prpić-Oršić
Faculty of Engineering, University of Rijeka, Vukovarska 58, 51000 Rijeka, Croatia
e-mail: mvalcic@riteh.hr

J. Prpić-Oršić
e-mail: jasnapo@riteh.hr

D. Vučinić
Department of Mechanical Engineering (MECH), Department of Electronics and Informatics (ETRO), Faculty of Engineering Sciences, Vrije Universiteit Brussel (VUB), Pleinlaan 2, 1050 Brussels, Belgium
e-mail: dean.vucinic@vub.be

D. Vucinic et al. (eds.), *Advances in Visualization and Optimization Techniques for Multidisciplinary Research*, Lecture Notes in Mechanical Engineering, https://doi.org/10.1007/978-981-13-9806-3_5

1 Introduction

Although the load on ships and marine structures due to wind does not represent the most important part of environmental loads, it still plays an important role in many aspects of exposed marine structure exploitation. Accurate estimation of forces and moments caused by wind represents a challenge because of its implications for various analysis related to ship stability, ship speed estimation, manoeuvring, station-keeping and berthing. Experimental research is still the best and most reliable approach. However, experiments are very demanding and expensive. It is necessary to have a wind tunnel at disposal and build a ship model which is lifelike as much as possible. Even then, the obtained results can be used only for that specific ship. Researchers realized quite early that it is necessary to systematically perform a series of experiments and then overcome the gap between them by some interpolation method.

Isherwood [1] collected the results of wind resistance experiments carried out at different laboratories and for a wide range of ship types. After analysing those results he proposed numerical expressions derived by multiple regression analysis in terms of coefficients for lateral and transverse wind forces as well as for the yawing moment. According to Isherwood, the drag coefficients can be estimated by polynomials depending on the derived polynomial coefficients and the basic characteristics of the hull shape above the water. The polynomial coefficients depend on ship type and wind direction relative to the ship.

Gould [2] presented a numerical procedure to determine the ahead force, side force and yawing moment of ships at anchor or in motion, caused by wind from any direction acting on the superstructures of ships. The procedure is based on the simplification of frontal and lateral projections of the superstructure above the waterline which are then subdivided into convenient elements recommended as "universal elements" in order to determine the effective wind speed and the lateral centre of pressure for a gradient wind. A series of ship models were tested in a wind tunnel in uniform and gradient velocity profiles. The tests were performed for a wide range of ship types and over a range of wind directions.

Blendermann [3–5] published a series of papers making systematic collection of wind load data derived from wind tunnel tests with uniform and non-uniform air flow on a scale model. Depending on a random shape of a deck cargo with regard to type and distribution, he suggested that wind loads on ships should be analysed as statistical data. With that assumption, Blendermann [5] carried out work that led to expressions for longitudinal and side force coefficients as well as coefficients for the yaw and roll moments in terms of wind angle of attack and as a function of the frontal and lateral projected areas. Blendermann [3] concluded that with respect to aerodynamic loading there are only three relevant basic ship shapes: multiform shapes (e.g. container ships, cargo vessels, fishing vessels, etc.), rectangular cubes (e.g. car carriers, ferries, passenger liners, etc.) and longitudinally unsymmetrical shapes (e.g. offshore supply vessels, tugs, etc.). The longitudinal force against the wind direction is the aerodynamic image of these shapes.

The Oil Companies International Marine Forum (OCIMF) report [6] presented coefficients and a procedure for computing wind and current loads on Very Large Crude Carriers (VLCC), i.e. tankers in the 150,000–500,000 DWT range with both prismatic and spherical tanks. However, the method could also be applied to smaller tankers with similar geometry. The non-dimensional wind force and moment coefficients based upon data obtained from wind tunnel tests conducted at the University of Michigan in the 1960s are presented for a moored vessel with load conditions ranging from ballasted to fully loaded.

Haddara and Guedes Soares [7] proposed a universal model for the estimation of the wind loads on ships using neural network techniques. Blendermann's experimental data for 19 ships were used to train the network and obtain the universal expression for wind load estimation, unconstrained with ship type. The data for a large tanker, which were not used in the training of the network, were used to test the ability of the network to predict wind loads for a new ship. They selected four methods (Isherwood, Blendermann, Gould and OCIMF) to determine and compare wind loads on tankers in the loaded and ballast conditions with the results obtained by the neural network approach.

Andersen [8] carried out an investigation of the influence of container configuration on the deck of a 9000+ TEU container ship on wind forces through a series of wind tunnel tests. She showed the influence of container configuration on longitudinal force.

Recently, computational fluid dynamics (CFD) is increasingly used in the assessment of the impact of wind on marine structures. Compared with wind tunnel testing, CFD has many advantages, of which the most obvious is lower cost and faster performance. Obviously, this doesn't mean that such tests can replace measurements in laboratory conditions, but they are very suitable for visualization of results and for preparatory calculations. However, the results can often be deceiving because they are affected by the mesh and input data selection as well as by the choice of the CFD method. Brizzolara and Rizzuto [9] used CFD methods to investigate the wind pressure field on superstructures of large commercial ships, in particular the suction area on the main deck caused by the presence of a negative pressure field.

Wnek and Guedes Soares [10] used a CFD code to analyse wind forces on a floating LNG platform and an LNG carrier. They compared obtained results with experimental ones and achieved a reasonable agreement. There are several other authors who have dealt with this kind of problem and a short overview can be found in Wnek and Guedes Soares [10].

The new approach to wind load estimation presented in this paper is based on elliptic Fourier features of a closed contour that can be used for describing ship frontal and lateral closed contours. This obtained set of data combined with the wind load coefficients data are used to train the approximating radial basis neural network. The new methodology is simple to use and can capture variability of transverse and longitudinal ship contour.

2 Theoretical Framework and Computational Background

The approach of wind load estimation method proposed by Valčić and Prpić-Oršić [11], presented in this chapter, consists of four basic parts. Thus, phases of image editing, pre-processing and data preparation for training, validation and testing of neural network should include the following steps:

(i) *Acquisition and pre-processing of vessel images*
At first, the capturing of real images, i.e. bow and side vessel views with their transformations into a digital representation are required. This can be done by digital cameras, but other acquisition techniques can also be used as well, e.g. scanning or editing of existing photos and project/technical documentation, editing of 2D or 3D CAD models, etc. The pre-processing phase is based on further digital editing of photos/figures which can include: discarding of unimportant image parts, overall image enhancement, grey scale conversion, image segmentation, etc. It is convenient to emphasize that the most important part of the pre-processing phase is image binarization, i.e. converting background colour to white and vessel colour to black or vice versa.

(ii) *Feature extraction of frontal and lateral projections*
This phase covers the procedure of obtaining the information about vessel outer contour from the vessel binarized image. For this purpose, the Freeman chain encoding method is used [12]. Further analysis includes mathematical description of frontal and lateral projected areas. Although several computational methods for describing different shapes have been introduced so far [13, 14], elliptic Fourier descriptors (EFDs) are of particular interest for contour determination of vessel frontal and lateral projected areas [15].

(iii) *Input and target data preparation for neural network training*
Regardless of the selected neural network or some other appropriate machine learning algorithm for solving the problem of multivariable nonlinear regression, it is necessary to prepare the input-output data set of parameters for training, validation and testing. The input set includes the parameters that are based on elliptic Fourier descriptors and the output dataset includes non-dimensional wind load coefficients.

(iv) *Validating, testing and retraining of deployed neural network*
In order to determine the mapping of elliptic Fourier descriptors to wind load coefficients, generalized regression neural network [16] has been used in this paper. Training and testing were conducted for several different vessel groups, i.e. for a group of car carriers, container ships and offshore supply vessels. All vessel figures and data required for training and testing have been taken from Blendermann [5].

For the purpose of this work, training and testing were conducted for several different vessel types, i.e. for a group of car carriers, a group of container ships and a group of offshore supply vessels. Moreover, in order to analyse reliability and stability of created neural network model, a simple sensitivity analysis was performed for the same container ship but with different container configurations, i.e. for various loading conditions. All figures and characteristics were adopted from Blendermann [5] and Andersen [8].

2.1 Digital Editing of Vessel Images

Photos and/or figures of vessels that will be used for further processing must be initially edited and simplified as much as possible in a way that none of the important parts of the superstructure significant for wind loads are omitted. On the other hand, since the methodology based on Fourier descriptors assumes closeness of contours of projected areas, very complex parts of the superstructure, such as lattice girders, narrow beams, thin antennas, etc., have not been taken into account in this paper due to simplification of the proposed estimation method.

In order to determine the contours of the frontal and lateral vessel projections, each processed image needs to be transformed into a binarized form, whereas for instance, the image background can be shown in white and the vessel projection in black, or vice versa. An example of simplified image binarization and contour extraction of lateral projected area above the waterline is shown in Fig. 1.

2.2 Freeman Chain Encoding of Vessel Contour

According to Freeman [12], the closed contour of an area can be written as a chain-code in the form:

$$V = a_1 a_2 a_3 \ldots a_K, \tag{1}$$

where $a_i \in \{0, 1, 2, 3, 4, 5, 6, 7\}$ is a chain link and K is the length of the chain, i.e. the number of chain links.

Each chain link a_i has its own length

$$|a_i| = 1 + \frac{\sqrt{2} - 1}{2}[1 - (-1)^{a_i}] \tag{2}$$

and orientation

$$\varphi_i = \frac{\pi}{4} a_i \tag{3}$$

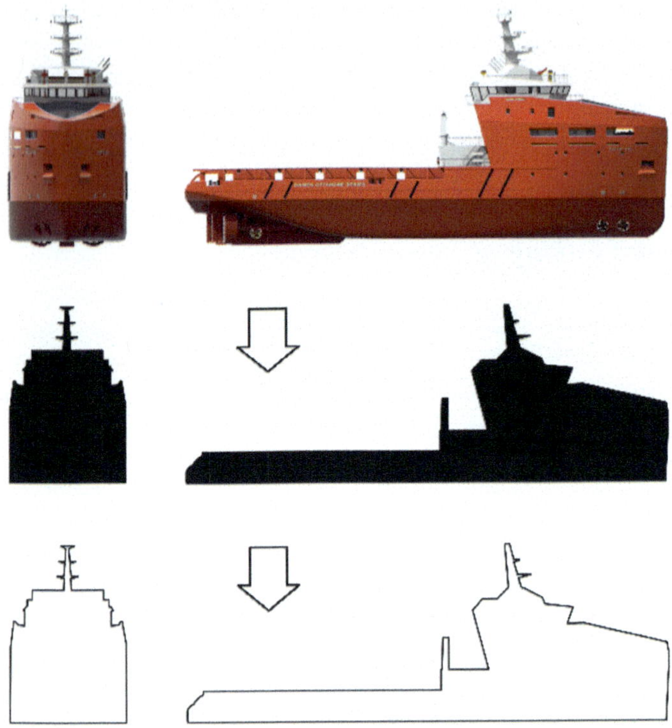

Fig. 1 Image binarization and contour extraction of lateral projected area above the waterline (PSV photo: Courtesy of Damen, https://www.damen.com/en)

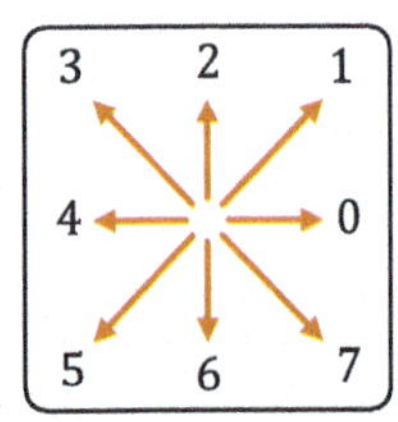

Fig. 2 Vector representation of components a_i of the chain V

in counter-clockwise direction from the positive X axis. Graphical representation of Freeman chain encoding is shown in Fig. 2.

Therefore, after the vessel image has been edited and binarized the chain encoding of frontal or lateral projected area can be performed from an arbitrary starting point. In Fig. 3 an example of chain encoding of lateral projected area procedure is presented, where the yellow square indicates the starting point and each grey square presents one image pixel. Encoding can be performed both in a clockwise or counter-clockwise direction, but for the purpose of this paper encoding

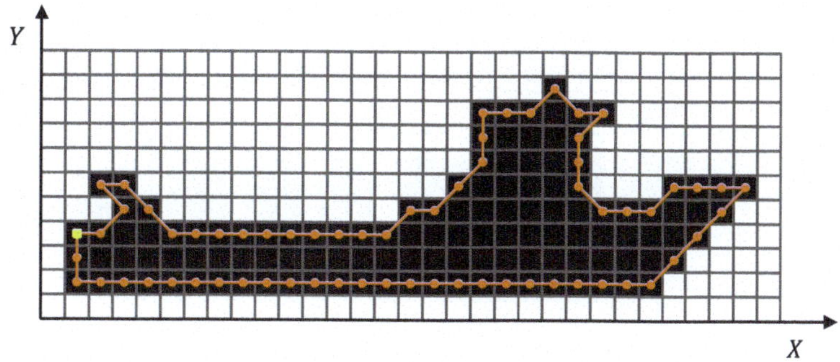

Fig. 3 Simplified contour extraction based on Freeman chain encoding for an offshore supply vessel

is done in a counter-clockwise direction. Thus, the chain code for the vessel in Fig. 3 has the following form.

$$\begin{aligned} V = {} & 660000000000000000000000011114445 \\ & 4432214354466554544444444434754. \end{aligned}$$

An example of a more complex lateral ship profile is shown in Fig. 4. In this case, the chain code has the following form.

$$\begin{aligned} V = {} & 66700000000000000000000000000000000000000000 \\ & 211144444445322244666532222446666532222446666 \\ & 532222446666532222446666532222446666532222224 \\ & 4366666665322224466665322224466665. \end{aligned}$$

Changes in the (x, y) projections of components a_i of the chain V, i.e. Δx_i and Δy_i of the chain in X-Y coordinate system (Figs. 3 and 4) can take values from the set $\{-1, 0, 1\}$ regarding the length and orientation of the chain link a_i (Fig. 2). This leads to more general expressions for evaluation of Δx_i and Δy_i.

$$\Delta x_i = \mathrm{sgn}(6 - a_i) \cdot \mathrm{sgn}(2 - a_i), \tag{4}$$

$$\Delta y_i = \mathrm{sgn}(4 - a_i) \cdot \mathrm{sgn}(a_i), \tag{5}$$

where

$$\mathrm{sgn}(x) = \begin{cases} -1, & x < 0 \\ 0, & x = 0\,. \\ 1, & x > 0 \end{cases} \tag{6}$$

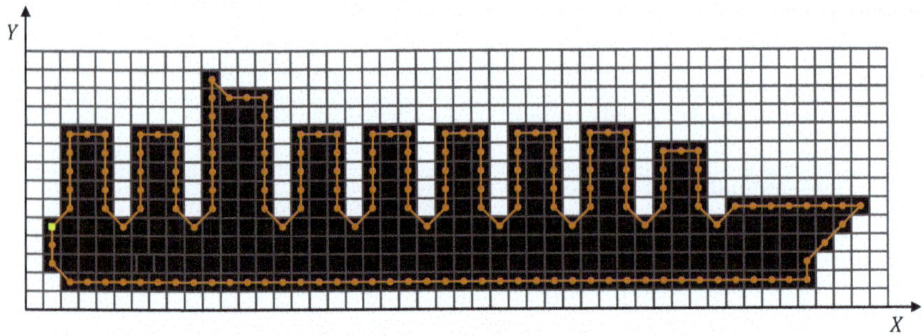

Fig. 4 Simplified contour extraction based on Freeman chain encoding for a container ship

Therefore, the (x, y) projections on the X and Y axis of the first p elements of the chain can be written as

$$x_p = \sum_{i=1}^{p} \Delta x_i, \tag{7}$$

$$y_p = \sum_{i=1}^{p} \Delta y_i. \tag{8}$$

2.3 Elliptic Fourier Features of a Closed Contour

Description of the contour can be achieved by letting the image plane be defined by the complex plane, i.e. each image pixel is represented by a complex number [17]. The first coordinate $x(t)$ represents the real part and the second coordinate $y(t)$ represents the imaginary part. Hence, a contour is defined as

$$u(t) = x(t) + jy(t), \tag{9}$$

where the parameter t is given by the arc-length parameterization and j is imaginary unit.

In order to determine the elliptic Fourier features of a closed contour, it is necessary to obtain the Fourier expansion of the contour in (9). The Fourier expansion can be performed by using the complex [17] or trigonometric representation [15]. The latter approach based on results of Kuhl and Giardina [15] is used in this paper.

If we assume that concatenation of the contour is performed with constant unit speed, then the time Δt_i required for transit over some arbitrary chain link a_i is equal to $|a_i|$. Hence, the time for p chain links is equal to

$$t_p = \sum_{i=1}^{p} \Delta t_i, \tag{10}$$

and the basic period of the complete chain code is equal to $T = t_K$.

This final conclusion about chain periodicity allows the use of Fourier analysis for approximation of a closed contour as a sum of elliptic harmonics.

The Fourier series expansion for the x and y projections of the chain code of the entire contour is defined in trigonometric matrix form as

$$\begin{bmatrix} x(t) \\ y(t) \end{bmatrix} = \begin{bmatrix} A_0 \\ C_0 \end{bmatrix} + \sum_{n=1}^{\infty} \begin{bmatrix} a_n & b_n \\ c_n & d_n \end{bmatrix} \begin{bmatrix} \cos(2\pi nt/T) \\ \sin(2\pi nt/T) \end{bmatrix}, \tag{11}$$

where $x(t)$ and $y(t)$ are piecewise linear and continuous functions, while the coefficients A_0, a_n, b_n, C_0, c_n and d_n are defined as follows

$$\begin{bmatrix} A_0 \\ C_0 \end{bmatrix} = \frac{1}{T} \int_0^T \begin{bmatrix} x(t) \\ y(t) \end{bmatrix} dt, \tag{12}$$

$$\begin{bmatrix} a_n \\ b_n \end{bmatrix} = \frac{2}{T} \int_0^T x(t) \begin{bmatrix} \cos(2\pi nt/T) \\ \sin(2\pi nt/T) \end{bmatrix} dt, \tag{13}$$

$$\begin{bmatrix} c_n \\ d_n \end{bmatrix} = \frac{2}{T} \int_0^T y(t) \begin{bmatrix} \cos(2\pi nt/T) \\ \sin(2\pi nt/T) \end{bmatrix} dt. \tag{14}$$

It is obvious that continuous-time Fourier coefficients cannot be used for contour extraction due to the image square-box quantization and the chain encoding nature. Therefore, Kuhl and Giardina [15] introduced discrete Fourier coefficients, i.e. elliptic Fourier descriptors (EFDs), for each harmonic $n = 1, 2, \ldots, N$ as

$$\begin{bmatrix} a_n \\ b_n \end{bmatrix} = \frac{T}{2\pi^2 n^2} \sum_{i=1}^{K} \frac{\Delta x_i}{\Delta t_i} \begin{bmatrix} \cos(2\pi n t_i/T) - \cos(2\pi n t_{i-1}/T) \\ \sin(2\pi n t_i/T) - \sin(2\pi n t_{i-1}/T) \end{bmatrix}, \tag{15}$$

$$\begin{bmatrix} c_n \\ d_n \end{bmatrix} = \frac{T}{2\pi^2 n^2} \sum_{i=1}^{K} \frac{\Delta y_i}{\Delta t_i} \begin{bmatrix} \cos(2\pi n t_i/T) - \cos(2\pi n t_{i-1}/T) \\ \sin(2\pi n t_i/T) - \sin(2\pi n t_{i-1}/T) \end{bmatrix}. \tag{16}$$

The inverse process allows determination of the closed contour based on N harmonics. Hence, the pixel coordinates of an i-th point of the contour can be expressed as

$$\begin{bmatrix} X_i \\ Y_i \end{bmatrix} = \begin{bmatrix} X_c \\ Y_c \end{bmatrix} + \sum_{n=1}^{N} \begin{bmatrix} a_n & b_n \\ c_n & d_n \end{bmatrix} \begin{bmatrix} \cos(2\pi n t_i/T) \\ \sin(2\pi n t_i/T) \end{bmatrix}, \tag{17}$$

where (X_c, Y_c) is the centre of gravity of the contour which can be calculated according to

$$\begin{bmatrix} X_c \\ Y_c \end{bmatrix} = \frac{1}{T} \sum_{i=1}^{K} \begin{bmatrix} \Delta x_i/(2\Delta t_i) & \xi_i \\ \Delta y_i/(2\Delta t_i) & \delta_i \end{bmatrix} \begin{bmatrix} t_i^2 - t_{i-1}^2 \\ t_i - t_{i-1} \end{bmatrix}, \tag{18}$$

$$\begin{bmatrix} \xi_i \\ \delta_i \end{bmatrix} = \sum_{m=1}^{i-1} \begin{bmatrix} \Delta x_m \\ \Delta y_m \end{bmatrix} - \frac{1}{\Delta t_i} \begin{bmatrix} \Delta x_i \\ \Delta y_i \end{bmatrix} \sum_{m=1}^{i-1} \Delta t_m \tag{19}$$

and $\xi_1 = \delta_1 = 0$.

According to (11) and (17), the truncated Fourier approximation of a closed contour can be expressed as

$$\begin{bmatrix} x(t) \\ y(t) \end{bmatrix} = \begin{bmatrix} X_c \\ Y_c \end{bmatrix} + \sum_{n=1}^{N} \begin{bmatrix} a_n & b_n \\ c_n & d_n \end{bmatrix} \begin{bmatrix} \cos(2\pi n t/T) \\ \sin(2\pi n t/T) \end{bmatrix}. \tag{20}$$

Kuhl and Giardina [15] interpreted each summand in (20) as a parameterization with parameter $t \in [0, 2\pi]$ of an ellipse that represents a visualization of the n-th Fourier coefficients. This approach explains why Fourier descriptors (15) and (16) are often referred to as elliptic.

Depending on the contour representation, the calculated Fourier descriptors can have different values with respect to different geometric transformations such as:

- phase shift of a starting point in the procedure of Freeman chain encoding,
- rotation,
- scaling,
- translation.

Several simple rules of normalization should be introduced in a way that Fourier coefficients (15) and (16) can be transformed into a form of invariant descriptors.

Fourier descriptors can be made invariant regarding the random or arbitrary selection of the starting point by introducing a phase shift, i.e. the angular rotation θ from the first semi-major axis. This phase shift can be calculated according to the following expression.

$$\theta = \frac{1}{2} \arctan \frac{2(a_1 b_1 + c_1 d_1)}{a_1^2 + c_1^2 - b_1^2 - d_1^2}, \tag{21}$$

where (a_1, b_1, c_1, d_1) presents a set of first harmonics and $0 \leq \theta < \pi$.

Now, each harmonic can be rotated by the angle $n\theta$ according to

$$\begin{bmatrix} a_n^{\theta} & b_n^{\theta} \\ c_n^{\theta} & d_n^{\theta} \end{bmatrix} = \begin{bmatrix} a_n & b_n \\ c_n & d_n \end{bmatrix} \cdot \mathbf{T}_n^{\theta} \tag{22}$$

until alignment with the first semi-major axis of the original ellipsis is achieved, where $\mathbf{T}_n^{\theta}$ is a transformation matrix defined as

$$\mathbf{T}_n^{\theta} = \begin{bmatrix} \cos(n\theta) & -\sin(n\theta) \\ \sin(n\theta) & \cos(n\theta) \end{bmatrix}. \tag{23}$$

Furthermore, the invariance of descriptors regarding rotation can be achieved if the semi-major axes of the first harmonics of each analysed photo/figure have been aligned to the original X axis. This can be done by additional harmonic rotation according to

$$\begin{bmatrix} a_n^{\phi} & b_n^{\phi} \\ c_n^{\phi} & d_n^{\phi} \end{bmatrix} = \begin{bmatrix} a_n^{\theta} & b_n^{\theta} \\ c_n^{\theta} & d_n^{\theta} \end{bmatrix} \cdot \mathbf{R}_n^{\phi} \tag{24}$$

where matrix $\mathbf{R}_n^{\phi}$ is defined as

$$\mathbf{R}_n^{\phi} = \begin{bmatrix} \cos\phi & \sin\phi \\ -\sin\phi & \cos\phi \end{bmatrix}. \tag{25}$$

Transformation matrix $\mathbf{R}_n^{\phi}$ is expressed with respect to the rotation angle $\phi, 0 \le \phi \le 2\pi$ that is obtained from (22) for $n = 1$ as

$$\phi = \arctan \frac{c_1^{\theta}}{a_1^{\theta}}. \tag{26}$$

Finally, the invariance of descriptors regarding scaling is obtained by dividing transformed descriptors (24) by the length of the semi-major axis L of the first harmonic.

$$L = \sqrt{(c_1^{\theta})^2 + (a_1^{\theta})^2}, \tag{27}$$

So the final transformed descriptors $(a_n^*, b_n^*, c_n^*, d_n^*)$ normalized with respect to the starting point, rotation and scaling can be calculated according to

$$\begin{bmatrix} a_n^* & b_n^* \\ c_n^* & d_n^* \end{bmatrix} = \frac{1}{L} \begin{bmatrix} a_n^{\phi} & b_n^{\phi} \\ c_n^{\phi} & d_n^{\phi} \end{bmatrix}. \tag{28}$$

The first harmonics of the normalized descriptors $(a_n^*, b_n^*, c_n^*, d_n^*)$ are characterized by relations

$$
\begin{aligned}
a_1^* &= 1.0 \\
b_1^* &= 0.0 \\
c_1^* &= 0.0 \\
|d_1^*| &< 1.0.
\end{aligned}
$$

If the invariance of descriptors to translation is required, the bias terms X_c and Y_c in (18) should be neglected.

2.4 Generalized Regression Neural Network

The generalized regression neural network (GRNN) is one of the most commonly used neural networks (NN) based on radial basis functions (RBF) with the aim of approximation of multivariable continuous functions. A brief period of time required for the creation and training of GRNN makes this network particularly suitable for solving any non-linear regression problem in engineering. Moreover, it is shown that even with sparse data in a multidimensional measurement space the algorithm provides smooth transitions from one observed value to another [16].

The GRNN is a one-pass learning algorithm with a highly parallel structure. This type of network is based on radial basis function and it usually consists of four layers. The input layer has n neurons, both the first hidden radial basis layer and the second hidden special linear layer have m neurons, while the output layer has only one neuron. Schematic representation of the GRNN structure [18] is shown in Fig. 5.

Once the input matrix $\mathbf{P} \in \mathbb{R}^{m \times n}$ and the target vector $\mathbf{T} \in \mathbb{R}^{m}$ are presented to the network, weight coefficient matrix $\mathbf{IW} \in \mathbb{R}^{m \times n}$ of the radial basis layer and weight coefficient matrix $\mathbf{LW} \in \mathbb{R}^{m \times m}$ of the special linear layer are formed. In this way, m neurons are created in each hidden layer.

If a certain vector $\mathbf{p} \in \mathbb{R}^{n}$ is presented as input vector to the GRNN, vector $\mathbf{n}^1 \in \mathbb{R}^{m}$ with components

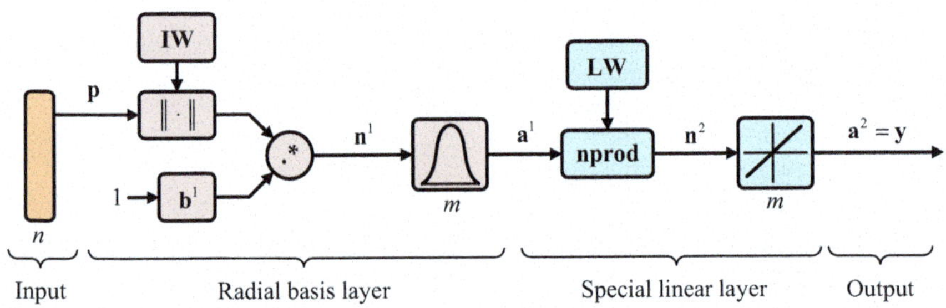

Fig. 5 Structure of a generalized regression neural network [18]

$$n_i^1 = \left\|\mathbf{IW}^i - \mathbf{p}\right\|. * b_i^1 \tag{29}$$

is formed in the radial basis layer. These components present Euclidean distance $\| \cdot \|$ between input vector $\mathbf{p}$ and all samples from the training set ($\mathbf{IW}$) multiplied by bias vector $\mathbf{b}^1 \in \mathbb{R}^m$. The term (29) presents an operation with respect to the i-th row of the matrix $\mathbf{IW}$ and bias vector $\mathbf{b}^1$, where $i = 1, \ldots, m$, and .* is dot product (element-wise multiplication).

If for a certain neuron $b = 1$ then the output value from the activation function will be equal to 0.5 for all samples which are distant 0.833 from the weighting vector of that neuron. By increasing the bias b, distance between the sample and the weighting vector will be increased as well and vice versa. The higher value of bias b entails increased spread of the sample. Spread s as a measure of spreading is directly related to the bias b, according to the expression

$$b = 0.833/s. \tag{30}$$

Output from the first hidden layer is a vector $\mathbf{a}^1 \in \mathbb{R}^m$ whose components

$$a_i^1 = \text{radbas}(n_i^1) \tag{31}$$

are a distance measure of the input vector $\mathbf{p}$ from each training sample, whereby applies: if a certain value is closer to one, the input vector $\mathbf{p}$ is closer to the associated weight vector, i.e. to a sample from the training set, and if it is closer to zero then it is farther away. Radial basis function *radbas* from (31) is defined as

$$\text{radbas}\,(x) = \exp\,(-x^2). \tag{32}$$

The components of the vector $\mathbf{a}^1$ are input data to the second linear layer in which the approximated value of the function y is determined. Components of the vector $\mathbf{n}^2 \in \mathbb{R}^m$ are determined by normalization weighting function *nprod* according to the expression

$$n_i^2 = \frac{\mathbf{a}^1. * \mathbf{LW}^i}{\sum_{i=1}^m a_i^1} \tag{33}$$

where $\mathbf{LW}^i \in \mathbb{R}^m$ is the i-th row of the linear layer weights.

The resulting vector $\mathbf{n}^2$ is finally mapped to the output layer of GRNN network by the activation function of special linear layer (*purelin*) according to

$$\mathbf{a}^2 = \mathbf{n}^2 \tag{34}$$

which ultimately gives the value of the output vector $\mathbf{y} = \mathbf{a}^2 \in \mathbb{R}^m$.

The described procedure is very similar in the case when the target vector is substituted with a target matrix $\mathbf{T} \in \mathbb{R}^{m \times r}$ where $r > 1$ represents the number of target parameters, i.e. the number of neurons in the output layer. This is related to multi-input and multi-output (MIMO) mapping. Further information and more details about the GRNN can be found in Specht [16] and Beale et al. [18].

3 Wind Load Estimation Based on Elliptic Fourier Descriptors and Generalized Regression Neural Network

3.1 Reference Frames and Notation

Geographic reference frames are illustrated in Fig. 6, where x_n and y_n denote North and East axis of North-East-Down (NED) reference frame $\{n\}$, while x_b and y_b denote surge and sway axis of body-reference frame $\{b\}$, respectively [19]. The wind speed V_w and wind direction β_w are measured by anemometers with respect to $\{b\}$, but considering the wind sensors are frequently coupled to the navigation system, in particular with gyro-compass that measures heading (ψ), results are usually provided in $\{n\}$.

The wind angle of attack γ_w is defined in $\{b\}$ with respect to x_b axis in a counter-clockwise direction. Although this is a commonly used notation in many textbooks and papers, e.g. Fossen [19] and Blendermann [5], the notation by which the wind angle of attack is denoted as α_w and is defined in $\{b\}$ with respect to x_b, but in clockwise direction, can also be applied. In that case, a simple transformation can be used.

$$\gamma_w = 2\pi - \alpha_w. \tag{35}$$

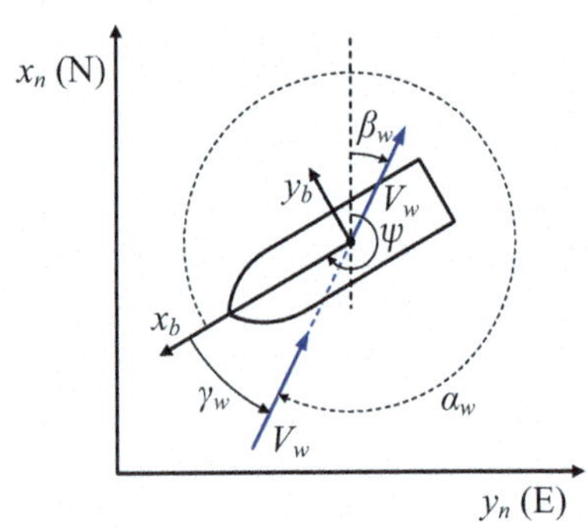

Fig. 6 Reference frames and notation [19]

3.2 Relative Wind Speed and Direction

Wind is defined as the movement of air relative to the surface of the Earth. Mathematical models of wind forces and moments are mostly used in ship motion control systems to improve the performance and robustness of the system in extreme conditions.

When the air flows over the ocean surface from any direction a natural boundary layer is formed. This means that the wind velocity at the surface is zero and increases with higher altitude. The local wind field caused by the movement of the ship does not have a boundary layer and is homogenous as illustrated in Fig. 7.

The actual wind field encountered by the part of the ship above the water surface is thus a combination of the wind field with a boundary layer and the homogenous wind field caused by the ship's forward speed. The relative wind direction γ_{rw} is found by adding the ship's wind field and the relative wind vectors as shown in Fig. 8, where u and v represent ship velocities in surge and sway, respectively, U is resulting ship speed, u_w and v_w are components of resulting absolute wind speed V_w, and γ_w is absolute wind direction.

The resulting wind field thus depends on the ship's speed and heading relative to the wind direction and the natural wind speed and direction. The relative wind speed V_{rw} is found using components of the ship speed U and components of the wind speed V_w according to

$$V_{rw} = \sqrt{u_{rw}^2 + v_{rw}^2}, \tag{36}$$

where

$$u_{rw} = u - u_w, \tag{37}$$

$$v_{rw} = v - v_w. \tag{38}$$

The relative wind angle of attack γ_{rw} can be calculated using the following relation.

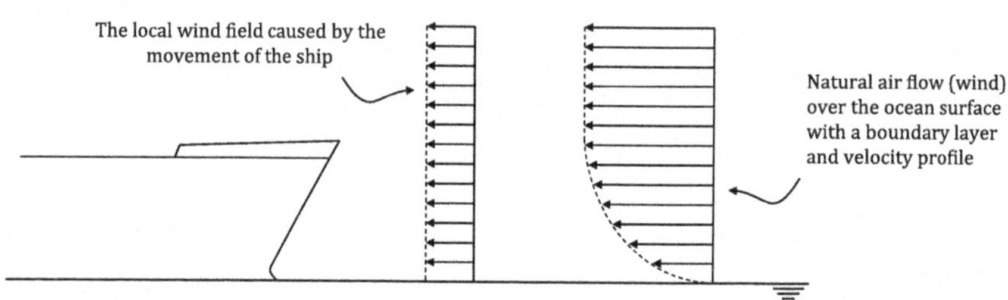

Fig. 7 The local and natural wind field near the vessel [8]

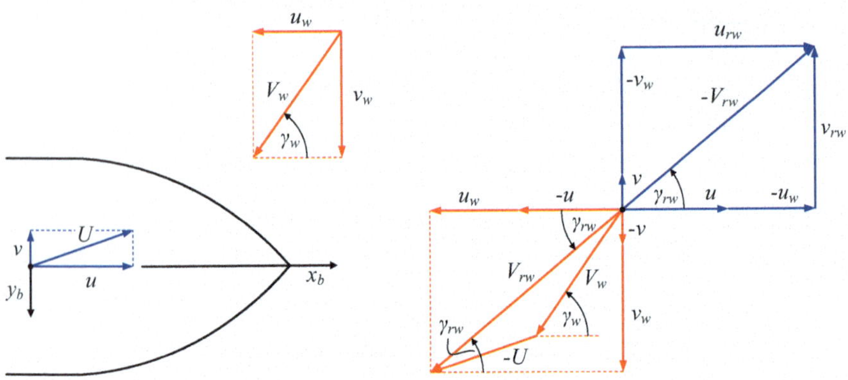

Fig. 8 Relative wind speed and direction

$$\gamma_{rw} = -\text{atan2}\left(v_{rw}, u_{rw}\right) = -\text{atan2}\left(\frac{v_{rw}}{u_{rw}}\right) = -\text{atan2}\left(\frac{v - v_w}{u - u_w}\right) \tag{39}$$

where *atan2* function, whose range is $\left[-\frac{\pi}{2}, \frac{\pi}{2}\right]$, can be expressed as follows.

$$\text{atan2}(y, x) = \begin{cases} \arctan\left(\frac{y}{x}\right), & \text{for } x > 0 \\ \arctan\left(\frac{y}{x}\right) + \pi, & \text{for } x < 0 \text{ and } y \geq 0 \\ \arctan\left(\frac{y}{x}\right) - \pi, & \text{for } x < 0 \text{ and } y < 0 \\ \frac{\pi}{2}, & \text{for } x = 0 \text{ and } y > 0 \\ -\frac{\pi}{2}, & \text{for } x = 0 \text{ and } y < 0 \\ \text{undefined} & \text{for } x = 0 \text{ and } y = 0. \end{cases} \tag{40}$$

3.3 Wind Loads on Ships and Marine Objects at Rest

Forces and moments in the horizontal plane with respect to the wind, i.e. apparent wind if the ship is moving, are the drag force *D*, positive in the wind direction, and the cross force *C*, positive to the right when facing into the wind. Only the forces in the horizontal plane are considered. The vertical wind force and the pitching moment are unimportant for the behaviour of conventional ships, though they might be for new and special ship types [5]. Due to simplicity reasons, resulting horizontal wind force is usually decomposed into two components, i.e. into longitudinal force (X_w) and transverse force (Y_w) as shown in Fig. 9.

The wind forces on a ship generally influence the ship by increasing the resistance experienced by the ship while under way. The longitudinal force generally constitutes the largest part of the total wind induced resistance. Due to the ship's forward speed the relative wind direction and thereby the longitudinal force is in most cases directed opposite the ship's heading [8].

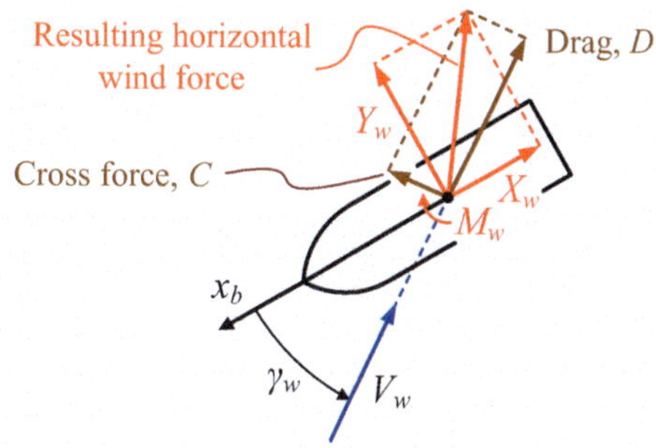

Fig. 9 Wind forces and moments in the horizontal plane [5]

The transverse force causes yaw, drift and deviation from the intended course, which can cause added resistance in two ways. The ship's heading is not aligned with the course, which gives greater resistance. The drift relative to the intended course means that the rudder angle must be increased. Increased rudder angle will also cause increased resistance [8].

For a marine vessel at rest, i.e. with zero forward speed $(u = v = 0)$, the vector of wind loads

$$\boldsymbol{\tau}_w = [X_w \quad Y_w \quad M_w]^T \tag{41}$$

in horizontal plane can be expressed in terms of the non-dimensional wind load coefficients $C_X(\gamma_w)$, $C_Y(\gamma_w)$ and $C_N(\gamma_w)$, according to

$$\begin{bmatrix} X_w \\ Y_w \\ M_w \end{bmatrix} = \frac{1}{2}\rho_a V_w^2 \begin{bmatrix} C_X(\gamma_w)A_F \\ C_Y(\gamma_w)A_L \\ C_N(\gamma_w)A_L L_{oa} \end{bmatrix}, \tag{42}$$

where X_w and Y_w are longitudinal and transverse wind forces, respectively, M_w is yawing wind moment, ρ_a is air density, A_F and A_L are the vessel's frontal and lateral projected areas above the water line, respectively, L_{oa} is the vessel's length over all, and γ_w is wind angle of attack.

It is important to emphasize that if the vessel is moving at a forward speed U then (42) should be redefined in terms of relative wind speed and relative angle of attack. The relative wind speed originating from the forward speed is a uniform field, while the meteorological wind comes with a boundary layer profile, which is neglected in this case.

Non-dimensional wind load coefficients are usually computed numerically or can be determined experimentally in wind tunnels using appropriate scaled vessel models and other necessary equipment. Based on (42), they can be expressed relative to the bow as

$$\begin{bmatrix} C_X(\gamma_w) \\ C_Y(\gamma_w) \\ C_N(\gamma_w) \end{bmatrix} = \frac{2}{\rho_a V_w^2} \begin{bmatrix} X_w/A_F \\ Y_w/A_L \\ M_w/(A_L L_{oa}) \end{bmatrix}. \tag{43}$$

The non-dimensional wind coefficients $C_X(\gamma_w)$, $C_Y(\gamma_w)$ and $C_N(\gamma_w)$ are usually computed using $h = 10$ m as reference height above the sea surface. To convert the non-dimensional wind coefficients to a different reference height, the ratio between the dynamic pressures at the two heights are usually used [19].

Except the experimental approach in wind tunnels, the values of wind load coefficients can be determined by means of various semi-empirical methods where the ones proposed by Isherwood [1], Gould [2], Blendermann [3–5] and OCIMF [6] are the most commonly used in industry.

In the following section a new approach to wind load estimation is presented which is based on the mathematical determination of the outer contours of frontal and lateral projected areas described in the previous sections [11]. In comparison with similar approaches, e.g. Haddara and Guedes Soares [7], this one includes all aspects of the variability of the above-water transversal and lateral ship profile with exception of very complex parts of the superstructure and hull equipment, particularly in cases where closeness of the contour cannot be achieved in a simple way. Multivariable nonlinear regression is conducted by means of the generalized regression neural networks.

3.4 A Novel Method for Wind Load Estimation Based on Elliptic Fourier Descriptors and Generalized Regression Neural Network

The proposed method, graphically presented in Fig. 10, can be subdivided into two characteristic parts:

(i) Model deployment based on EFDs and GRNN;
(ii) Additional independent testing and application of deployed NN model.

Therefore, in the first part of the proposed method, it is necessary to build an appropriate knowledge data-base for NN training, validation and testing. For this purpose, all available vessel images should be collected and edited in accordance to the criteria defined in Sect. 2.1. It is important to emphasize that the vessel's length (L_{oa}) and beam (B), if necessary, can be used for image scaling while frontal and lateral projected areas (A_F, A_L) can be used as control variables. The resulting binarized images of frontal and lateral projected areas should be prepared so the Freeman chain encoding can be performed in the simplest way possible.

Afterwards, the obtained chain-codes are used to calculate EFDs of frontal and lateral projected areas according to (15) and (16). Hence, we need to distinguish two ordered quadruples of EFDs, i.e. an ordered quadruple (a_n, b_n, c_n, d_n) of EFDs

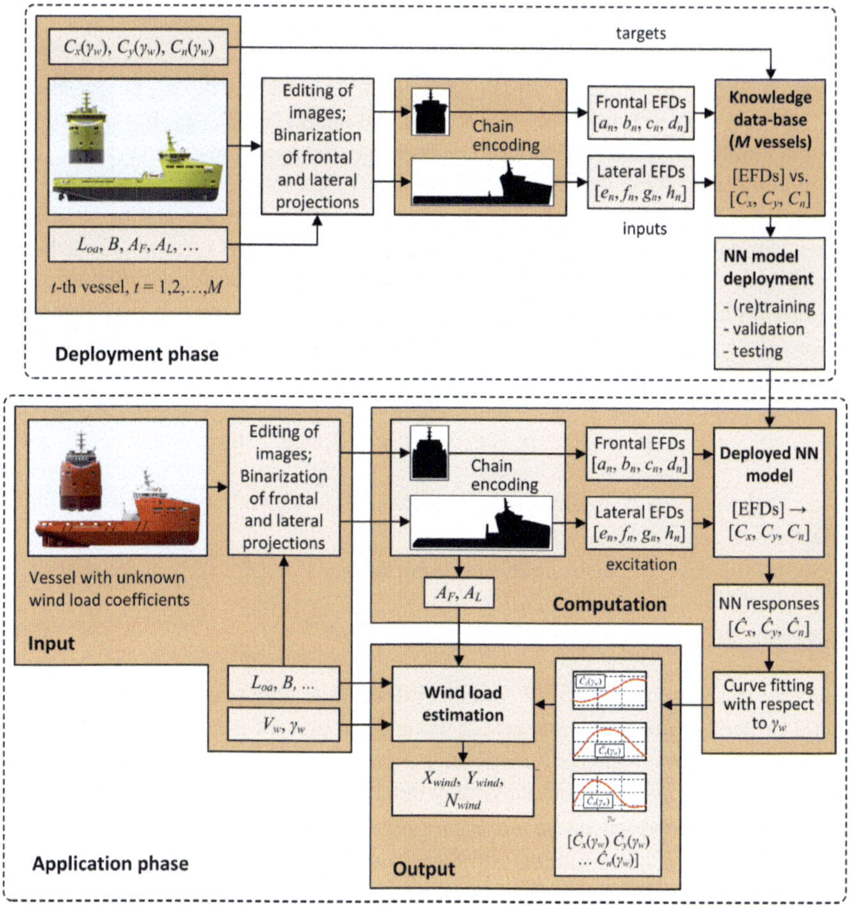

Fig. 10 Framework of proposed wind load estimation method based on elliptic Fourier descriptors and generalized regression neural network (PSV photos: Courtesy of Damen, https://www.damen.com/en)

for describing frontal projected area and an ordered quadruple (e_n, f_n, g_n, h_n) of EFDs for describing lateral projected area, where $n = 1, 2, \ldots, N$. Both of these quadruples are calculated using Freeman chain encoding and elliptic Fourier descriptors (15) and (16) for images of frontal and lateral vessel profiles.

Calculated EFDs are the foundation of a knowledge database since they represent input parameters which contain all the required information about frontal and lateral vessel profile. In order to use them properly within the construction of a neural network input matrix, it is necessary to rearrange them as follows. So, let

$$\mathbf{x}_n^t = [x_1^t,\ x_2^t, \ldots,\ x_N^t] \in \mathbb{R}^N \tag{44}$$

denote a vector of arbitrary EFD where $\mathbf{x}_n \in \{\mathbf{a}_n,\ \mathbf{b}_n,\ \mathbf{c}_n,\ \mathbf{d}_n,\ \mathbf{e}_n,\ \mathbf{f}_n,\ \mathbf{g}_n,\ \mathbf{h}_n\}$ is defined according to frontal $(\mathbf{a}_n,\ \mathbf{b}_n,\ \mathbf{c}_n,\ \mathbf{d}_n)$ and lateral $(\mathbf{e}_n,\ \mathbf{f}_n,\ \mathbf{g}_n,\ \mathbf{h}_n)$ quadruples, N is the number of terms in Fourier expansion, i.e. the length of each vector $\mathbf{x}_n$, M is the number of vessels, $n = 1, 2, \ldots, N$ and $t = 1, 2, \ldots, M$.

With the appropriate number of harmonics N, close contour can be described very accurately. Figure 11 shows the ship contour (blue line) versus contour based on different numbers (N = 1, 2, 5, 10, 20 and 50) of harmonics (red line). It is obvious that the chosen contours can be very well captured with only 50 harmonics.

Therefore, EFDs that describe frontal and lateral contours of an arbitrary t-th vessel can be aggregated as a row vector $\mathbf{p}^t \in \mathbb{R}^{8N}$

$$\mathbf{p}^t = \begin{bmatrix} \mathbf{a}_n^t & \mathbf{b}_n^t & \mathbf{c}_n^t & \mathbf{d}_n^t & \mathbf{e}_n^t & \mathbf{f}_n^t & \mathbf{g}_n^t & \mathbf{h}_n^t \end{bmatrix}, \tag{45}$$

where each vector in $\mathbf{p}^t$ is additionally expressed by its own harmonics according to (44).

So, in this case, the neural network input matrix $\mathbf{P} \in \mathbb{R}^{M \times (8N)}$ can be written as

$$\mathbf{P} = \begin{bmatrix} \mathbf{p}^1 & \mathbf{p}^2 & \cdots & \mathbf{p}^M \end{bmatrix}^T, \tag{46}$$

where a t-th row of the matrix $\mathbf{P}$ contains information about contours of frontal and lateral projected areas of a t-th vessel, columns represent harmonics of EFDs and superscript T denotes matrix transpose.

On the other hand, accompanying wind load coefficients (43) for M vessels should also be reorganized in an appropriate way. Hence, let $\mathbf{C}_l^t \in \mathbb{R}^{(180/\Delta\gamma + 1)}$ denote a vector of an arbitrary non-dimensional wind load coefficient of a t-th vessel, where $\mathbf{C}_l \in \{\mathbf{C}_X,\ \mathbf{C}_Y,\ \mathbf{C}_N\}$, $t = 1, 2, \ldots, M$, and $\Delta\gamma$ stands for the difference between two adjacent wind angles of attack, i.e.

$$\Delta\gamma = \gamma_{k+1} - \gamma_k, \tag{47}$$

where $k = 1, 2, \ldots, (180/\Delta\gamma)$ and $\Delta\gamma \in \{1^\circ,\ 2^\circ,\ 3^\circ,\ 4^\circ,\ 5^\circ,\ 10^\circ\}$.

Values of $\Delta\gamma$ from the previously defined set have been chosen so that the common condition of $(180/\Delta\gamma) \in \mathbb{N}$ is fulfilled. Once the difference between two adjacent wind angles of attack has been selected, it should be kept constant for all the vessels used during GRNN training, validation and testing. Moreover, for the purpose of this paper, only symmetrical vessel models have been used so the assumption $\gamma \in [0^\circ, 180^\circ]$ is sufficient. If required, this assumption can easily be extended to $\gamma \in [0^\circ, 360^\circ]$.

Now, wind load coefficients in longitudinal, transverse and vertical directions of an arbitrary t-th vessel can be aggregated as a row vector $\mathbf{t}^t \in \mathbb{R}^{3(180/\Delta\gamma + 1)}$ in the following form

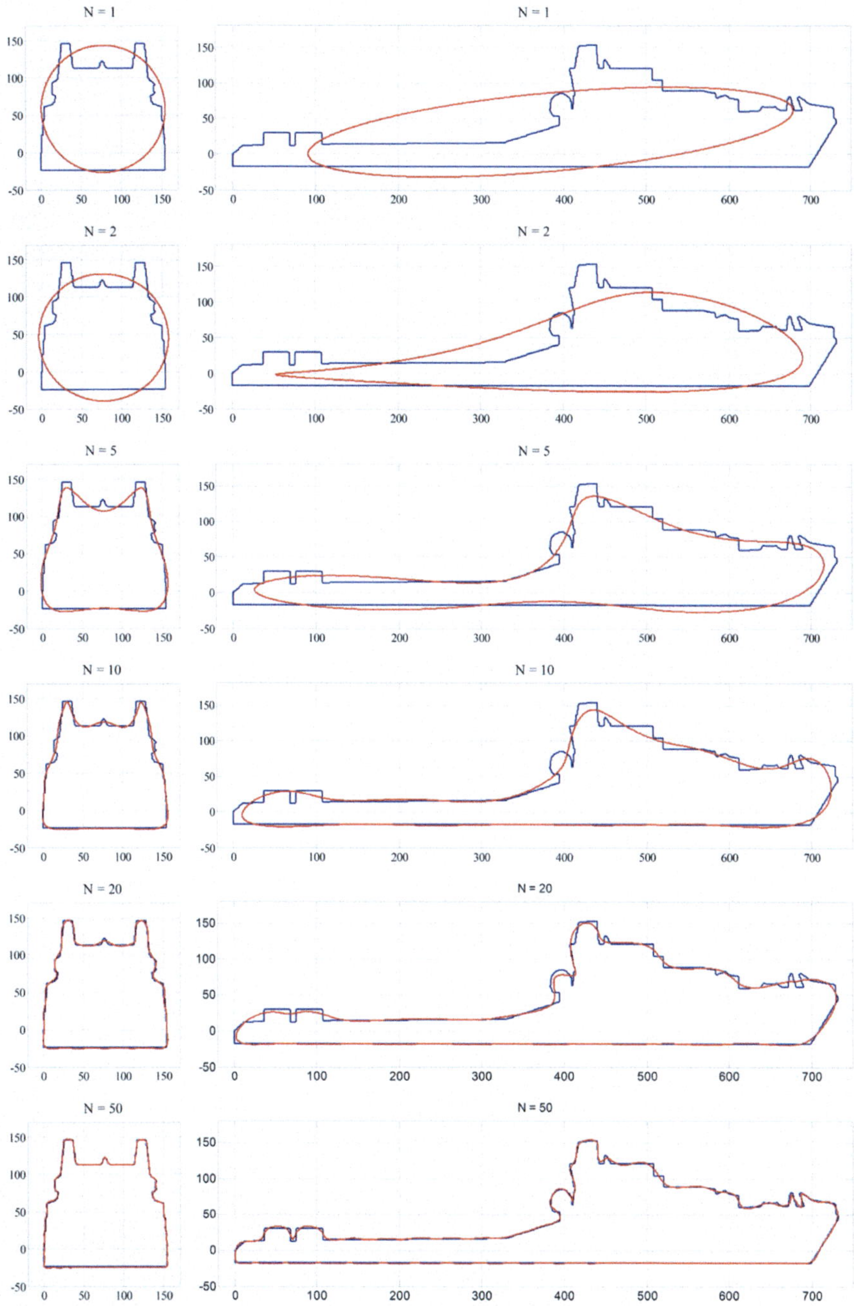

Fig. 11 Contour capturing with different number of harmonics N

$$\mathbf{t}^t = [\, \mathbf{C}_X^t \quad \mathbf{C}_Y^t \quad \mathbf{C}_N^t \,], \tag{48}$$

and thus, the neural network target matrix $\mathbf{T} \in \mathbb{R}^{M \times 3(180/\Delta\gamma + 1)}$ can be written as

$$\mathbf{T} = [\mathbf{t}^1 \quad \mathbf{t}^2 \quad \cdots \quad \mathbf{t}^M]^{\mathrm{T}}, \tag{49}$$

where a t-th row of the matrix $\mathbf{T}$ contains information about wind load coefficients with respect to frontal and lateral projected areas of a t-th vessel and columns represent wind load coefficients for all analysed vessels in longitudinal, transverse and vertical directions, respectively.

Due to reasons of simplicity, only one sub-matrix $\mathbf{C}_l^t$ $(t = 1, 2, \ldots, M;\ l = X, Y, N)$ of target matrix $\mathbf{T}$ can be used for neural network training, but in that case three networks are required for covering wind loads in a horizontal plane. A similar conclusion can be drawn for the input matrix $\mathbf{P}$ and the possibility of its separation with respect to frontal and lateral EFDs. On the other hand, it is important to emphasize that this separation is not recommended due to the evident loss of information which will undoubtedly affect the performance of the neural network during both training and testing.

Once the training, validation and testing of the GRNN is completed in a manner conventional in machine learning, the deployed GRNN model can be used in the forthcoming applications, i.e. in the case when the estimation of non-dimensional wind load coefficients of some new vessel is required. This deployed GRNN model presents a multivariable nonlinear mapping

$$[\mathrm{EFDs}] \overset{\mathrm{GRNN}}{\longrightarrow} [C_X, C_Y, C_N] \tag{50}$$

i.e. the mapping of frontal and lateral contours described by EFDs to wind load coefficients.

Thus, for any new vessel it is necessary to acquire images of frontal and lateral projections by means of digital camera or based on available project documentation. These images should again be properly edited and binarized according to the previously described procedure so the frontal and lateral EFDs of a new vessel can be aggregated in the form of a vector $\mathbf{p}^{new} \in \mathbb{R}^{8N}$

$$\mathbf{p}^{new} = [\, \mathbf{a}_n^{new} \quad \mathbf{b}_n^{new} \quad \mathbf{c}_n^{new} \quad \mathbf{d}_n^{new} \quad \mathbf{e}_n^{new} \quad \mathbf{f}_n^{new} \quad \mathbf{g}_n^{new} \quad \mathbf{h}_n^{new} \,], \tag{51}$$

which represents an excitation of the deployed GRNN model (50).

The resulting GRNN response or output is also presented as a vector, but in the form $\hat{\mathbf{o}}^{new} \in \mathbb{R}^{3(180/\Delta\gamma + 1)}$ which yields

$$\hat{\mathbf{o}}^{new} = [\, \hat{\mathbf{C}}_X^{new} \quad \hat{\mathbf{C}}_Y^{new} \quad \hat{\mathbf{C}}_N^{new} \,]. \tag{52}$$

Vector (52) can easily be transformed in a more suitable matrix $\hat{\mathbf{c}}^{new} \in \mathbb{R}^{(180/\Delta\gamma+1)\times 3}$ as

$$\hat{\mathbf{c}}^{new} = [\,(\hat{\mathbf{C}}_X^{new})^{\mathrm{T}} \quad (\hat{\mathbf{C}}_Y^{new})^{\mathrm{T}} \quad (\hat{\mathbf{C}}_N^{new})^{\mathrm{T}}\,], \tag{53}$$

where rows represent the values of ordered triples $(\hat{C}_X^{new}, \hat{C}_Y^{new}, \hat{C}_N^{new})$ with respect to corresponding wind angles of attack that are selected according to (47).

The matrix data in (53) can be used for additional curve-fitting in order to obtain more applicable expressions $\hat{C}_X(\gamma_w)$, $\hat{C}_Y(\gamma_w)$ and $\hat{C}_N(\gamma_w)$ for the estimation of wind load coefficients in terms of the wind angle of attack γ_w. For this purpose, the sequence of simpler GRNN models

$$\left.\begin{array}{l} [\gamma_w] \overset{\mathrm{GRNN}_1}{\longrightarrow} [\hat{C}_X(\gamma_w)] \\ [\gamma_w] \overset{\mathrm{GRNN}_2}{\longrightarrow} [\hat{C}_Y(\gamma_w)] \\ [\gamma_w] \overset{\mathrm{GRNN}_3}{\longrightarrow} [\hat{C}_N(\gamma_w)] \end{array}\right\} \tag{54}$$

in longitudinal, transverse and vertical directions should be derived or some other curve-fitting approach can be used as well, e.g. Fourier expansion, nonlinear polynomial fitting, spline fitting, etc. In this way, the deployed GRNN model, i.e. the responses generated with it, can be transformed in a very simple model for the estimation of non-dimensional wind load coefficients.

Finally, according to (42), the estimation of wind forces and moment can be easily performed by using previously deployed models (54), frontal and lateral projected areas (A_F, A_L) and filtered low frequency anemometer data (V_w, γ_w). It is also important to highlight the capability of this approach, according to which it is possible to easily calculate the frontal and lateral projected areas if they are not known in advance. This calculation can be performed during the determination of EFDs in (51).

If the wind load coefficients (C_X, C_Y, C_N) are already known for some arbitrary new vessel, they can be used either for additional independent testing in the application phase or for retraining, i.e. updating of deployed NN model in deployment phase in order to increase reliability, accuracy and overall model performance.

4 Results and Discussion

4.1 Numerical Examples and Assessment of Findings

The developed program using MATLAB R2015b software [20] has been run for three groups of vessels based on data provided in Blendermann [5]. Thus, a group of six offshore supply vessels, a group of four car carriers and a group of four

Table 1 Characteristics of offshore supply vessels [5]

#t	Contours of front and lateral projected areas	L_{oa} (m)	A_F (m^2)	A_L (m^2)	Vessel code
1	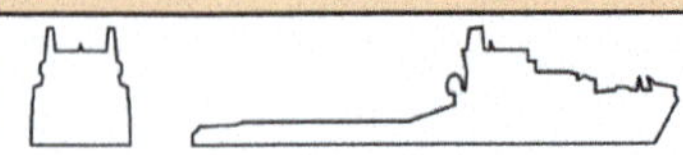	61.95	137.48	336.80	SUP0101BN
2	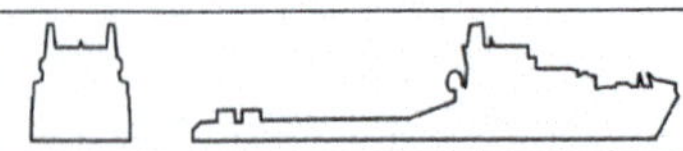	61.95	137.48	344.16	SUP0102BN
3	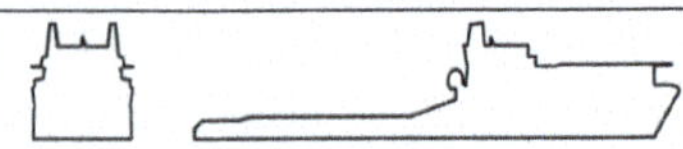	61.95	139.78	329.50	SUP0106BN
4		61.00	110.12	260.33	SUP0107BN
5	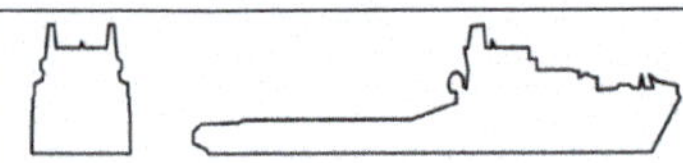	61.95	152.56	402.42	SUP0109BN
6		61.00	124.96	326.13	SUP0111BN

Table 2 Characteristics of car carriers [5]

#t	Contours of front and lateral projected areas	L_{oa} (m)	A_F (m^2)	A_L (m^2)	Vessel code
1	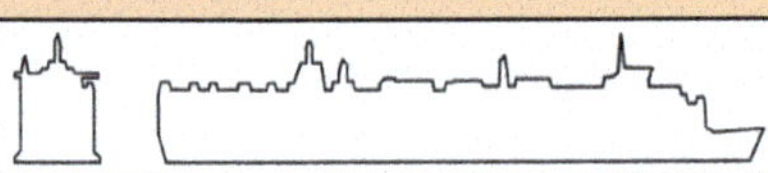	190.65	653.98	4256.55	CAR0101BN
2	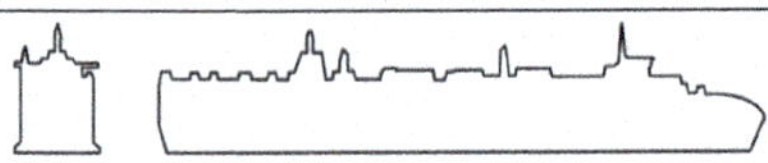	190.65	653.98	4378.28	CAR0102BN
3	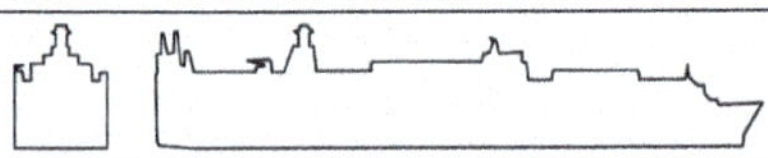	195.00	811.68	4584.16	CAR0201BN
4	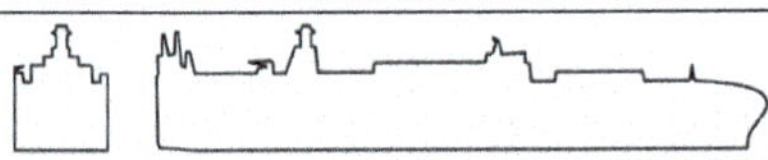	195.00	811.68	4693.94	CAR0202BN

containers ships were used, and the main characteristics of these vessels are shown in Tables 1, 2 and 3, respectively.

The training of GRNN was performed according to the following procedure. In general, any particular group of M vessels can be written as a set $S = \{S_1, S_2, \ldots, S_M\}$. Therefore, if we choose some arbitrary t-th vessel $S_t \in S$ for testing, all other vessels from the set $S \backslash \{S_t\}$ are used for GRNN training. If we vary testing vessel S_t according to $t \in \{1, 2, \ldots, M\}$, we can perform training and testing of M independent cases where each of them is denoted as Case #t, $t = 1, 2, \ldots, M$.

Table 3 Characteristics of container ships [5]

#t	Contours of front and lateral projected areas	L_{oa} (m)	A_F (m^2)	A_L (m^2)	Vessel code
1	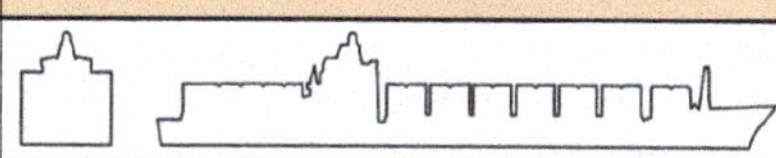	210.75	801.97	3751.11	CON0101BN
2	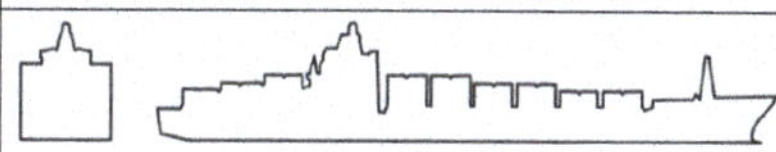	210.75	857.06	3773.51	CON0107BN
3	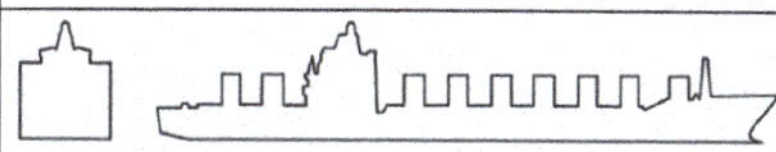	210.75	857.04	3500.2	CON0110BN
4	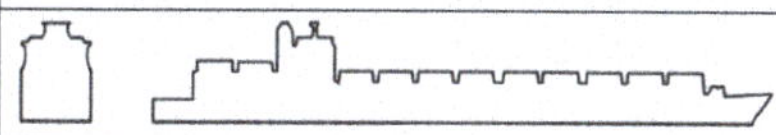	216.40	674.06	3562.72	CON0201BN

In other words, Case #*t* indicates a scenario where *t*-th vessel $S_t \in S$ is used for the evaluation of GRNN responses, i.e. the obtained estimated values of wind load coefficients of the vessel S_t, while EFDs and wind load coefficients of all other vessels from the set $S \backslash \{S_t\}$ are used as inputs and targets during the training phase of GRNN. In that way, the wind load coefficients $\hat{\mathbf{C}}_X^{new}$, $\hat{\mathbf{C}}_Y^{new}$ and $\hat{\mathbf{C}}_N^{new}$ obtained as GRNN estimates can be easily compared with the wind load coefficients $\mathbf{C}_X^t$, $\mathbf{C}_Y^t$ and $\mathbf{C}_N^t$ provided by Blendermann [5] for the same *t*-th ship. This also applies to results presented in Figs. 12, 13, 14 and Table 4.

For evaluation of GRNN responses, normalized root mean squared error (NRMSE) performance index was used wherein the normalization has been performed with respect to the range of target values. According to (47), (48) and (52), NRMSE is defined as follows.

$$NRMSE_l(\hat{\mathbf{C}}_l^{new}, \mathbf{C}_l^t) = \frac{1}{\max(\mathbf{C}_l^t) - \min(\mathbf{C}_l^t)} \sqrt{\frac{1}{180/\Delta\gamma + 1}(\hat{\mathbf{C}}_l^{new} - \mathbf{C}_l^t)(\hat{\mathbf{C}}_l^{new} - \mathbf{C}_l^t)^{\mathrm{T}}} \tag{55}$$

where $l \in \{X, Y, N\}$.

In this way, performance index NRMSE presents a percentage of the target wind load coefficient corresponding to each case, where lower values indicate less residual variance. NRMSE based results of performance evaluation of GRNN estimates for all three groups of analysed vessels, particularly with respect to each individual Case #*t* ($t = 1, 2, \ldots, M$), are presented in Table 4.

In Fig. 12 the non-dimensional wind load coefficients $C_X(\gamma_w)$, $C_Y(\gamma_w)$ and $C_N(\gamma_w)$ obtained by GRNN (red dashed lines) are compared with those of Blendermann (blue solid lines). As already mentioned, the comparison obtained for the *t*-th ship is denoted as Case #*t*. Results show good agreement, especially for

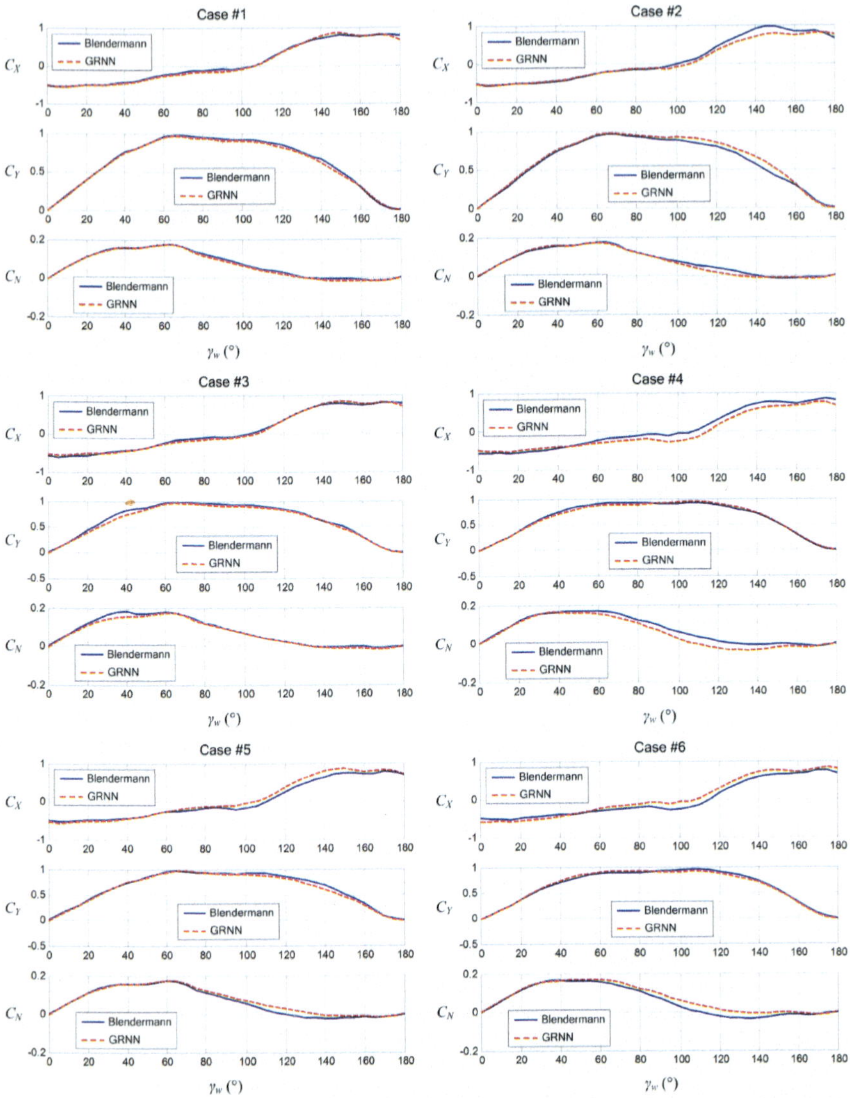

Fig. 12 Comparison of wind load coefficients for a group of offshore supply vessels

Cases #1 and #3. There are some slight deviations in Cases #2 and #4, but from the engineering application point of view, results are quite satisfactory.

Similar comparison is shown in Figs. 13 and 14 for the group of car carriers and the group of container ships, respectively. While the results for the car carrier group of vessels are excellent, there are some significant deviations in the group of container ships, particularly for wind load coefficients in longitudinal directions for Cases #1, #2 and #3. It is quite obvious that the reason for this lies in the fact that

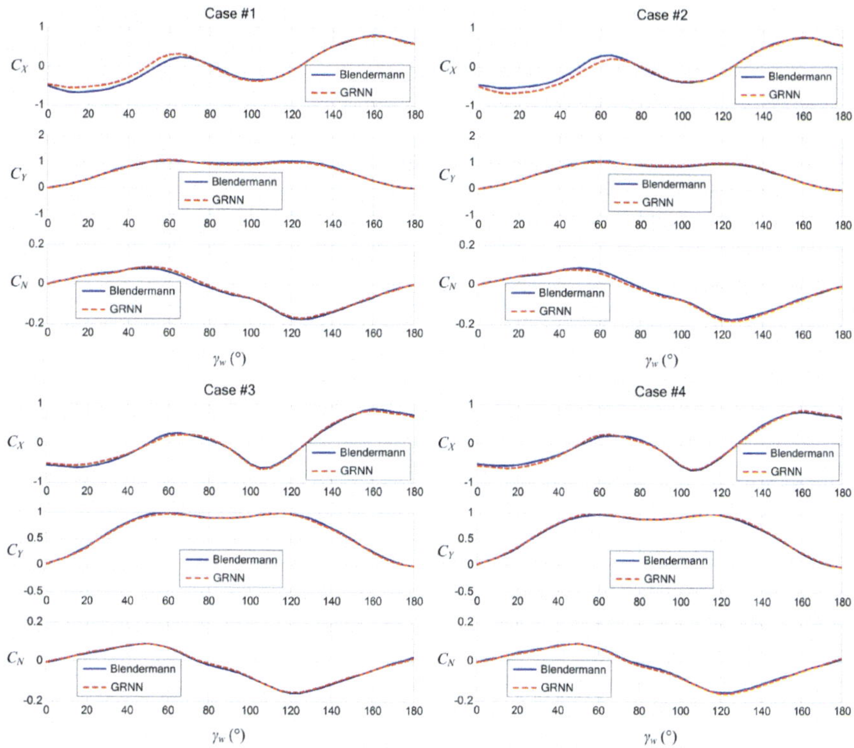

Fig. 13 Comparison of wind load coefficients for a group of car carriers

$C_X(\gamma_w)$ of the container ship depends significantly upon loading conditions while at the same time training of the GRNN for this group of vessels was performed using only three container ships at a time. Although this approach also takes into account the variability of ship lateral profile, i.e. the variability in loading conditions while the training in longitudinal direction is performed, some additional improvement should be introduced. However, at this stage, it is reasonable to expect much better GRNN responses after retraining of GRNN performed with an increased number of container ships with various loading conditions.

Values of performance index (NRMSE) for all 14 analysed cases are presented in Table 4. The results clearly indicate that GRNN performance is practically equable for offshore supply vessels and car carriers while the results for container ships are somewhat worse, mainly due to the aforementioned problem of relationship between the loading conditions, i.e. the variability of ship lateral profile, and wind load coefficients in longitudinal directions. At the same time, it can be seen that GRNN responses are the most accurate for wind loads in vertical direction, i.e. for yawing moment estimation, while the least accurate are for the wind loads in longitudinal directions. Nevertheless, despite the small available database of only 4–6 vessels, results indicate very good agreement. That is, of course, in part

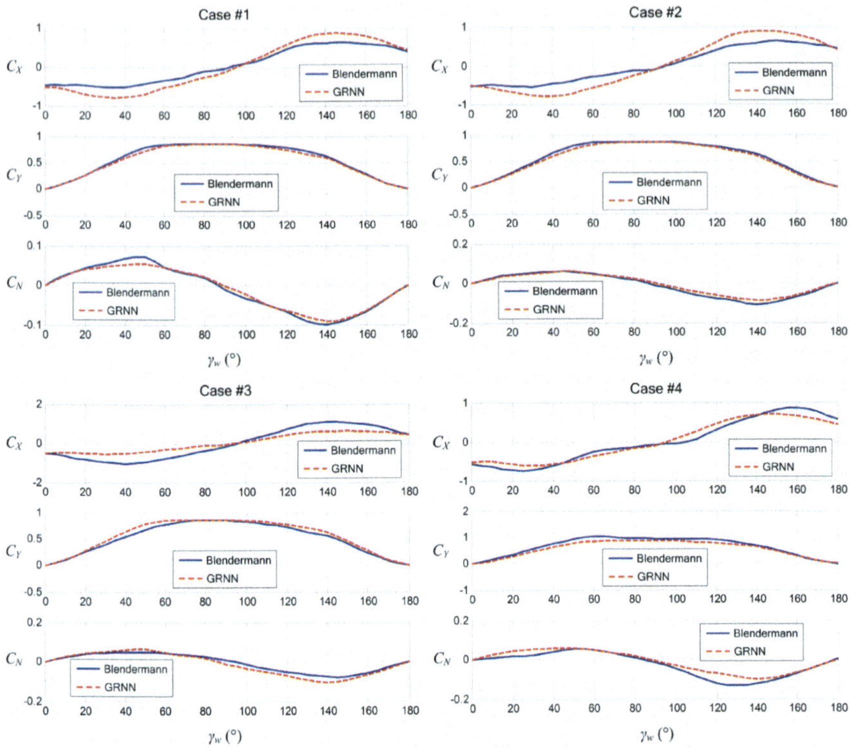

Fig. 14 Comparison of wind load coefficients for a group of container ships

Table 4 Performance evaluation of GRNN estimates

Vessel type	Case	$NRMSE_X(\hat{\mathbf{C}}_X^{new}, \mathbf{C}_X^t)$	$NRMSE_Y(\hat{\mathbf{C}}_Y^{new}, \mathbf{C}_Y^t)$	$NRMSE_N(\hat{\mathbf{C}}_N^{new}, \mathbf{C}_N^t)$
Offshore supply vessels	#1	0.024422	0.019019	0.023452
	#2	0.053642	0.046583	0.046095
	#3	0.025530	0.033853	0.046202
	#4	0.071526	0.028642	0.110108
	#5	0.064673	0.032129	0.074079
	#6	0.079620	0.027299	0.102332
Car carriers	#1	0.052502	0.018751	0.021599
	#2	0.058427	0.018279	0.021608
	#3	0.025494	0.013530	0.013311
	#4	0.025747	0.013881	0.013637
Container ships	#1	0.148562	0.032557	0.041508
	#2	0.176991	0.034352	0.062090
	#3	0.164506	0.063636	0.118085
	#4	0.078484	0.094968	0.118881

because ships are of similar form, but this does not diminish the effectiveness of the method, especially with respect to the number of vessels used for training.

Considering that GRNN can capture very accurately any nonlinearity developed GRNN model is highly dependent on the accuracy of experimental (target) data that are used for GRNN training. Results are also dependent on the quality of images used for the extraction of frontal and lateral contours, but not too much from the image resolution point of view. Image details, i.e. parts of the hull, superstructure and hull equipment are much more important. Therefore, omitting image details due to the previously mentioned image simplification should be performed with caution. High quality pictures with a lot of negligible details are not necessary because increasing the number of terms in Fourier expansions makes sense only to a certain extent. The use of too many harmonics in Fourier expansion will consequently cause overfitting problems during the training process of neural networks which might result in degradation of model performance with respect to generality.

Finally, the regression methods introduced so far for the estimation of wind load coefficients are mainly based on basic vessel characteristics such as various linear parameters (e.g. length, width, draft, other distances among specific points of interest), planar parameters (e.g. frontal and lateral projected areas) as well as on some of the corresponding ratios of these variables. However, the fact remains that for a selected set of these parameters there can be many vessels with different values of wind load coefficients. On the other hand, an approach based on EFDs takes into account the complete contour variability of the frontal and lateral projections. This is particularly important when it is necessary to combine the information of both frontal and lateral contours in order to determine $C_X(\gamma_w)$. Namely, this is a very common case when several ships have very similar frontal projections and significantly different values of $C_X(\gamma_w)$. The reason for this usually lies in geometrical differences of lateral profiles which can easily be taken into account by the approach proposed in this paper. However, to achieve this goal, it is necessary to have a sufficiently large database of vessels for neural network training.

4.2 *Sensitivity Analysis*

In order to investigate stability and reliability of created NN model, a simple sensitivity analysis has been performed as well [21]. For this reason, the developed

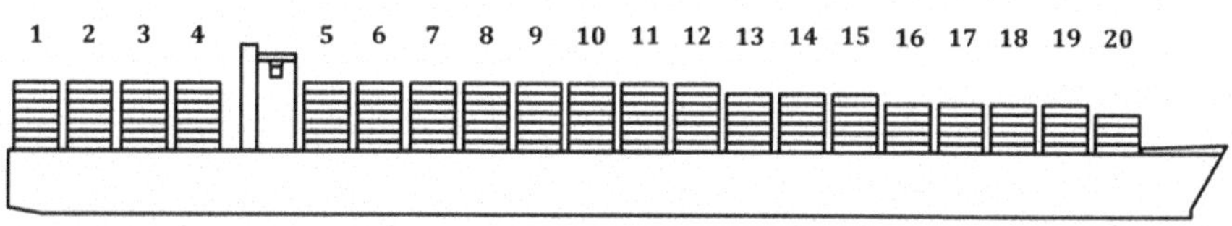

Fig. 15 Container ship with 20 container bays [8]

Matlab program has been run for the 9000+ container ship shown in Fig. 15 with the main characteristics: L_{oa} = 340 m, L_{pp} = 320 m, B = 45 m [8].

The 17 different container configurations on the deck analysed by Andersen [8] are shown in Table 5, where only the first 14 configurations were used in this study. Hence, the GRNN was trained with 13 arbitrary configurations at a time and the responses are compared with respect to remaining 14th vessel. In other words, the wind coefficients of 14 available configurations of containers obtained as GRNN responses are compared with corresponding wind coefficients that were obtained experimentally.

In Fig. 16, wind load coefficients $C_X(\gamma_w)$, $C_Y(\gamma_w)$ and $C_N(\gamma_w)$, obtained by GRNN (red dashed line) and Andersen (blue solid line), are compared. The comparison conducted for the i-th vessel is denoted as Conf. #i, where, due to simplicity reasons, only selected cases are shown, i.e. for i = 1, 2, 3, 8, 9, 14. Results show good agreement, especially for Conf. #1, Conf. #9 and Conf. #14.

The wind load coefficients $\hat{\mathbf{C}}_X^{new}$, $\hat{\mathbf{C}}_Y^{new}$ and $\hat{\mathbf{C}}_N^{new}$ obtained as GRNN estimates can be easily compared with the wind load coefficients $\mathbf{C}_X^t$, $\mathbf{C}_Y^t$ and $\mathbf{C}_N^t$ provided by Andersen [8] for the same t-th ship. For evaluation of GRNN responses in this sensitivity analysis, mean squared error (MSE) performance index was used which is defined as follows.

$$MSE_l(\hat{\mathbf{C}}_l^{new}, \mathbf{C}_l^t) = \frac{1}{180/\Delta\gamma + 1}(\hat{\mathbf{C}}_l^{new} - \mathbf{C}_l^t)(\hat{\mathbf{C}}_l^{new} - \mathbf{C}_l^t)^T \tag{56}$$

where $l \in \{X, Y, N\}$.

Table 5 Containership configurations [8]

#	Frontal and lateral configuration	#	Frontal and lateral configuration
1		10	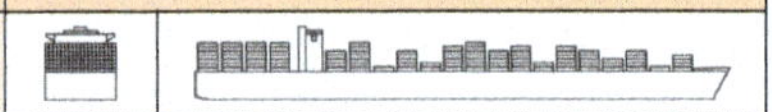
2		11	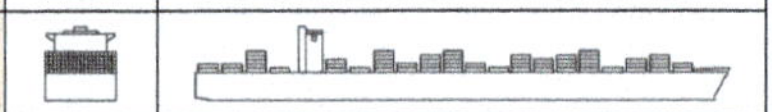
3	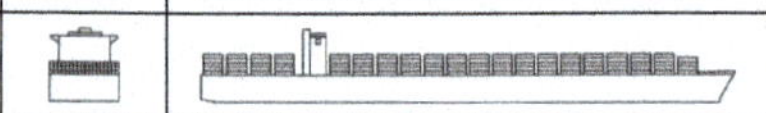	12	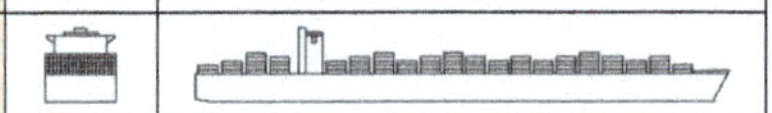
4	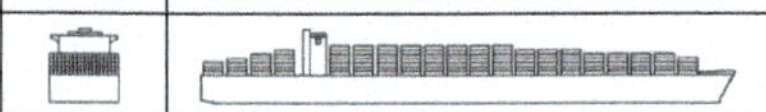	13	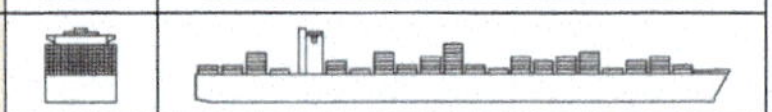
5	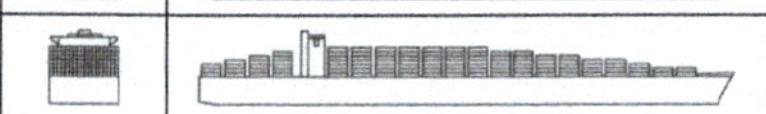	14	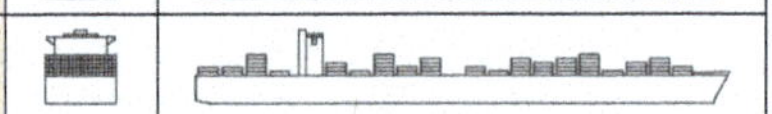
6	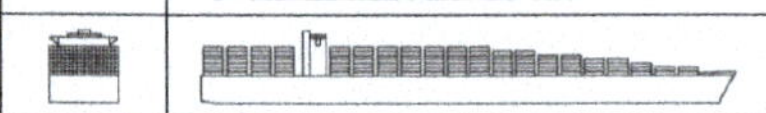	15	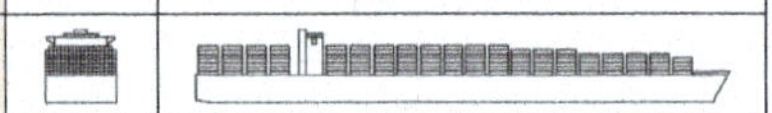
7	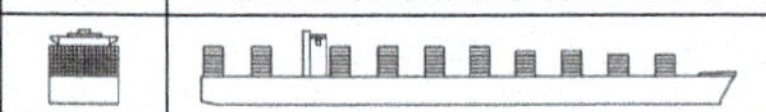	16	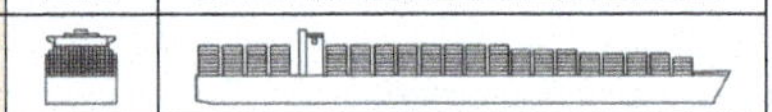
8	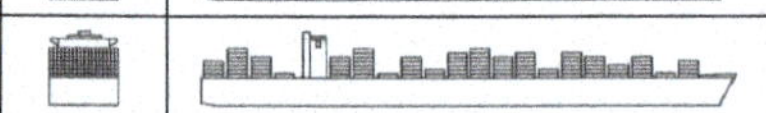	17	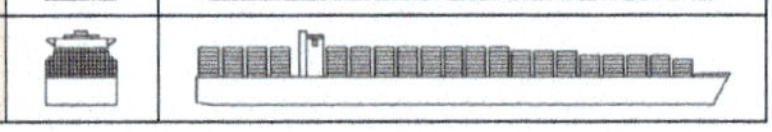
9	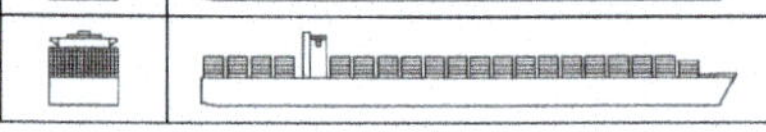		

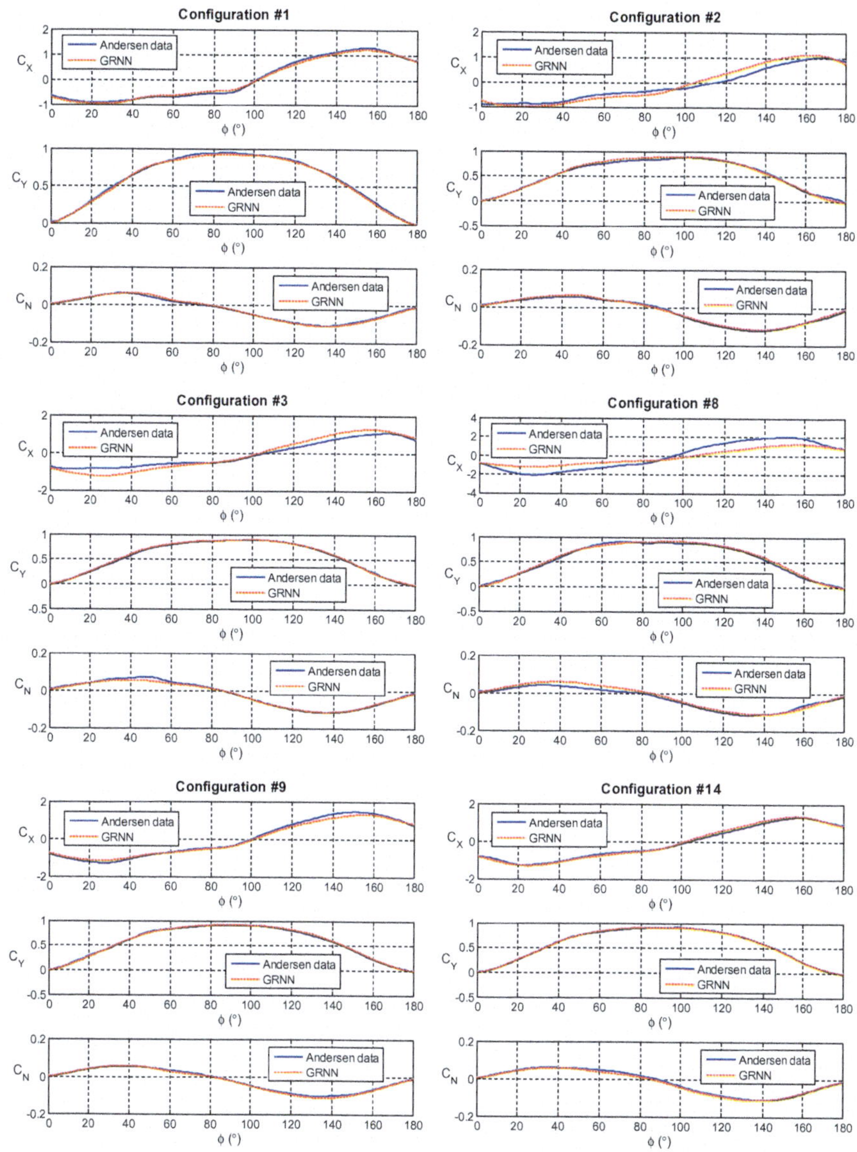

Fig. 16 Comparison of wind coefficients

Results based on MSE performance evaluation of GRNN estimates for all 14 analysed configurations are presented in Table 6. For all cases the deviations are within the following ranges: $MSE_X < 0.05$ (except for the Config. #8), $MSE_Y < 0.001$ and $MSE_N < 0.001$. The results clearly indicate that GRNN performance is practically equable for the variability of ship lateral profile caused by different container configurations and that results indicate very good agreement.

Table 6 Performance evaluation of GRNN estimates for sensitivity analysis

Conf.	$MSE_X(\hat{\mathbf{C}}_X^{new}, \mathbf{C}_X^t)$	$MSE_Y(\hat{\mathbf{C}}_Y^{new}, \mathbf{C}_Y^t)$	$MSE_N(\hat{\mathbf{C}}_N^{new}, \mathbf{C}_N^t)$
#1	0.0034	0.0002	0.0000
#2	0.0286	0.0007	0.0000
#3	0.0476	0.0003	0.0000
#4	0.0141	0.0002	0.0000
#5	0.0216	0.0002	0.0002
#6	0.0360	0.0003	0.0000
#7	0.0123	0.0007	0.0003
#8	0.4064	0.0009	0.0002
#9	0.0095	0.0002	0.0000
#10	0.0240	0.0002	0.0001
#11	0.0071	0.0000	0.0000
#12	0.0412	0.0001	0.0000
#13	0.0147	0.0003	0.0001
#14	0.0063	0.0001	0.0000

The lateral variability of containers configuration is represented as percentage of mutual differences between two configurations. This percentage is estimated on the basis of the highest possible difference (100% diversity), and that is the difference between an empty and completely loaded ship. Percentage of difference for configurations i and j, Δ_{ij}, is calculated as

$$\Delta_{ij} = \sum_{k=1}^{20} \frac{(c_{ik} - c_{jk})^2}{980} \cdot 100\%, \quad i,j = 1, \ldots, 14, \tag{57}$$

where c_{ik} and c_{jk} are the numbers of container rows at the k-th bay for the i-th and j-th container configuration respectively.

Figure 17 represents experimental values of wind load coefficients (left) as well as numerical results (right) obtained by GRNN. Results are presented for configuration #1, as reference configuration, which is compared with configurations which are, based on Δ_{ij} values, 5, 10, 15, 33 and 55% different (configurations 4, 10, 9, 14, 2, respectively).

Relative deviations of wind load coefficients for container configuration #1 versus configurations #4, #10, #9, #14 and #2 are presented in Table 7 and shown in Fig. 18. The deviations are calculated only for the wind directions where the maximum deviations of the non-dimensional wind load coefficients $C_X(\gamma_w)$, $C_Y(\gamma_w)$ and $C_N(\gamma_w)$ were encountered, i.e. for 30°, 40° and 70°. One can notice that the results for longitudinal and transverse wind load coefficients do not variate much with the variation of container configurations while the results for yawing moment wind load coefficient vary up to 125%.

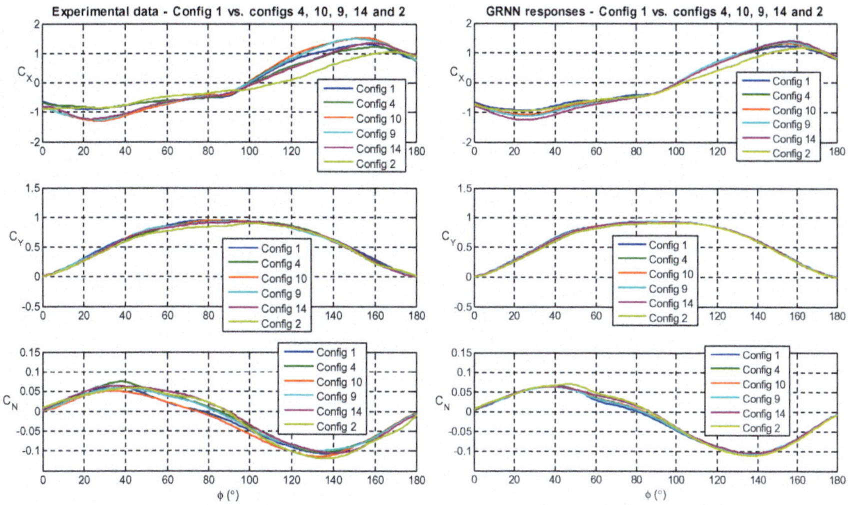

Fig. 17 Experimental data [8] and numerical values (GRNN responses) for Config. #1 versus configurations #4, #10, #9, #14 and #2

Table 7 Relative deviations of wind load coefficients

Conf. #1 versus	Conf. #4 (5%)	Conf. #10 (10%)	Conf. #9 (15%)	Conf. #14 (33%)	Conf. #2 (55%)
$\delta C_X(30^\circ)$	0.1620	0.1506	0.2034	0.3324	0.0154
$\delta C_Y(40^\circ)$	−0.0196	−0.0205	−0.0237	−0.0397	−0.0915
$\delta C_N(70^\circ)$	0.4583	0.3904	0.3154	0.9743	1.2575

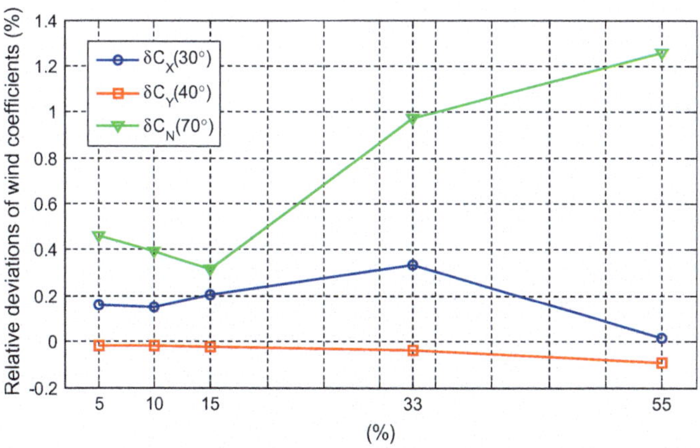

Fig. 18 Relative deviations of wind load coefficients

4.3 Limitations and Avenues of Future Research

The simplicity of approach presented in this seminar is primarily manifested by the fact that to estimate wind load coefficients of a new vessel it is sufficient to take appropriate photos by digital camera or to use project documentation in order to obtain binarized images of frontal and lateral projections. However, it is also important to point out that at the same time one must take into account the draft of the vessel in analysed images. Hence, estimated wind load coefficients based on analysed images will be in relation to the current draft of the vessel, i.e. in relation to the loading conditions of the vessel, as pointed out similarly in Blendermann [5].

Although the obtained results are somewhat limited due to the small number of vessels used in GRNN training, this is not a general limitation of the proposed model. As stated previously, this can also be easily solved by increasing the number of vessels in the input database. On the other hand, since EFDs presented in this paper can deal only with closed contours, the problem of how to manage more complicated longitudinal and lateral profile geometries that may include, for instance, complex helideck constructions (vessels and marine structures in offshore industry), cranes (e.g. offshore and platform supply vessels, heavy-lift crane vessels), truss structures and derricks (e.g. pipe-laying vessels, drill ships and drill rigs), etc. remains open.

So, since EFDs have some limitations with respect to complex lateral area geometries, particular attention should be focused on how to manage lateral area geometries having one or more holes within the profile. In addition to that, further research should also be aimed at sensitivity analysis as well as uncertainties and reliability analysis. The selection of the neural network type, i.e. machine learning algorithm for nonlinear multivariate regression, and the number of harmonics could also affect the results, so they should also be analysed further. Therefore, future research should include the following steps:

- investigate and present conclusions and recommendations of image editing procedure with respect to image quality, omitted image details and number of terms in Fourier expansion;
- determine the optimum, or at least recommended, number of terms in Fourier expansion with respect to contour complexity of frontal and lateral projected areas;
- determine the optimum, or at least recommended, number of vessels required for GRNN training;
- improve the proposed methodology based on elliptic Fourier features of closed contours so the holes within the profiles can be managed as well;
- investigate the application possibilities of other nonlinear multivariable regression algorithms for mapping of EFDs to wind load coefficients, emphasize their advantages and disadvantages, and propose the most convenient algorithm for this purpose.

5 Conclusion

A novel method for estimating the wind loads on ships and marine structures is presented. The method is based on elliptic Fourier descriptors which are used for ship frontal and lateral closed contour representation. The generalized regression radial basis neural network was trained with the elliptic Fourier descriptors of closed contours as inputs and wind load data derived from wind tunnel tests as targets. The methodology was tested on three groups of ships: offshore supply vessels, car carriers and container ships.

This approach takes into account all aspects of the variability of the above-water frontal and lateral ship profile. It is very suitable for assessing wind loads on marine structures wherever we have a wind load database for a group of similar vessels. Thus, the method could be of great use in hydrodynamic basins that have a large database obtained experimentally. In addition, the data required for network training can be obtained by CFD as well. In this way, the cheaper and faster calculation can bridge the gap between ship shapes for which calculations or experiments have already been made.

The advantage of the method is also in the ease of input data preparation wherein it is sufficient to extract or to draw the contour of the transversal and longitudinal above water projected areas in any suitable computer program for photo editing and to save them into one of the well-known formats, like bmp, jpg, tiff, png, etc. In such a way, input data can be easily prepared for uploading and further computational analysis.

For the purpose of this study, GRNN training was conducted with a modest knowledge database consisting of no more than five vessels at a time. Therefore, it is reasonable to conclude that GRNN performance can be significantly improved by increasing the number of available vessels for neural network training. Nevertheless, the obtained results are very satisfactory and promising, particularly from the engineering application point of view and with respect to possibilities within future research.

Moreover, a simple sensitivity analysis was also performed in order to investigate the sensitivity and stability of the proposed neural network model. Hence, the wind load data derived from wind tunnel tests for 14 different container configurations of 9000+ container ship were used. The neural network is tested on mentioned 14 container configurations in a way the network is trained on arbitrary 13 configurations, and the 14th one is used for testing. Sensitivity analysis was performed by varying container configuration by 5, 10, 15, 33 and 55%. It shows that model is quite stable and that deviations of non-dimensional wind load coefficients due to container configuration change is rather small as expected.

Finally, it is important to emphasize that this approach also presents an extension of application capabilities of elliptic Fourier descriptors from the usual pattern recognition and classification problems to problems of very complex nonlinear multivariable approximations of multi-input-multi-output functions.

Acknowledgements This work has been fully supported by the Croatian Science Foundation under the project IP-2018-01-3739 and by the University of Rijeka under the project numbers uniri-tehnic-18-18 and uniri-tehnic-18-266.

References

1. Isherwood RM (1972) Wind resistance of merchant ships. Transcr R Inst Nav Arch 115: 327–338
2. Gould RWF (1982) The estimation of wind loads on ship superstructures. The royal institution of naval architects, Report RINA-MTM-8, No 8, pp 1–34
3. Blendermann W (1994) Parameter identification of wind loads on ships. J Wind Eng Ind Aerod 51:339–351
4. Blendermann W (1995) Estimation of wind loads on ships in wind with a strong gradient. In: Proceedings of the 14th international conference on offshore mechanics and arctic engineering (OMAE), New York, pp 271–277
5. Blendermann W (1996) Wind loading of ships—collected data from wind tunnel tests in uniform flow. Report No. 574, Institut für Schiffbau der Universität Hamburg, Hamburg, Germany
6. OCIMF (1994) Prediction of wind loads and current loads on VLCCs, 2nd edn. Witherby and Co., London, UK
7. Haddara MR, Guedes Soares C (1999) Wind loads on marine structures. Mar Struct 12(3): 199–209
8. Andersen IMV (2013) Wind loads on post-panamax containership. Ocean Eng 58:115–134
9. Brizzolara S, Rizzuto E (2006) Wind heeling moments on very large ships. Some insights through CFD results. In: Proceedings of the 9th international conference on stability of ships and ocean vehicles, Rio de Janeiro, Brasil, pp 781–793
10. Wnek AD, Guedes Soares C (2015) CFD assessment of the wind loads on an LNG carrier and floating platform models. Ocean Eng 97:30–36
11. Valčić M, Prpić-Oršić J (2016) Hybrid method for estimating wind loads on ships based on elliptic Fourier analysis and radial basis neural networks. Ocean Eng 122:227–240
12. Freeman H (1974) Computer processing of line-drawing images. Comput Surv 6(1):57–97
13. Nixon MS, Aguado AS (2008) Feature extraction and image processing, 2nd edn. Academic Press (Elsevier), London, UK
14. Pavlidis T (1982) Algorithms for graphics and image processing. Springer, Berlin, Heidelberg, Germany
15. Kuhl FP, Giardina CR (1982) Elliptic fourier features of a closed contour. Comput Vision Graph 18:236–258
16. Specht DF (1991) A general regression neural network. IEEE T Neural Netw 2(6):568–576
17. Granlund GH (1972) Fourier preprocessing for hand print character recognition. IEEE T Comput 21:195–201
18. Beale MH, Hagan MT, Demuth HB (2010) Neural network toolbox: user's guide. The MathWorks Inc, Natick, MA
19. Fossen TI (2011) Handbook of marine craft hydrodynamics and motion control. Wiley, Chichester, UK
20. MathWorks (2016) MATLAB—the language of technical computing. Available online https://www.mathworks.com/
21. Prpić-Oršić J, Valčić M (2016) Sensitivity analysis of wind load estimation method based on elliptic Fourier descriptors. In: Proceedings of the 3rd international conference on maritime technology and engineering (MARTECH 2016). Lisbon, Portugal, pp 151–160

Tensairity, an Extra-Light Weight Structure for Airships

Anna Suñol and Dean Vučinić

Abstract Heavy payload airships and non-conventional shape airships require a rigid keel to maintain their shape under loading. The keel increases the structural weight, decreasing the payload capacity of the airship. A potential solution to maintain a light structural weight is to apply Tensairity as structural elements for an airship keel. Tensairity is a novel lightweight structure, which consists of a reinforced inflatable beam. A first demonstrator was publicly presented in 2002 by Pedretti, Luchsinger et al., and consisted of a car bridge. Since then, the concept is under continuous development, principally in the civil engineering domain. In this chapter, the potential of applying Tensairity as structural elements for an airships' keel has been assessed by comparing the weight of a Tensairity frame to the weight of an I-beam frame. It is significant to mention that we can find in literature the comparison of Tensairity beams and conventional beams weight, but the differences on loading conditions, type of structural element (frame instead of beam) and scale do not allow a correlation. After developing the necessary numerical model, a large Tensairity frame has been sized. We show that we can obtain an airship weight, which is eight times lower dimensioned against stress, strain and buckling. However, the buckling results are preliminary and should be validated experimentally. Considering the mentioned conclusions, we have identified Tensairity as a novel technology with large potential for the Lighter Than Air (LTA) vehicles application, making possible hull sizes and geometries, which are not possible to be constructed nowadays.

Keywords Tensairity · Lighter than air (LTA) vehicles · Lightweight structure

---

A. Suñol (✉)
3D Printing R&D Engineer at Idneo Technologies, Barcelona, Spain
e-mail: anna.sunyol.jimenez@gmail.com

D. Vučinić
Vrije Universiteit Brussel (VUB), Pleinlaan 2, 1050 Brussels, Belgium
e-mail: dean.vucinic@vub.be

D. Vucinic et al. (eds.), *Advances in Visualization and Optimization Techniques for Multidisciplinary Research*, Lecture Notes in Mechanical Engineering, https://doi.org/10.1007/978-981-13-9806-3_6

Nomenclature

b	Strut section width
e	Fabric thickness
E	Young modulus
E_{fabric}	Fabric Young's modulus
E_{strut}	Strut Young's modulus
h	Strut section height
I_{fabric}	Fabric moment of inertia
I_{strut}	Strut moment of inertia
I_z	Beam moment of inertia
k_b	Buckling constant for Tensairity frames
l	Column's length
n_{xx}	Membrane tension on the x-axis
n_{yy}	Membrane tension on the x-axis
p	Pressure
P_{buckling}	Euler critical load
p_{cr}	Critical overpressure
p_{dist}	Distributed pressure
q	Distributed load
q_{buckling}	Euler critical distributed load
r	Section radius
R	Frame radius
S	Beam's cross-section area
S_{fabric}	Fabric's cross-section area
S_{struts}	Strut's cross-section area
y	Distance to the beam section centre
y_{in}	Vertical distance of the inlet mid-section to the origin
$y+$	Dimensionless wall distance
ν	Poisson's ratio
ρ	Fluid density
θ	Thrust angle

1 Introduction

A current limitation in airship design is that heavy payload airships and non-conventional shape airships require a rigid keel to maintain its shape under loading. According to its hull configuration, airships are classified in three structural groups: (a) non-rigid airships, (b) semi-rigid airships and (c) rigid airships. Figure 1 shows the main structural characteristics of the three types of airships.

Non-rigid airships, or blimps, are those LTAs of which the shape is maintained solely by the envelope overpressure: the differential pressure between the inner lifting gas and the outer atmosphere pushes the envelope to the desired shape,

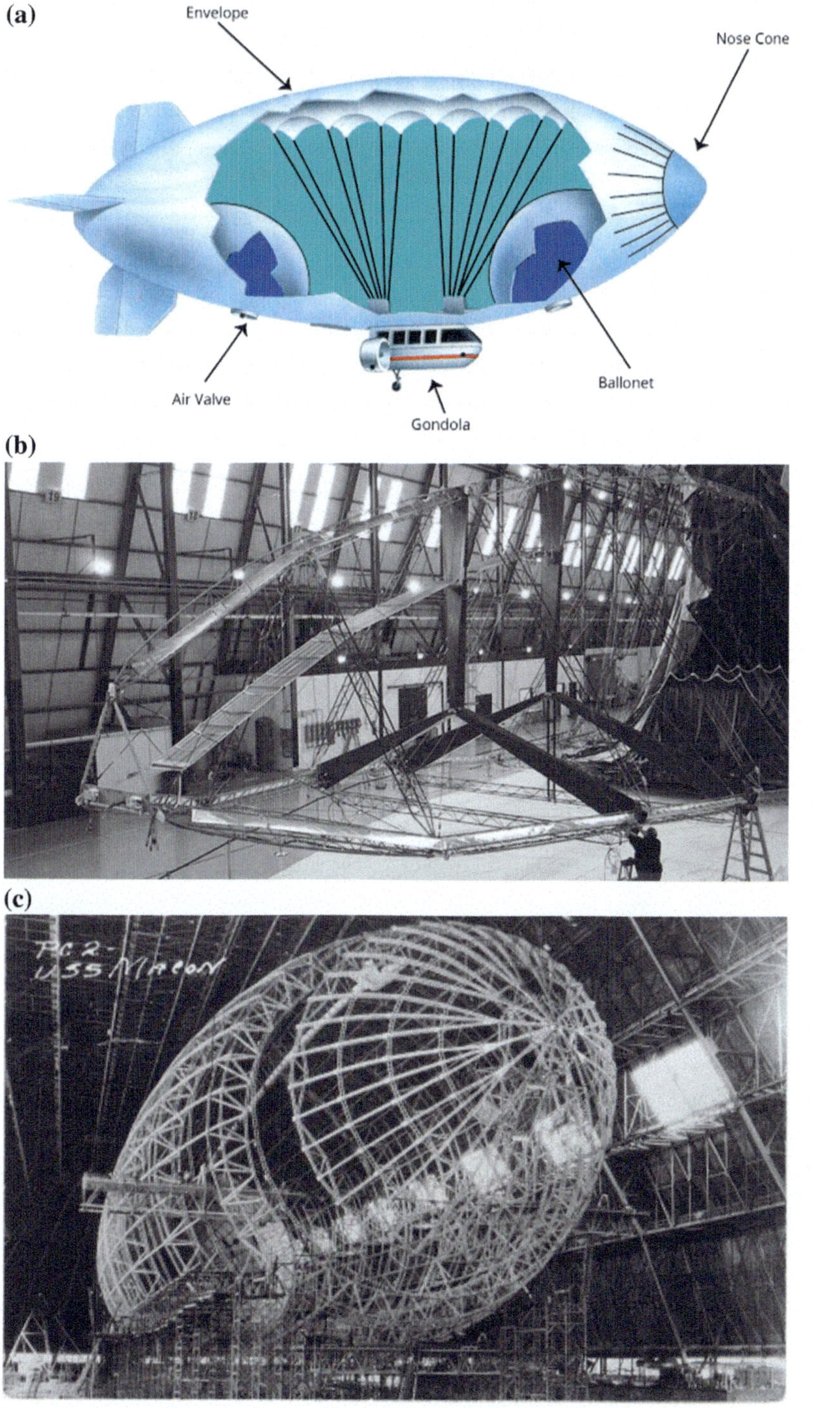

Fig. 1 Airships are classified in **a** non-rigid, **b** semi-rigid and **c** rigid airships. *Source* Ref. [3]

counteracting all other involved forces. Simple structures exist in the critical areas (as the nose cone), where otherwise the fabric would need to be thicker to withstand the loads only by itself. Loads are transmitted to the envelope by a system of cables and distributed by a suspension system (adjustable catenary cable system). Due to the absence of a rigid structure, non-rigid airships are light, easy to design, build and maintain. However, as the volume of the airship increases, drawbacks related to manufacturing and maximum load bearing of the membrane appear. Consequently, non-rigid airships are suitable for small airships [1].

Semi-rigid structures consist both of a rigid keel and a catenary suspension system that plays a much-reduced role in comparison to the non-rigid structures. The rigid keel distributes the gondola weight along the entire length of the airship and eliminates the main function of the catenary curtain, although the envelope still acts as a primary structure. Thus, the envelope carries part of the aerodynamic bending loads and, consequently, is still partially responsible of maintaining the desired shape by its overpressure [1].

Rigid airships are supported by a framework, which carries all the loads and maintains by itself the shape of the airship. In this type of airships, the purpose of the envelope is to retain the buoyant gas, and thus the strength requirement is much lower. Rigid airships allow gas compartmentalization, which increases safety. Rigid airships are not appropriate for sizes below one million cubic feet (which is around 28,300 m^3) volume. Its major benefit is that there is no size restriction due to the envelope fabric strength, which allows for larger volumes' dimensions [2].

According to the presented classification, the principal variable to decide the most appropriate structure for an airship is its size. Figure 2 shows the length of several airships that have been in use, without considering the projects currently under development. The conclusion is that the biggest airships that have ever flown were rigid. The data in Fig. 2 has been extracted from [2–7].

Besides the airship size, the airship shape also influences the need of structural elements. As internal overpressure is needed on non-rigid and semi-rigid airships, the envelope becomes a pressure vessel made by a material, incapable of withstanding bending moments and compression. Depending on the shape, form-defining elements are needed to force the membrane to follow the desired shape [8]. A clear example is the lenticular airship, as Fig. 3.

A rigid structure inherently leads to a higher weight of the structure and consequently to a lower payload capacity. As an example, comparing the Goodyear GZ

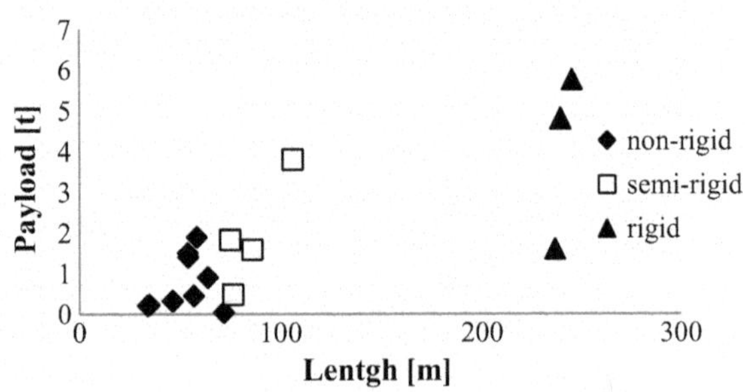

Fig. 2 Large and/or heavy payload LTA vehicles that have ever flown until now have a rigid keel

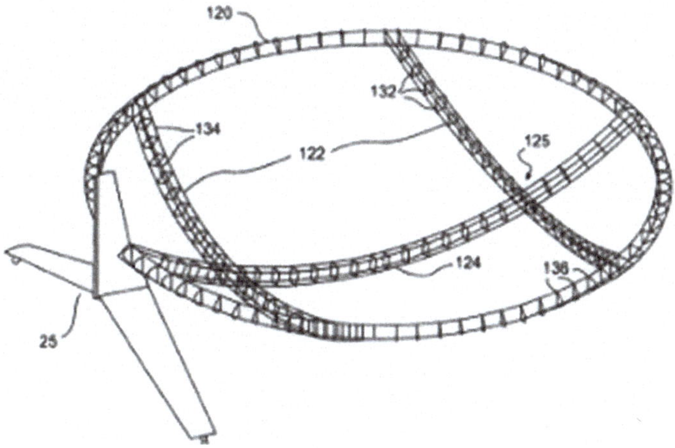

Fig. 3 Hulls which are not a pressure vessel require form-defining elements to maintain the desired shape. *Source* Ref. [8]

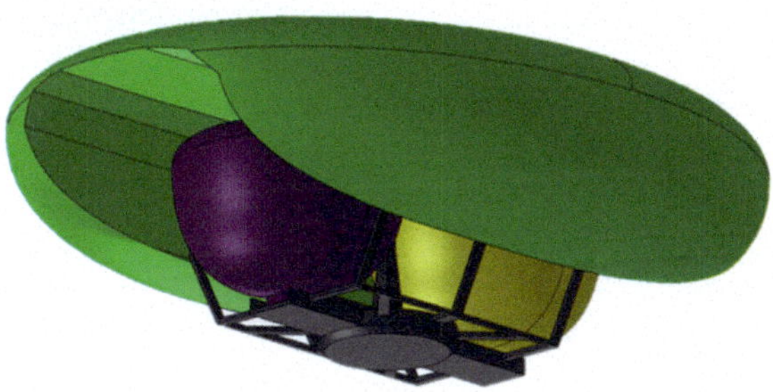

Fig. 4 The studied docking operation consists on the approach and connection of two airships at stratospheric altitude

20A (non-rigid airship) and Goodyear NT (semi-rigid airship), the ratio payload/ total weight is approximately 20% higher for a non-rigid airship than for a semi-rigid airship. This fact has motivated us to assess the potentiality of applying an extra-light weight structure called Tensairity as structural elements for airship keels. The principal objective of this paper is to assess the feasibility and the weight-reduction potential of Tensairity frames as a structure for keels of large and/ or non-conventional airships against stress and strain.

1.1 The MAAT Airships

The European project MAAT (Multi-body Advanced Airship for Transport) is a project from the European Seventh Framework Programme (FP7) for Transport including Aeronautics [9], and targets at designing a new concept of transport system. MAAT is based on a modular concept design, which is composed of two different airships that have the possibility to join, respectively called cruiser and feeder (Fig. 4). The advantage of an airship system compared to a single LTA aircraft is that only small airships land on ground based airports, which simplifies ground operations, while the big airship remains at the cruising altitude and only lands for maintenance purposes.

Following this concept, the feeders perform linking operations between cruisers and ground based airport hubs. They are designed to lift up and down from the ground to the interception altitude, where they join the cruiser in a unique MAAT transportation system. The cruiser remains in flight at an economical altitude and speeds, and it has static hovering capabilities to simplify the feeder docking operations. This economical altitude is between 13 and 16 km [9]. The flight altitude has been selected to maximize the energetic efficiency. One of the requirements of the MAAT project is the cancellation of any energy cost by an adequate photovoltaic (PV) energy source combined with hydrogen cells [10]. In [11], a thorough study of the photovoltaic productivity at high altitude is presented. As a conclusion, in [11] it has been demonstrated that the efficiency of the photovoltaic cells is at least twice at the targeted flight altitude than at sea level (for identical latitudes).

The MAAT Consortium has developed the conceptual design of the two airships, which in both cases is radically different to conventional airships. The MAAT system is breakdown in one cruiser and two feeders. The research performed in this paper focuses on the cruiser.

1.2 The Tensairity Concept

Tensairity is a relatively new lightweight structure concept based on the Tensegrity concept: it is based on the functional separation of its elements, aiming to maximize its efficiency. In tensegrity, some elements are only able to withstand tension and the others are only optimal on withstanding compression, which leads to the most efficient structure in terms of weight. However, designing stable tensegrity structures is not straightforward. In tensegrity structures, the interaction of compression and tension components makes the nodes stay in the state of balance (self-equilibrium state). Consequently, a tensegrity structure cannot be arbitrarily designed, but must ensure that the connections and pre-stress of the tension elements ensure such self-equilibrium state (this part of the design is called form-finding). The novelty of Tensairity is that the different elements are connected with a pre-stressed membrane, providing rigidity to the system. In other words, Tensairity overtakes the tensegrity principle of constructive separation of tension and compression in cables and struts, and a third element is added: low pressured air for stabilization [12].

As example, a Tensairity beam of 10 m span, slenderness γ of 10 and a distributed load of 2 kN/m is by factor of 6 lighter than a steel HEB profile beam (when designed for similar loading conditions) [13]. The larger the slenderness, the larger this factor will be. Summarizing, the basic Tensairity beam consists of three major parts: cables, which solely withstand tension; a strut, conceived to withstand compression; and a pressurized membrane, which pretensions the cables and stabilizes the compression element against buckling. The different elements existing in a Tensairity beam are represented in Fig. 5. Tensairity is a relatively new concept which is still under continuous study, and that has never been applied on the LTA sector. In order to assess the potential on such sector, we have researched a Tensairity frame. The researched frame is made of an inflated membrane reinforced with two struts, one located at the inner radius of the frame and the other at the outer radius. We have selected the type of reinforcements (two struts instead of a combination of cables and struts) and their position due to the internal forces distribution, which combine traction and compression loads along both the frontal and rear part of the frame. Future optimization of the concept would substitute the struts for cables in the positions where only traction is expected (which have been identified in this research).

Tensairity has already been successfully applied in different civil applications, such as bridges, roofs or hangars [12, 14]. Figure 6 show examples of such applications. Considering the existing applications and its fundamental mechanics, we have identified it as a potential option for LTA structures. A possible drawback of their implementation on airships is that Tensairity structures require a pressure control system due to the gas leakage. Although LTA vehicles already require such subsystem, the dependence of a stable structure on it decreases the vehicle's safety. It is also significant to mention that the weight reduction is less pronounced if dimensioned against 3D loads, since in that case stiff elements are required in those areas where the stress is high. In addition, the application of a Tensairity keel is such a novel concept that would involve an exhaustive design to ensure safety: definition and analysis of all the loading conditions, analysis of failure modes, redundancy of some elements, or the use of higher safety factors due to the use of fabric instead of metal.

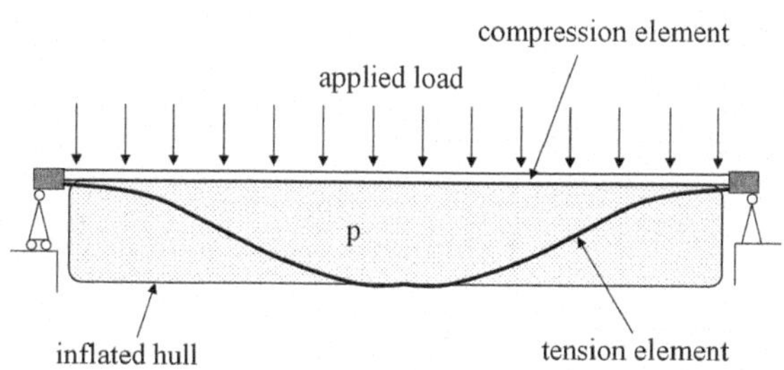

Fig. 5 Tensairity beam consisting of a compression element (strut), a tension element (cable), and a stabilizing element (membrane, inflated hull). *Source* Ref. [13]

Fig. 6 Tensairity has already been applied successfully as long span beams on a roof and on a bridge, portable hangars and kites [12, 14]

2 Fundamentals of Tensairity Structures

In a Tensairity structure, the elements responsible of withstanding tension and compression are tight to an inflated air beam, reinforcing it. The pressure inside the air beam must be higher than the external pressure, since sufficient pre-stress is needed to provide stability to the air beam. The air beam can be modelled as a thin walled pressure cylindrical vessel, and consequently the stress on the fabric can be computed from Eqs. (1) and (2). The assumptions to apply the thin walled pressure vessel model are:

- the wall is assumed to be very thin compared to the other dimensions (thickness/radius << 1)
- the geometry and the loading are cylindrically symmetric
- the internal pressure is uniform and positive
- end effects are neglected.

In Eqs. (1) and (2), n_{xx} refers to the membrane tension on the x-axis, R_x to the radius in the x-axis, p to the internal overpressure, σ_{xx} to the stress on the x-direction and e to the membrane thickness. The subscript y refers to the y-axis.

$$\frac{n_{xx}}{R_x} + \frac{n_{yy}}{R_y} = p \tag{1}$$

$$\sigma_{xx} = \frac{n_{xx}}{e} \tag{2}$$

A standard air beam reaches its limit load for practical applications when wrinkling occurs at the compression side. When the tension becomes zero, the membrane cannot take any compressive forces. By connecting a stiff element on this compressed side, the load bearing behaviour is increased because the compressive forces are now taken by this stiff element [15]. If cables are connected on areas where tension is occurring, the cables will withstand such forces. Thus, by adding a compression element and cables to an inflated tube, a structural system is obtained. The strut and cables accommodate the applied load and the air tube transfers forces between both and stabilizes the compression element against buckling. In this way, the load carrying capacity is increased compared to a traditional air beam, whereas the pressure inside the air tube can be lowered. The result is a modified air beam with the same load bearing capacity as a steel beam but with a drastically reduced weight [13]. In [13], Tensairity is described with a combination of beam theory and membrane theory. However, it is not obvious if the described modelling can be applied to a frame, and if it is valid for dimensions one factor larger (which is the case in large airships). They consider a Tensairity beam similar to the one shown in Fig. 5, under a distributed load q applied on the strut. Applying beam theory, they deduce the total force of the cable, which is shown in Eq. (3).

$$T = \frac{qL\gamma}{8} \tag{3}$$

Compared to existing Tensairity structures, the principal difference of a frame is its lack of a point where the compression and the traction element merge. In addition, the type of loading conditions and internal forces distribution are also different. Related to the influence of a larger scale, the validity of the assumptions found in literature was also unknown. More specifically, we have found in literature analytical models to estimate the force withstood by the traction and the compression elements [16, 17]. Equation (3) does not consider the fabric mechanical contribution, assuming that it can be neglected since the fabric Young modulus is typically one factor smaller than the Young modulus of the other Tensairity components. In other words, although the experimental and numerical modelling did include the fabric contribution, the analytical modelling neglected it. In [18], an analytical and numerical analysis is performed to understand the validity of such hypothesis.

Since a structure must be sized against stress, strain and stability, it is necessary also to understand also the mechanical performance of Tensairity structures on buckling. According to literature, Tensairity structures have a larger weight reduction potential at large span since, contrary to conventional structures, its buckling load is independent from it. More specifically, as the membrane continuously supports the element carrying compression, buckling can be avoided if the proper internal pressure and inertia are chosen [13]. In conventional columns, the Euler critical load P_{buckling}. decreases with the square of the column length l, as shown in Eq. (4). In this equation E is the material's Young modulus for a standard beam made of a homogeneous material. Such dependence on the span implies the necessity of either increasing the number of supports to decrease the length or increasing the column inertia I_z. Both solutions lead to an increase of material and, consequently, of weight. On the contrary, Tensairity can be described as a beam connected to a continuous elastic support. The elements, which are prone to buckle, are the struts, which are tightly connected to the fabric. The fabric acts as a continuous support of the strut, restricting the struts deflection and, consequently, reducing the buckling load. The stiffness of this support is determined by the fabric stress, which depends on the internal overpressure p. Equation (5) describes the buckling load for a column on a continuous support, particularized for Tensairity columns [13], and shows that the buckling force of columns with a continuous support is independent of the length. In Sect. 6, these equations are reformulated for a Tensairity frame.

$$P_{\text{buckling}} = \pi^2 \frac{EI_z}{l^2} \tag{4}$$

$$P_{\text{bucklingtensairiry}} = 2\sqrt{\pi p E_{\text{strut}} I_{\text{strut}}} \tag{5}$$

Extending those equations to our case study require extensive research, since it is necessary both to adapt Eq. (5) to a frame and to understand the scalability of the problem. An attempt has been performed in this paper where Eq. (5) has been adapted to a Tensairity frame and where the buckling load of small scale Tensairity frames has been computed numerically. We have also highlighted that if the equation developed in Sect. 6 is valid at any scale, the weight reduction improvement would be much more significant than for stress and strain. More specifically, the weight reduction would be of 8 times for a frame connected to the keel by 12 points, taking into account stress, strain and stability. However, the results are preliminary, since they have not been compared to experimental data. In addition, the numerical model at large scales has not been developed, and the validity of the developed expression at larger scales has not been investigated. Consequently, the analysis of the buckling load must be further researched. More details of the buckling model and the weight comparison can be found in Sect. 6.

3 Design Requirements: Geometry, Loading Conditions and Dimensioning Criteria

In this section, we detail the geometry, the loading conditions, the dimensioning criteria and the simplifications that we have considered to dimension a Tensairity frame that would be applied in a large and non-conventional hull. In next sections, the weight of the Tensairity frame has been compared to the weight of an I-beam frame and a truss frame, all of them withstanding the same loading conditions. It is significant to mention that the Tensairity frame has not been optimized, since the scope of this research is a preliminary evaluation of the advantages of applying Tensairity in airships. Besides the comparison to the I-beam frame, a non-reinforced inflated beam has been modelled, aiming to assess if a regular inflated beam performs worse than a Tensairity frame at identical overpressure. Regarding the numerical model, it is identical to the one described in Sect. 4.

3.1 Geometry

We have studied an elliptical frame airship of 266 m span and 330 m length (the dimensions of the cruiser airship), represented in Fig. 7 and specified in Table 1. Preliminary investigations evidence that such envelope geometry could not be designed as a non-rigid airship. Consequently, it has been identified as a scenario where a Tensairity keel could be an optimal light-weight solution. We have not considered the breakdown of the air beam structure in multiple compartments, which might be necessary for safety reasons.

3.2 Loads and Boundary Conditions

We have studied a frame with a constant distributed load. We have dimensioned the frame under 2 types of boundary conditions: the first one constrained at the same direction as the load only at 2 points (fundamental case study) and the second one constrained at 12 points (frame attached to the keel). The constraint that simulates the action of the propellers is only in the flight direction, while the constraints that simulate the connections to the keel is in all the directions except for the flight

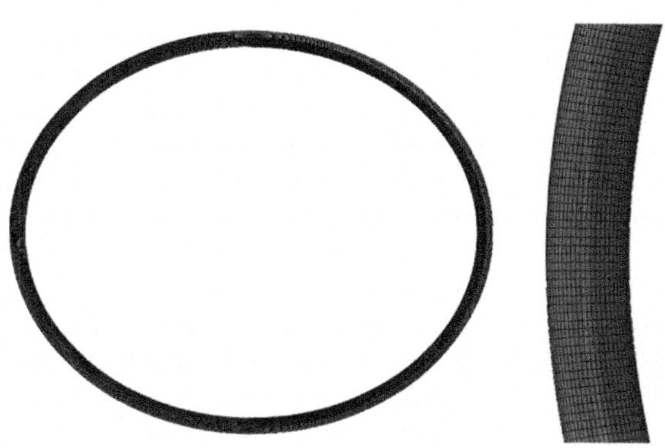

Fig. 7 Elliptical airship Tensairity frame and details of the mesh

Table 1 The dimensions of the ellipse correspond to the dimensions of the cruiser airship

Dimension	Value
Length of the ellipse (m)	330
Width of the ellipse (m)	266

direction. This is a simplification of the real conditions, in which the horizontal frames are the principals responsible of withstanding the drag force and the other frames are the principal responsible of withstanding the pitch moment and the gravity load. Consequently, from those 12 points, 2 are constrained in the 6 degrees of freedom and the other 10 are constrained in all the directions except for the flight direction.

The uniform distributed load is a simplification of the real pressure distribution. In the first case, the value of the distributed load is 100 N/m, while in next second we increase it to 300 N/m (150 kN in total). The reason is that the analysis performed in the fundamental case does not include the interaction with the other members of the keel, which would increase the stiffness of the frame. Consequently, and according to the obtained results, designing a frame that withstands 300 N/m for the first case would be challenging. For the second case, 300 N/m is approximately 1/3 of the total drag. Consequently, 2 more horizontal frames (similarly to Fig. 10) would be sufficient to withstand the totality of the drag force. Besides increasing the value of the distributed load, the second case includes the self-weight, modelled using the gravity load. Regarding internal overpressure, we have considered a uniform pressure distribution, which is a simplification that neglects the temperature variations or the influence of altitude.

It is significant to mention that there are still several simplifications: the use of a continuous volume of air along the frame, the assumption of a homogeneous distribution both for the drag load and the internal overpressure, and the idealization of the other members of the keel, which neglects their deformation in all the directions except for the one parallel to the drag force. In a detailed design, those factors should also be considered.

3.3 Dimensioning Criteria

We have dimensioned the Tensairity frame considering stress and strain, while a preliminary analysis of its stability is discussed in Sect. 6. We have considered 0.01 as maximum allowable deformation. Regarding the stress, and considering 6061 T451 aluminium for the struts, their maximum allowable stress is 95 MPa (the yield stress is 145 MPa [19] and the safety factor is 1.5). Regarding the fabric, we have considered the fabric material described in Sect. 4. In Sect. 4, we discuss the effects of changing the fabric material. According to FAA-ADC or the German LFLS [20], the structural design criteria for the envelope is to not exceed a maximum stress larger than the yield stress, applying a security factor of 4 to account for fatigue.

Considering such security factor, 40 MPa has been assumed a conventional value of maximum allowable fabric stress [21]. Considering a 5 m radius section and applying Eq. (1) and (2), the selected overpressure has been 5000 Pa (50 mbar). 50 mbar is a value significantly lower than conventional Tensairity beams, inflated with a few hundreds of mbar. This difference lies on the large beam section radius selected.

4 Numerical Model

This section describes the numerical model, which is based on the modelling described by Pedretti et al. [12]. In order to model a Tensairity frame, the commercial software package ABAQUS has been used, similar approach described in [12]. ABAQUS is especially appropriate for simulating non-linear analysis, which is the case for the orthotropic material considered [22].

It is significant to highlight that the most significant deformations after inflation are not perpendicular to the cross-section but in the frame radial direction. As a consequence, at large scale Tensairity frames the stiff elements cannot be added before inflation: the pre-stress would be excessive. The reason lays on the large deformations involved in the torus inflation and on the interaction fabric-struts, which are two materials with more than one order of magnitude difference in their elastic modulus. As a result, the stress on the struts after inflation is excessively large. In the studied test case, the struts stress after inflation would be approximately 155 MPa. Partial struts pre-stress might be desirable to avoid compression regions and substitute the struts for cables, which are a lighter solution. As this is part of an optimization process, in the studied case the struts have been attached after inflation.

Due to the complexity of simulating the fabric material deformation, a dynamic analysis has been required to model the Tensairity beam inflation and loading. The internal overpressure and the drag loading have been modelled in two subsequent transient steps. Thus, inertial forces are also considered during the analysis. In order to avoid large inertial forces due to the large dimensions, which are not of interest since the aim is to model the steady loaded case, a sufficiently large time step has been assumed. In this case, and after different attempts, the time step imposed is 30 s. Moreover, the loads are applied only during the first half of the step, allowing the system to stabilize [22].

The fabric has been modelled as SHELL elements, while the struts have been modelled as BEAM elements. If the thickness of the membrane is small, SHELL and MEMBRANE elements provide similar results [23]. We have decided to use SHELL elements since this type of element is more robust in terms of numerical computation, and can be applied to cases where the MEMBRANE elements cannot. This is important since in order to model the large scale Tensairity frame we had to attach the struts on the air beam after inflation, and that operation is only supported by ABAQUS if SHELL elements are used instead. The rigid elements have been modelled as SHELL elements, applying aluminium 6061 T451 as material. The decision of using SHELL elements instead of BEAM elements relies on the convenience of defining a surface to be tight to the air beam. This approach is similar to the one described in [12]. The mesh is generated after 4 m uniform seeding discretization and imposing quad-dominated elements with sweep meshing algorithm, resulting in 240 S4R elements for the internal strut and 368 for the external strut. The air beam has been modelled similar to the small circular frame test case. A 1 m seeding has been considered in the frame axis direction, while for the cross-section the seeding is 0.6 m. After discretization, 29,920 S4R elements have been created. Figure 8 shows a detail of the mesh. The surfaces in contact have been tight using the surface-to-surface contact interaction, imposing zero slip and not allowing separation once they are in contact. According to [23], the air beam must be modelled as the master surface due to its larger size compared to the struts. This is a robust approach to simulate the interaction between two surfaces [23], and has already been used in [12]. The two surfaces have been tight after the inflation step. An explicit numerical scheme has been applied.

We have selected aluminium 6061 T451 as material for the stiff elements, and a fabric nylon material already applied in Tensairity prototypes for the fabrics [24]. Table 2 shows the mechanical properties of the considered fabric, extracted from its datasheet (Riverseal 2006). In Table 2, E stands for Young modulus, ν is the Poisson's ratio and G is the Shear modulus. The subscript 1 stands for the warp direction, 2 for the weft direction and 3 for the orthogonal direction. It is beyond the scope of this paper to perform an extensive research on fabric materials, which should take into account requirements as maximum bearing stress, minimum weight and gas permeability. However, we have performed a comparison of some common structural fabrics to understand the influence of the fabric material properties on the conclusions extracted in this section.

Fabric materials can be modelled with different techniques, depending on the desired level of fidelity. Some of the most common techniques are the (a) Linear elastic orthotropic, (b) Hyperelastic orthotropic, (c) Hybrid constructions, or

Fig. 8 The mesh was created applying a uniform seeding of 0.6 m

Table 2 Mechanical properties of the Tensairity fabric

Density (kg m^3)	Thickness (mm)	E_1 (GPa)	E_2 (GPa)	E_3 (GPa)
1400	0.6	1.1	1.1	0.1
ν_{12} (–)	ν_{13} (–)	ν_{23} (-)	G_{12} (MPa)	G_{13} (MPa)
0.31	0.2	0.2	30	10

(d) applying a user defined material law specially developed for woven fabrics. The Tensairity membrane has been modelled as SHELL elements with orthotropic material properties, which is common in FEA of airships [25]. The fabric has been oriented placing the material principal directions parallel to the torus axis and the radial direction, in order to avoid non-desired shear stresses that would deform the membrane from its desired shape.

The modelling of the boundary conditions and the loads is similar to the ones described in [12]. The air beam overpressure has been modelled as a constant pressure force on the internal surface. The distributed load has been modelled as a line force on the frontal part of the frame.

4.1 *The Membrane Material: Discussion of Its Influence*

In order to assess the potential of Tensairity frames we have considered aluminium for the struts and a nylon fabric for the membrane (Sect. 3). As already mentioned, it is out of the scope of this paper to select the optimal fabric material for the Tensairity frame. This is a complex task, since the selection of the fabric should take into account many requirements, the most significant of which are surface weight, tensile strength and Young modulus, flammability, permeability to the buoyant gas, price and durability. In addition, the characteristics of a structural membrane do not only depend on the material of its yarns but also in the weave pattern, the number of yarns per centimetre, the coating and the finishing. The combination of those parameters will determine the surface weight and the mechanical and physical properties of the membrane. Consequently, to adapt to the different market requirements, there are many variants of commercial membranes, and there is also the option of designing a new solution to optimize the fabric to the case study. We have identified the selection of an appropriate fabric as a future next step to perform a more accurate research on Tensairity frames for large scale airships.

In this section, we discuss the effect of using another fabric on the conclusions extracted in this paper. The materials used for architectural membranes generally consist of a woven fabric coated with a polymeric resin, as shown in Fig. 9. As shown in Table 3, we have compared the mechanical properties of the most common yarn materials used in the aerospace and the architecture sectors: nylon, polyester, glass fibber, Vectran and Kevlar. We have extracted the data shown in Table 3 from [2] and [26]. Although the number of yarns per cm, weave pattern and coating will also determine the mechanical performance of the membrane, the

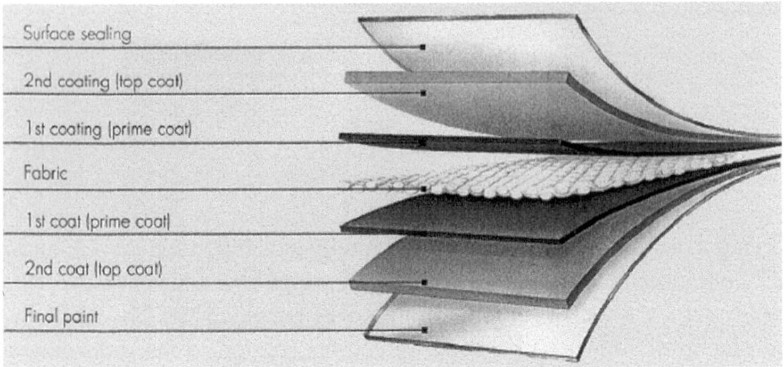

Fig. 9 Structural membranes are made by a fabric material coated with a polymeric resin [26]

Table 3 The two materials normally applied in airships are Nylon and PVC coated polyester fabric, but current state of the art shows the existence of more advanced fabric materials

Yarn material	Application	Density (kg m^3)	Tensile strength (GPa)	Young modulus (GPa)
Nylon	Airships	1140	0.8	5
Polyester	Airships/ architecture	1380	1.0	12
Glass-fibber	Architecture	2600	2.4	72
LCP (Vectran)	Space	1400	3.28	104
High modulus aramid (Kevlar)	Space	1450	3.32	160

values shown in Table 3 already indicate an order of magnitude of the differences that can be expected by changing the fabric material. Nylon and PVC coated polyester fabrics are commonly applied in airship envelopes [2]. In the architecture domain, the most commonly used materials are PVC coated polyester fabrics and PTFE coated glass fabrics, and of these two PVC coated polyester fabrics is more widely used due to the compromise of cost, performance and durability [26]. Finally, LCP fabrics (Vectran) and high modulus aramid (Kevlar) are high performance fabrics that have been used in space, applied on airbag landings on Mar and as inflatable decelerators for atmospheric entries [27]. We have only focused on the two first types of materials, since they are the ones commonly applied in airships. However, an insight extracted from Table 3 is that, focusing only on the requirement of minimum weight for maximum tensile strength and Young modulus, the use of the more advanced fabric materials (as Vectran or Kevlar) would allow to obtain a weight reduction more significant than the one that we have obtained. We have extracted such conclusion by comparing density, tensile strength and Young's modulus of those materials.

Regarding the results of strain, stress and stability, a different density would only affect on the inertial forces. Since the analysis is static, a different density would not provide different results than the ones discussed. On the opposite, a different Young modulus would affect on the results as shown in Eq. (6) [18].

One of the conclusions extracted at [18] is that the mechanical contribution of the fabric can be neglected in the modelling for small scale Tensairity frames, while at large scales (i.e. 100 m of radius) ignoring the fabric leads to inaccuracies on the prediction of the struts stress of 25%. If we would have conducted such study using PVC coated polyester fabric instead of nylon, the ratio $\frac{E_{\text{fabric}}}{E_{\text{strut}}}$ would increase, and consequently such percentage would also increase according to Eq. (6).

$$\sigma_{\text{strut}} = -\frac{M_z y \mathrm{E}_{\text{strut}}}{2E_{\text{strut}} I_{\text{strut}} + E_{\text{fabric}} I_{\text{fabric}}} + \frac{N \mathrm{E}_{\text{strut}}}{2E_{\text{strut}} S_{\text{strut}} + E_{\text{fabric}} S_{\text{fabric}}} \tag{6}$$

Regarding maximum stress criteria, we have used the tensile strength to estimate the internal overpressure. If we would have used PVC-polyester instead of nylon, the maximum internal overpressure or alternatively the internal radius r could also increase proportionally (according to Eqs. 1 and 2). In addition, since the Young modulus is approximately the double (i.e. polyester is stiffer than the nylon fabric), at identical conditions the deformation would be smaller. On the contrary, the density of polyester is 20% higher than the one of nylon. Taking into account that in a Tensairity frames the struts are the principal elements that provide stiffness and not the membrane (as mentioned, the case study shown in [18] indicate that the stress on the struts considering and neglecting the fabric differ on a 25%), it is not straightforward to deduce which fabric material is more weight-efficient.

As a general conclusion of the comparison of the fabric materials, the conclusions related to the analytical model and the analysis of the different design parameters would not vary. On the opposite, the weight of the Tensairity frame would be influenced by the selection of the fabric, reason why we have identified it as a task to be performed in future research. Similarly, the composite materials could replace the aluminium struts, and thus reducing their weight too.

5 Assessment of the Tensairity Potential Under Static Load (Stress and Deformations)

The objective of this section is to assess the potential of Tensairity as a structural element for airship keels. The aim is to deduce if, in the studied geometry, size and loading conditions, Tensairity potentially allows for a significant weight reduction, and thus its further study is justified. The designed frame is the horizontal frame, as shown in Fig. 10. As we expected the potential use of Tensairity for large and non-conventional airships is tested for an ellipsoidal airship of 266 m span.

Following this line, we have studied a fundamental case: we have dimensioned a large-scale Tensairity frame and compared its weight to the one of an I-beam frame

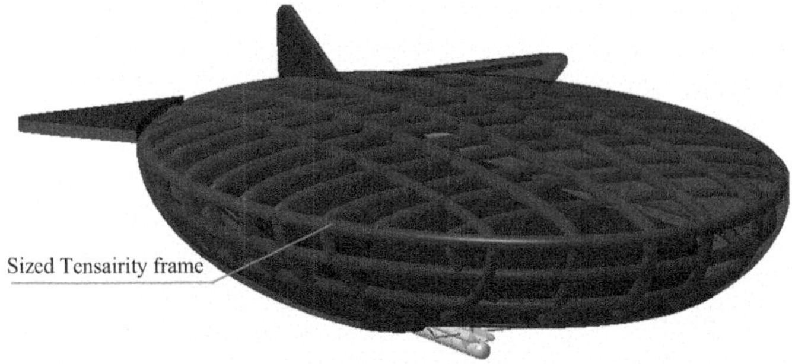

Fig. 10 The considered Tensairity frame is elliptical and must withstand the effect of frontal drag, which is a distributed load. This is similar to the test case studied in the previous section

and of a truss frame, all of them sized for identical span and loading conditions. Moreover, we have evaluated a non-reinforced air beam under similar conditions. We have performed such analysis under two conditions: considering only the frame without the interaction of the keel, and considering also the connections to the keel. The first condition, which is a fundamental case study that analyses the performance of the frame under a frontal distributed load, allows understanding the potential of Tensairity on large scale-frames. The second condition, which includes the self-weight load and the constraints that result from the interaction with other members of the keel, allows understanding the potential of Tensairity frames as part of an airship keel. For the fundamental case, (i.e. without the constraints of the keel), the value of the distribute load is only 100 N/m, since the absence of connections to the keel implies a low rigidity of the structure. In next section, the loading increases to 300 N/m, which is approximately 1/3 times the total drag of the airship. We have considered such value a realistic loading condition, since the keel shown in Fig. 10 has two more horizontal frames to withstand the drag load.

Initially, we have dimensioned the Tensairity frame against stress (the maximum allowable stress in the struts is the yield stress, using a safety factor of 1.5) and deformation (the maximum allowable deformation was 0.01) for the first condition. We have found that the first limitation is the deformation since, due to the lack of connections to the keel, the stiffness of the geometry is low. According to the results shown in this section, if we pursue identical deformation, for the first case the Tensairity frame is four times lighter than the I-beam frame and 8 times lighter than the truss frame. In addition, for identical overpressure, geometry and loading conditions, the non-reinforced inflatable beam fails.

For the second case, we have considered the Tensairity frame dimensioned for the first case to estimate the minimum number of connections to the keel, which has resulted in 12 (i.e. 6 frames). However, this might not be the most weight-efficient distribution, which must be determined during the design of the airship's keel. After adding the loads of gravity and drag, the maximum stress and the maximum displacement are both below the maximum allowable values. We have also analysed the rigid frames connected by the same number of points to the keel. For the I-beam frame, the maximum stress exceeds the yield stress due to its self-weight. It would require 24 points of connection, which means that the keel would require 12 frames more (increasing the global structural weight). For the truss frame, the displacements and stress are 1.54 m (which equals the maximum allowable deformation) and 54 MPa (below the maximum allowable stress). Consequently, for the considered loading conditions, the truss frame would require identical connections to the keel to the Tensairity frame.

5.1 Fundamental Case Study: Potential of Tensairity on Large Structures

The results presented in this section show that, considering only stress and deformation, a large-scale Tensairity frame without connections to the keel would be more than 4 times lighter than a rigid frame. The stress is below the maximum allowable stress, which is a consequence of the studied structure (if the considered torus included radial reinforcements to restrict the deformations, the stiffness of the structure would increase, turning into a more weight-efficient structure). Regarding the comparison to an air beam inflated at equal pressure and without struts, the air beam does not withstand the considered load, deforming excessively (deformation larger than 0.36 after applying the 78% of the load).

The maximum Von Mises stress on the fabric after inflation, σ_{VMmax}, is 37 MPa, which results from 21 MPa in the warp direction (σ_{xx}) and 44 MPa in the weft direction (σ_{yy}). As a validation of the results obtained after inflation, we have compared such value with the one predicted by Eqs. (1) and (2). We have obtained the 3 stresses (Von Mises, stress in warp direction and in weft direction) from the numerical calculation, and they satisfy Eq. (7). Equations (1) and (2) are also satisfied, since considering that, at the position of maximum stress, the radius in the warp and the weft direction are 160.7 m and 5.3 m after inflation respectively, they predict and overpressure of 51 mbar. Such value is only a deviation of 2% compared to the applied overpressure.

$$\sigma_{VMmax} = \sigma_{xx}^2 + \sigma_{yy}^2 - \sigma_{xx}\sigma_{yy} \tag{7}$$

The maximum displacement after inflation is approximately 1.85 m, as shown in Fig. 11. With the described geometry and loading conditions, struts of 300 mm width and 3 mm height are sufficient to ensure a stress lower than the maximum allowable stress and a deformation lower than 1%. As the considered test case is a single elliptical frame, without considering the interaction with other structural elements keel, the frame stiffness is low, and large deformations appear. Thus, the most restrictive criteria to dimension the frame is the maximum deformation instead of the maximum stress, both for the Tensairity frame and for the I-beam frame (Figs. 12, 13 and 14).

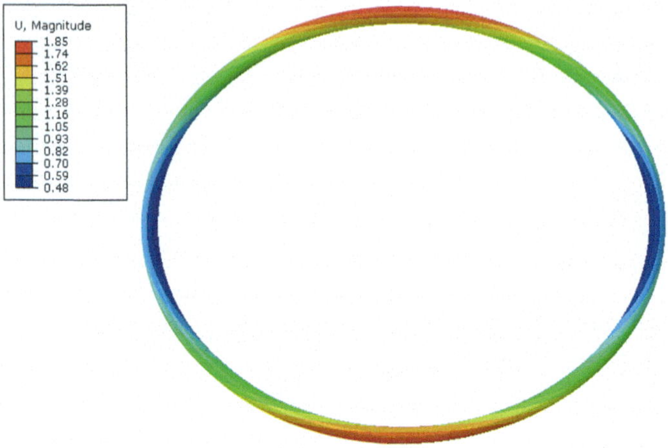

Fig. 11 The maximum displacement after inflation is approximately 1.85 m

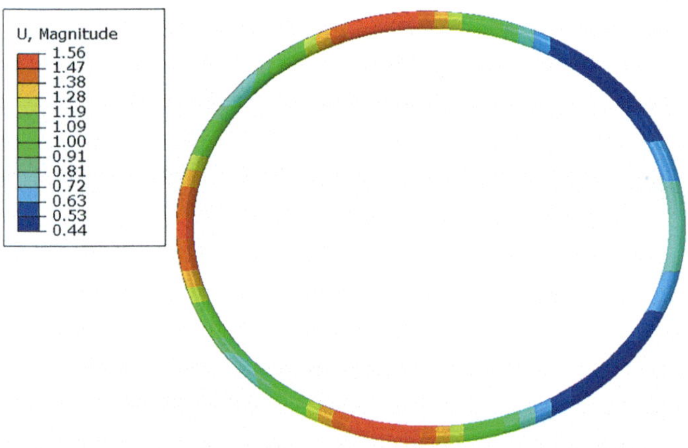

Fig. 12 The maximum displacement after loading is approximately 1.5 m

Table 4 shows the geometrical dimensions of the rigid frames (I-beam and truss). The maximum stress and displacement for the considered Tensairity frame are shown in Table 5. According to the results in Table 5, the maximum Von Mises stress is 38 MPa on the fabric and 20 MPa on the struts, both values below their respective maximum allowable stress. The maximum deformation is 0.01, as targeted. In addition, Table 5 shows a mass more than 4 times lower of the Tensairity frame than the lightest rigid frame for identical deformation, and 8 times lower for the truss frame.

For the designed Tensairity frame, 7.7 tones are due to aluminium, while 24.9 tones are due to the fabric material. Lighter weights might be obtained by a more

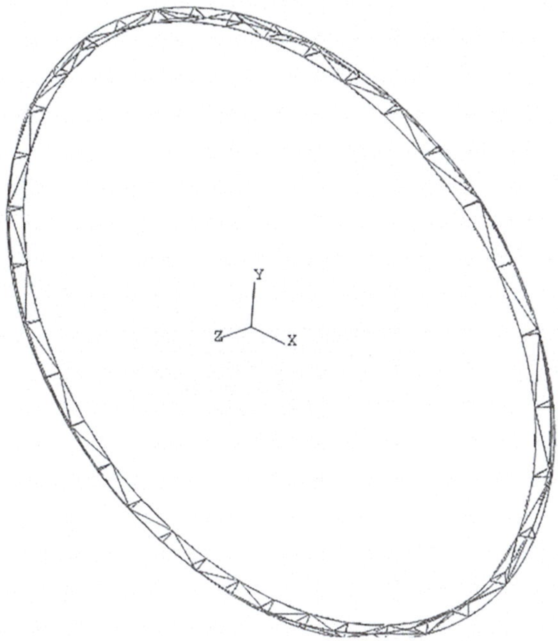

Fig. 13 The considered truss frame is heavier than the I-beam frame (fundamental case study, only stress and strain considered)

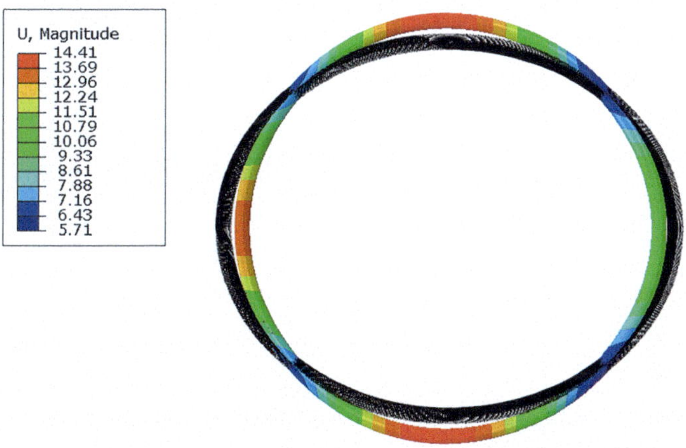

Fig. 14 A conventional air beam without reinforcements does not withstand the considered load

detailed material selection, based on lighter fabrics and composite materials for the struts. Moreover, according to literature, the weight difference between rigid and Tensairity structures might be larger when the buckling load is also considered. This is because selecting a sufficiently large overpressure and inertia allows

Table 4 Geometrical dimensions of the I-profile frame and of the truss frame circular cross-section

I-profile				Truss
Base (mm)	Height (mm)	Base thickness (mm)	Height thickness (mm)	Cross-section radius (mm)
300	1500	50	21.5	75

Table 5 The obtained preliminary results show a weight reduction of more than 4 times if dimensioned against stress and stiffness

	Max. Von Mises stress (MPa)	Max. displacement (m)	Mass (t)	Max. deformation (–)
Tensairity	20	1.56	32.6	0.01
I-beam	11.6	1.56	152	0.01
Truss	50.2	1.39	264.6	0.01

dimensioning the structure for yield stress criteria instead of for buckling. We have performed a first attempt in this sense, and it is described further in this section.

We have also compared an air beam without reinforcements (i.e. without struts) to the Tensairity frame, applying identical value of overpressure and identical geometric parameters. In such case, the computation could not be finished due to the appearance of excessive deformations (a deformation of 0.36 after applying 78% of the load). If an air beam must withstand the same loading conditions, higher overpressure must be applied. As already mentioned, this would lead to a challenging research on fabrics with sufficiently large breaking stress.

5.2 The Frame as Part of the Keel: Potential of Tensairity Applied on Airships

In this section, we discuss the potential of a Tensairity frame that is part of an airship's keel. Aiming to check the feasibility of the dimensioned Tensairity frame, we have studied the mechanical performance of the frame against its self-weight. With such objective, we have performed a numerical simulation in ABAQUS to check the stress and deformations that appear on the structure under the gravity load and the drag force.

The first step has been to estimate the minimum number of constraints required to withstand the frame's weight. Such constraints represent the interaction with the other structural members of the keel, which has not been designed yet and are out of the scope of this paper. In the previous section, the modelling included only two constraints. Those two points represent the action of the propellers, which generate a total horizontal force of value equal to the distributed load (drag force). In other

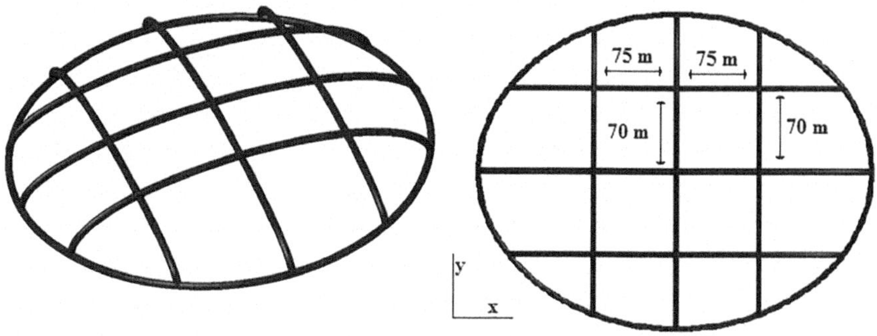

Fig. 15 The dimensioned frame requires to be connected to the keel by at least 12 connection points

words, the two points are not representing a constraint in the direction pointing to the ground (*z*-direction). On the opposite, the connections to the keel restrict the displacement in all the directions, although for this preliminary analysis we do not consider the constraint on the flight direction. Such simplification is made based on the assumption that in the considered keel, the dimensioned frame is responsible of withstanding principally the drag force, while the other frames are responsible of withstanding principally the weights and the aerodynamic pitch moment.

After estimating the minimum number of constraints in 12 (6 frames, 3 parallel the *x*-direction and 3 parallel to the *y*-direction), we have performed a second simulation including both the drag and the self-weight. According to the obtained results, the displacements decrease compared to the case where the self-weight and the interaction with the keel were not considered. The maximum stress on the struts increases from 20 to 57 MPa, but it is still below the maximum allowable stress. Figure 15 shows an isometric view and a top view of the frames that constraint the horizontal frame, as well as its position.

The modelling of the struts, of the membrane and of their interaction is identical to the one used previously. Regarding the load, we have simulated the self-weight with a gravity load. We have modelled the gravity load applying an acceleration of 9.8 m/s^2 towards the ground (z-direction). For the estimation of the minimum number of constraints, we have suppressed the drag load. The second simulation includes the two loads. Similarly to the modelling described in Sect. 4, we have simulated the inflation, attachment of the struts, self-weight and the drag load in subsequent steps, each of them of 30 s of duration except for the attachment of the struts, of 0.25 s.

The maximum allowable stress on the struts and on the membrane, and the maximum allowable deformation are identical to the ones considered for the drag force: 40 MPa on the membrane, 95 MPa on the struts and maximum displacement of 1.65 m. Connecting the dimensioned frame to 6 other frames, 3 parallel to the *x*-axis and 3 parallel to the *y*-axis, and considering only the self-weight, the maximum stress on the membrane is 38 MPa, the maximum stress on the struts is 50 MPa, and the maximum displacement is 0.91 m. Figure 16 shows the obtained displacements.

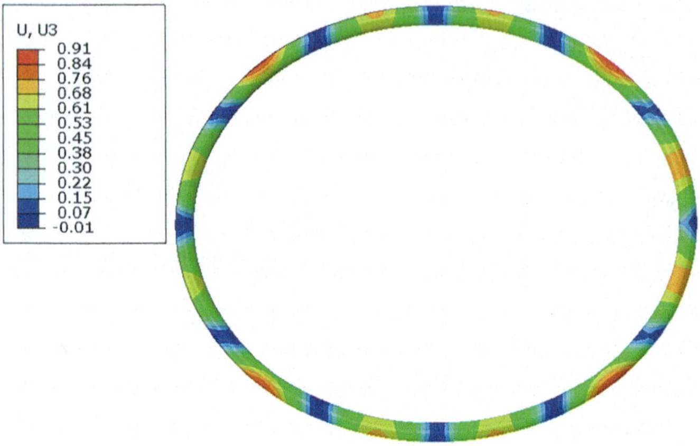

Fig. 16 If the considered frame is connected to the keel at 12 points, the maximum displacement is 0.91 m

As a conclusion of the first part of this analysis, the dimensioned Tensairity frame would require at least 12 points of connection to the keel (i.e. 6 frames). However, this might not be the most weight-efficient distribution, which must be determined during the design of the airship's keel. The design and dimensioning of the keel and its frame are not in the scope of this paper, and we have identified them as future research activities. As insight, it is significant to mention that the considered Tensairity frame has inner and outer reinforcements, but we have not reinforced it on the upper and lower sides. This is because in the keel shown in Fig. 15, the horizontal frames (as the one dimensioned) would withstand the drag forces, while the other frames would withstand the weights and the pitch moment of the airship. Consequently, the Tensairity frames perpendicular to the studied one would be made of the inflated membrane reinforced at the upper and lower sides.

After estimating the minimum number of connections to the keel, we have added a drag load of 300 N/m (150 kN in total). The maximum stress is 57.9 MPa and the maximum displacement is 1.48 m (see Fig. 17), of which 0.82 m are in the x-direction, as shown in Fig. 18. In other words, although the load is 3 times higher, the displacements reduce from 1.56 to 0.82 m, both in the x-direction. As mentioned in the previous section, the lack of radial reinforcements along the frame involves a low stiffness of the geometry, undergoing large displacements before reaching the maximum allowable stress. When constraining the structure in 12 points, even if the x-direction is not constraint, the structure stiffness increases. Another conclusion of the performed analysis is that the self-weight is a load that cannot be neglected in the design of such large structures, since the stress and displacement generated by the drag load and the self-weight are of similar scale.

We have also analysed the rigid frames under identical loading and boundary conditions. According to the obtained results, the stress due to the gravity load would exceed the yield stress for the I-beam profile, which implies that more

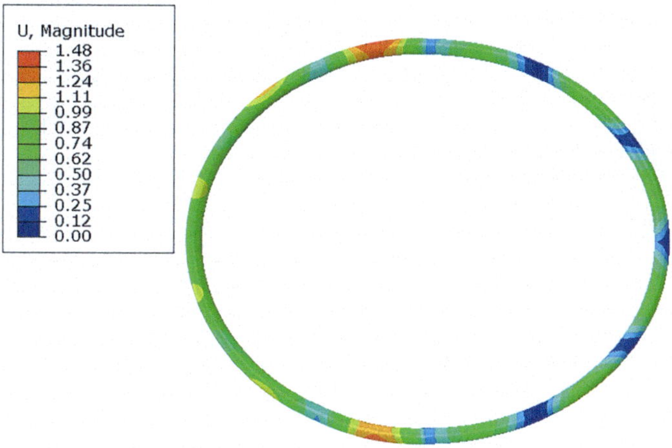

Fig. 17 The stress and displacements after loading show that the self-weight cannot be neglected in such large structures, since its contribution to the stress and strain is of similar magnitude to the one of the load [m]

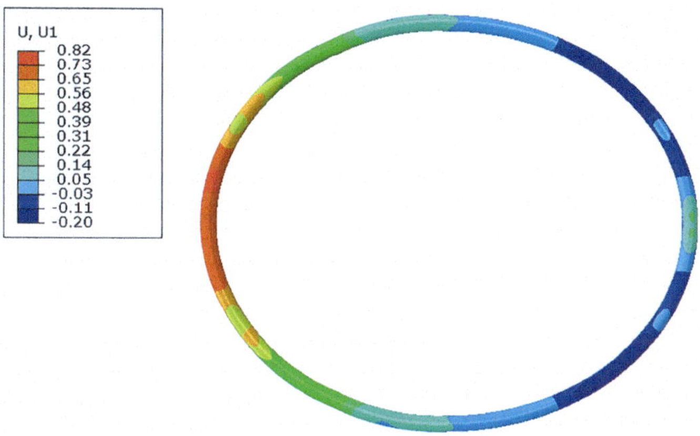

Fig. 18 Even increasing the load from 100 to 300 N/m, the displacements in the *x*-direction decrease due to the keel interaction [m]

connection points are required. More specifically, the minimum number of connections to the keel is 24, which implies an approximate distance between connections to be 40 m. Consequently, the keel would require 12 frames (6 more than the Tensairity frame), increasing the structural weight of the keel. Under such configuration, the maximum stress is 100 MPa (slightly above the maximum allowable stress) and the maximum displacement is 0.68 m. For the truss frame, the displacements and stress of the truss frame are 1.54 m and 54 MPa, both below the maximum allowable values (Fig. 19).

Consequently, the number of minimum connections to the keel is identical to the one of the Tensairity frame. As conclusion, considering only stress and strain, the

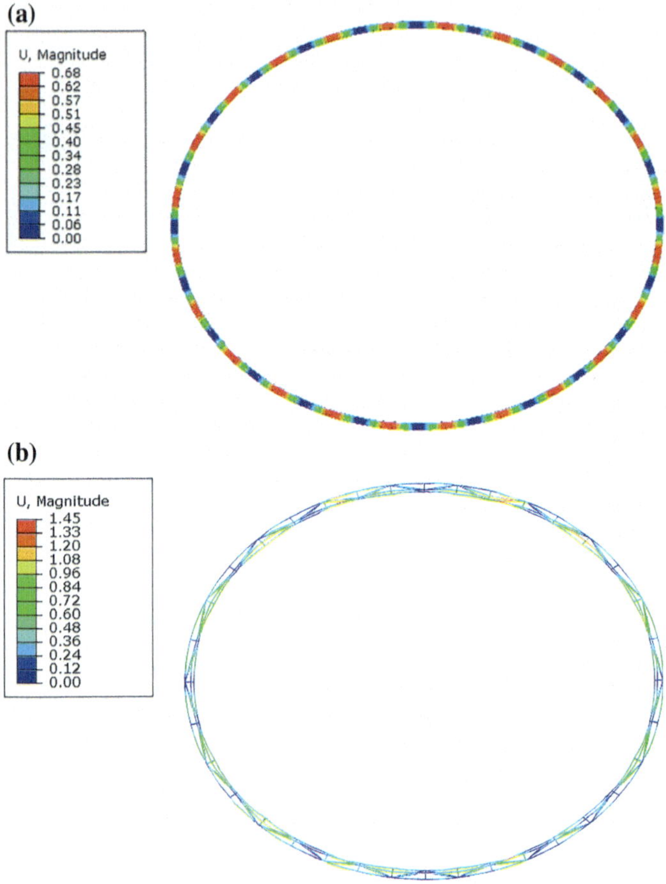

Fig. 19 The I-beam frame (**a**) would require the double of connections to the keel than the Tensairity and the truss frames (**b**), adding structural weight to the structure. The limitation for the I-beam frame is stress, while for the truss and Tensairity frames are the displacements [m]

lightest rigid frame is the truss frame, since it requires half the connections than the I-beam frame. As discussed previously, the truss frame is 8 times heavier than the Tensairity frame. Next section studies the stability of the frames.

6 Assessment of the Tensairity Potential Performance Against Buckling

After analysing the frames against stress and strain, we have assessed the buckling load for the three types of frames, under the two conditions. For the case without connections to the keel, the result of such assessment indicates that the designed Tensairity frame would not fail due to buckling for the considered load, while the I-beam and truss frames fail at a load one factor lower than the targeted one. For the case with 12 connections to the keel, the buckling load of the three frames is larger than the considered load. The largest buckling load is the one of the Tensairity frame. However, since the results have not been validated experimentally, the results are preliminary and we have identified this section as a future research task to be conducted.

As discussed in Sects. 2.2 and 2.3, for the considered geometry and loading conditions a Tensairity frame withstands compression forces. Consequently, estimating its buckling load is essential to assess its potential compared to conventional structures. Prior research on the buckling performance of Tensairity structures covers experimental analysis and the development of an analytical expression (Eq. 5, shown in Sect. 2) to determine the buckling load of Tensairity columns. To tackle Tensairity frames, it is necessary to develop analytical and numerical models, and this section is a first attempt in that direction.

We have analysed the literature to develop an analytical model that describes the buckling load of Tensairity frames. This model is a combination of the analytical model that describes buckling on Tensairity beams (Eq. 5) and a model that describes buckling on rigid circular frames (Eq. 10, which appears later in this section). Since the buckling behaviour of Tensairity frames has never been studied prior to this research, it was unknown if such modelling could describe the buckling on Tensairity frames. The good correlation of the obtained results points in that direction, as shown later in this section. It is significant to mention that there is no previous research of the validity of Eq. (5) on large-scale Tensairity structures, and that the results have not been validated experimentally. Consequently, those results are preliminary, and are a first step for a future research on the buckling performance of Tensairity frames.

The modelling applied previously cannot simulate the Tensairity frame buckling behaviour since the struts are attached after inflation. The commercial ABAQUS software does not support to perform buckling analysis after attaching the struts, and consequently we have only performed FEA simulations on small-scale Tensairity frames. With such methodology, the frame is modelled as a single part, modelling the struts as stringers and applying BEAM elements. Afterwards, we have combined the results from numerical simulations at small-scales with the analytical expression developed by us (an analytical expression that computes the buckling load of circular Tensairity frames, which is Eq. (10) and appears later in this section).

The developed analytical model includes a constant of, a priori, unknown value. We have estimated the value of the constant by computing numerically the buckling load of different small-scale Tensairity frames. More specifically, we have derived such value by obtaining the buckling load for different small-scale Tensairity frames, extracting the constant of our developed model to obtain identical buckling load. Since we have obtained virtually the same value for all the Tensairity frames, we have concluded that the analytical model describes the buckling performance of small-scale Tensairity frames. Finally, we have extracted the value of such constant for an elliptical frame, and applied the developed model to predict the buckling load of the large-scale Tensairity frame.

Simultaneously, we have computed with ABAQUS the buckling load for the I-beam and the truss frames designed in the previous section. This preliminary analysis shows that the advantages of Tensairity in terms of weight reduction are more relevant when considering buckling than when considering deformations: applying the procedure described above, the buckling load for the sized Tensairity structures results on 488 N/m. Comparing the buckling load with the considered load, in this scenario the Tensairity frame can be considered buckling-free. On the contrary, the buckling load for the considered I-beam profile and truss frames are 5.6 and 42.3 N/m respectively, which implies that such a large frame could not be designed in practice without reinforcing the frame with radial beams. If we consider also the connections to the keel, the buckling load of the Tensairity frame increases to 667 N/m, while the buckling load of the I-beam and the truss frames increase to 483 and 460 N/m respectively. If we increase the cross-section of the I-beam and the truss frames to withstand similar buckling load to the Tensairity frame, their weight would be approximately 8 times the one of the Tensairity frame (for the case where the constraints of the frame are considered). However, this would not be necessary, since their buckling load is higher than the considered load.

It is significant to highlight that, as already mentioned, it is not clear what the scalability of the developed expressions is, and the results must be considered preliminary. In addition, we have not considered yet the influence of self-weight on the buckling load, since the numerical analysis cover only small-scale frames.

6.1 Modelling Buckling on Tensairity Frames

One of the basic functions of the inflated membrane is to stabilize load-carrying compression elements against buckling by providing a continuous support. That is, the inflated membrane serves as an elastic foundation to the strut. The support of the inflated beam can be modelled as a continuous foundation of several equally spaced springs, as shown in Fig. 20 [28]. For a straight Tensairity beam, the value of the spring constant equals $p\pi$. Since the strut is continuously supported, its buckling load is independent of the column's length. On the contrary, as detailed in Section "2", in conventional columns the Euler critical load decreases with the square of the column length, (see Eq. 4). According to literature [16, 28, 29], the struts are the component prone to buckle since they are the principal components that withstand compression. Existing experimental research on Tensairity columns shows that the global buckling of Tensairity structures initiates with the local buckling of the strut. After the strut buckles, if the load increases further, the perturbation is transmitted to the rest of the structure, generating global buckling [28].

Since, according to literature, the buckling load of the Tensairity struts is independent of the span, Tensairity structures are suited for large span applications. This section describes the first attempt to estimate the buckling load of Tensairity frames.

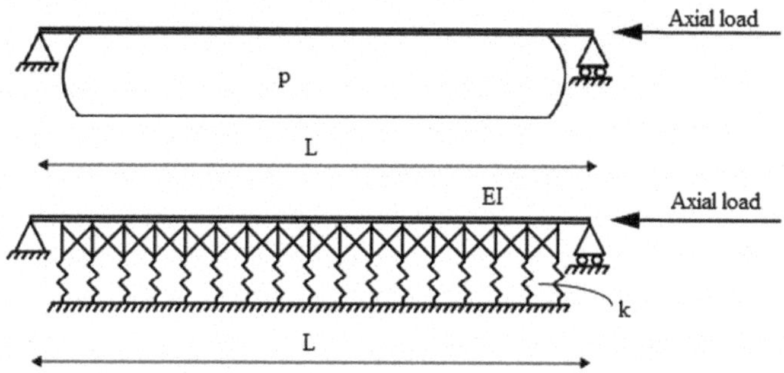

Fig. 20 The inflated membrane acts as a continuous support, which is equivalent to a continuous foundation of several springs. *Source* [28]

Regarding the development of an analytical model, the type of structure and loading conditions considered along the paper differ from those valid for Eq. (5): we are considering a frame with a frontal distributed load, while Eq. (5) considers a column with a punctual load of compression on the edge, similar to Fig. 20. Thus, an analogous expression is required. The value of developing such expression is that it allows computing the buckling load of large-scale Tensairity frames, which has not been possible numerically. However, it is significant to highlight that there is no previous research on the validity of Eq. (5), and consequently the obtained results are preliminary. As mentioned, the research shown in this section is a preliminary research on the buckling performance of Tensairity frames.

Aiming to develop an expression to compute the buckling load of Tensairity frames, we have considered the analytical expression that computes the buckling load of beams and columns under different types of constraints and loads (as long as they generate compression): they can all be expressed as Eq. (8). In Eq. (8), k_b is a

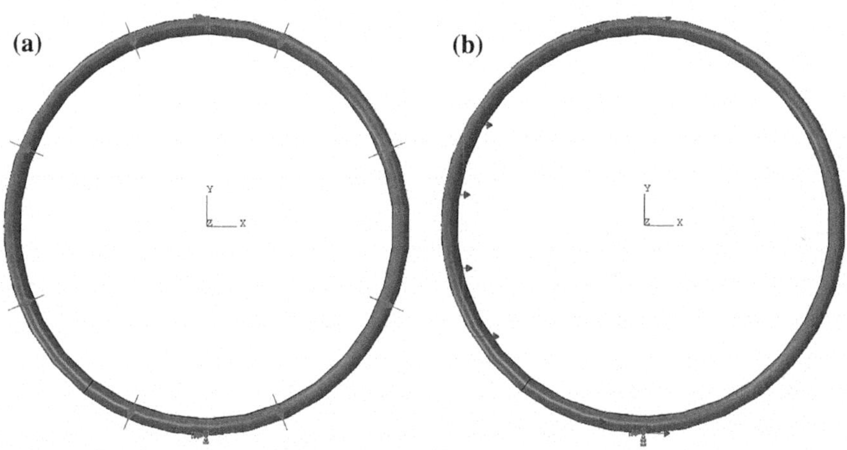

Fig. 21 We have modelled the buckling performance in two steps: **a** inflation and **b** buckling

constant that particularizes the expression according to the loading and boundary conditions.

Secondly, we have assumed that the Tensairity design parameters that affect the buckling behaviour of a Tensairity frame strut are identical to those for a Tensairity column, and affect in the same way. Those parameters are the internal overpressure, the Young modulus of the strut and the inertia of the strut, as shown in Eq. (5), in Sect. 2. We have analysed the validity of such assumption by comparing the deviation of the k_b constants obtained from the buckling loads predicted numerically.

Combining Eq. (8), with the Euler critical load for arches, Eq. (9) [30], analogously to Eq. (5), we obtain Eq. (10). Equation (10) assumes, similar to Eq. (5), for which the buckling load increases linearly with the square root of the internal overpressure and the inertia. This implies that the buckling load increases linearly with the cross-section radius r and with the struts section. In addition, since the case study is a distributed load instead of a punctual force, the buckling load decrease linearly with the Tensairity frame radius. In Eq. (9), E refers to the Young modulus, I to the beam cross-section inertia, R to the arch's radius and α to the angle along which the load is applied (being zero the centre). In addition, in Eq. (10) p stands for the internal overpressure, and includes a constant k_b to particularize the expression for the loading and boundary conditions. As already explained, this equation applies to the buckling of the strut, which in Tensairity structures is the element prone to buckle and is the element that initiates global buckling.

$$P_{\text{buckling}} = k_b \pi^2 \cdot \frac{EI_z}{l^2} \tag{8}$$

$$q_{\text{buckling}} = \frac{EI}{(2R)^3}\left(\frac{1}{\alpha^3}\right) \tag{9}$$

$$q_{\text{buckling}} = \frac{k_b \sqrt{\pi p E_{\text{strut}} I_{\text{strut}}}}{R} \tag{10}$$

The value of the constant k_b is unknown, and we have deduced it from the numerical analysis of different Tensairity frames. In other words, different Tensairity frames have been numerically investigated against buckling, and the values of the constant k_b have been compared. The objective is to elucidate the validity of Eq. (10) by comparing the variations of the constant k_b for different scenarios.

Concerning the numerical model, the inflation and loading have been simulated analogously to existing simulations of inflatable beams buckling [31]. It is significant to highlight that, prior to this research, this numerical approach has only been applied to conventional inflatable beams, and not to Tensairity structures. Following the procedure described in [31], we have applied the "General" analysis to simulate the inflation process followed by a "Buckling" analysis to simulate the buckling response on the distributed load. In [31], the accuracy of the commercial

ABAQUS software on predicting the buckling load of inflatable beams has been tested, comparing the results obtained experimentally, with ABAQUS, and applying a novel modelling to capture the variations on the fabric rigidity with internal overpressure. A significant conclusion obtained in [31] is that ABAQUS accuracy in the buckling load prediction can be compromised at normalized pressures lower than 4858. The normalized pressure is a non-dimensional parameter, and its value is obtained by dividing the internal overpressure by the critical pressure p_{cr}, defined as expressed in Eq. (11). The critical pressure is the external pressure at which a thin-walled circular cylinder buckles [32]. In Eq. (11), E stands for Young modulus, e for thickness, r for the beam's cross-section radius and υ for the Poisson's ratio. The subscript 1 refers to the warp direction, 2 for the weft direction and 3 for the orthogonal direction.

$$p_{cr} = \frac{E_1 \cdot e^3}{4r^3(1 - \upsilon_{13}\upsilon_{23})} \tag{11}$$

Regarding loads and boundary conditions, for this analysis we have not stabilized the system using SYMMETRY on the xy and the xz planes (Fig. 21). On the contrary, we have imposed no rotation on the two constraints at the y = ± 1 position. The other parameters of the numerical model are identical to the ones described in Sect. 4.

6.2 Buckling Results: Analysis of the Analytical Model Results, Extrapolation to Large Scale and Comparison to Rigid Frames

Table 6 shows the characteristics of the simulated Tensairity frames, as well as the obtained buckling load for the first mode and the value of k_b for that mode. The first buckling mode, shown in Fig. 22, shows that the component that buckles is the strut. In other words, the first mode is a local buckling, and not a global buckling of the whole frame. According to literature, if the load is further increased after the local buckling of the struts, the whole structure will buckle. Figure 23 shows the global buckling of the frame, initiated by the buckling mode shown in Fig. 22 and its symmetric mode. Other buckling modes are shown in the annex, showing other modes of local buckling of the internal and external struts.

Focusing on the first buckling mode, and averaging the results, the average value of k_b is 0.017, and the standard deviation is 0.0007, as shown in Table 6. Consequently, the repeatability is acceptable and suggest that Eq. (10) describes correctly the buckling performance of Tensairity frames. It is significant to mention that, even if for some of the studied cases the normalized pressures are lower than the mentioned value of 4858, we have not identified significant variations of the k_b (Table 7).

Table 6 Similar *kb* values (0.015–0.018) have been obtained for different configurations, which indicates that Eq. (10) describes the buckling behaviour of Tensairity frames

	R (m)	γ (–)	*P* (mbar)	*pn* (–)	*B* (mm)	*H* (mm)	$q_{buckling}$ (N/m)	*kb* (–)
Varying *p*	5	25	300	3879	25	2	694	0.016
	5	25	400	5172	25	2	827	0.016
	5	25	500	6465	25	2	961	0.017
	5	25	600	7758	25	2	1065	0.017
	5	25	700	9051	25	2	1086	0.016
	5	25	800	10,343	25	2	1109	0.015
Varying γ	5	25	300	3879	25	2	694	0.016
	5	10	300	60,606	25	2	1603	0.017
	5	8	200	78,914	25	2	1571	0.016
Varying R	3	15	800	10,343	25	2	1955	0.017
	5	15	300	17,957	25	2	1257	0.018
	7	15	300	49,275	25	2	1191	0.017
	9	15	300	104,727	25	2	1257	0.018
Varying b and h	5	11	400	60,712	10	2	1065	0.015
	5	11	300	45,534	25	2	1603	0.017

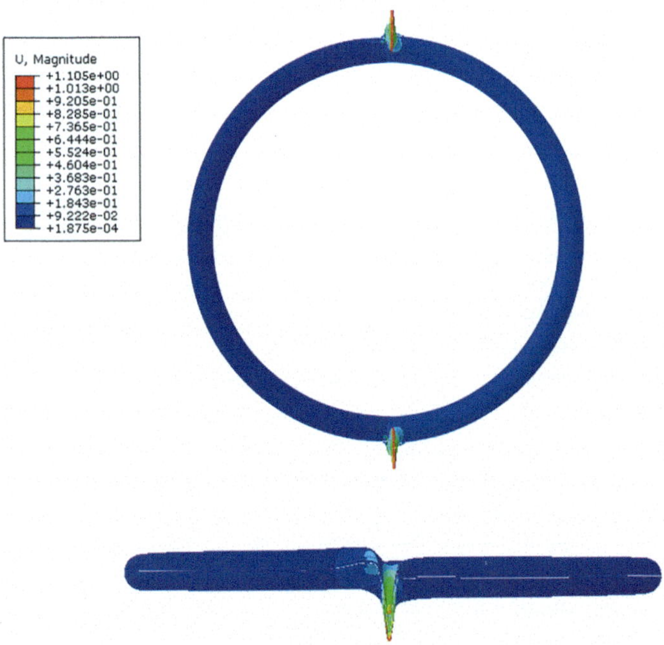

Fig. 22 The first buckling mode is the local buckling of the strut (k_b = 0.017)

Fig. 23 If the load is increased after the local buckling of the struts, the instabilities propagate, generating a global buckling of the frame ($k_b = 0.019$)

Table 7 The deviation of the $\boldsymbol{k}$ obtained at the different scenarios is low (0.015–0.018). $\boldsymbol{k_b}$ is the constant to be defined to on Eq. (10)

	k_b
Mean value	0.017
Standard deviation	0.0007

After concluding that Tensairity frames can be modelled using Eq. (10), we have obtained the constant k_b value of a small-scale elliptical frame of identical ratio length/span to the one described in Sect. 2. The dimensions and internal over-pressure of such small-scale Tensairity frame are shown in Table 8. The modelling in ABAQUS is identical to the one of the circular frame, already described. Similar to the circular frame, the first buckling mode is local buckling of the strut, as shown in Fig. 24. The buckling load is 490 N/m and the value of k_b is 0.013 (considering as R half the span of the ellipse, i.e. 5 m). Once the k_b value has been obtained, Eq. (10) has been applied to compute the buckling load of the Tensairity frame previously dimensioned against stress and strain. The main parameters of the large-scale frame are also shown in Table 8.

The results are shown in Table 9, which also includes the buckling load of the I-beam and the truss frames from Sect. 5. Table 9 shows that the mechanical performance of Tensairity, when compared to the I-beam and truss frames improves more the buckling than the strain. Applying Eq. (10), the buckling load for the sized Tensairity structures results on 488 N/m. Thus, the considered Tensairity frame can be considered buckling-free. On the contrary, the buckling load for the considered I-beam profile and truss frames are 5.6 and 42.3 N/m respectively, which implies that such a large frame could not be designed in practice without reinforcing the

Table 8 The small scale elliptical frame is of same ratio length/span and slenderness to the one in Sect. 3

	Length (m)	Span (m)	Radius r (m)	Strut width (mm)	Strut height (m)	Internal overpressure (mbar)
Large-scale	330	266	5	300	3	50
Small-scale	12.4	10	0.2	25	2	800

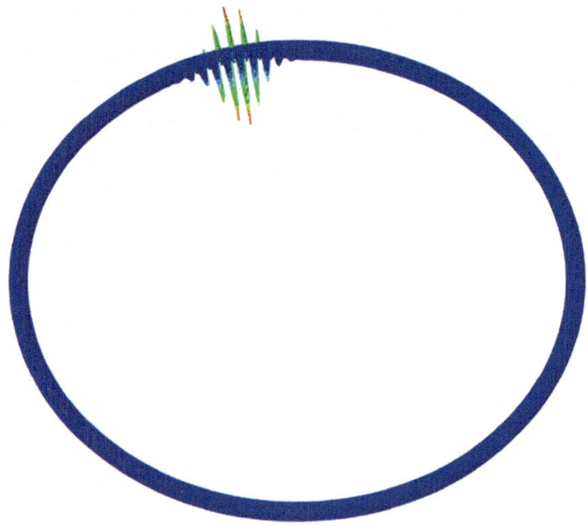

Fig. 24 The first buckling mode of an elliptical frame is also local buckling of the strut

Table 9 The Tensairity frame considered in the previous section is buckling free, while the rigid frames (I-beam and truss) could not be designed in practice without connections to the keel

Type of frame	Buckling load (N/m)
Tensairity	488.2
I-beam	5.6
Truss	42.3

frame with radial beams. Figure 25 shows the first buckling mode of the I-beam and the truss frames.

As mentioned, the buckling load of the rid frames depends on the length. More specifically, depends on the un-constraint length. Consequently, the attachments to the keel will decrease the buckling load. Although the comparison shown in Table 9 shows the large potential of Tensairity on large span structures, in a real airship structure the frame would be attached to the keel, and consequently the buckling load of the rigid frames would be lower. In order to assess the potential of Tensairity under more realistic conditions, we have repeated the analysis using the constraints described in Sect. 4 (Fig. 15). On the small-scale Tensairity frame, we have positioned those constraints proportionally, at 2.63 and 2.81 m instead of 70

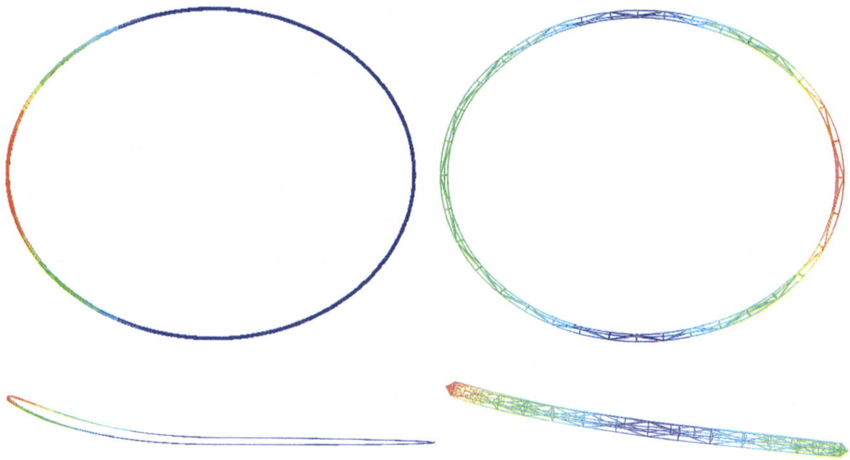

Fig. 25 First buckling mode of the I-beam and the truss frames

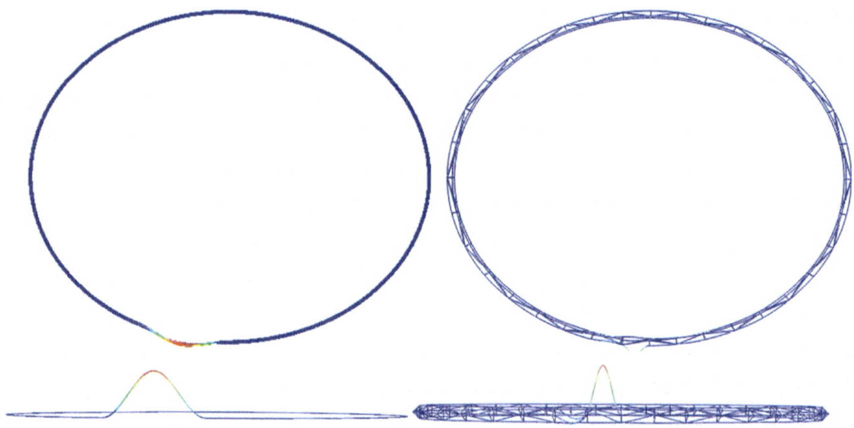

Fig. 26 First buckling mode of the I-beam and the truss frames (frame attached to the keel)

and 75 m respectively (see Fig. 15). We have repeated the simulation in ABAQUS, obtaining a buckling load of 670 N/m (which implies that the value of k_b is 0.018). Assuming the same value of k_b for the large-scale Tensairity frame, the value of the buckling load would be 667 N/m. In this case, and as shown in Table 9, the difference of buckling load between the Tensairity and the I-beam and truss frames are much less significant (the buckling load of the Tensairity frame is approximately 1.4 times the buckling load of the other frames). Figure 26 shows the first buckling mode of the I-beam and the truss frames.

If we increase the section of the I-beam profile and the truss sections pursuing identical buckling load to the Tensairity frame, the weight reduction is the double

Table 10 The difference on buckling load between the different types of frames are significantly reduced when considering the connections to the keel

Type of frame	Buckling load (N/m)
Tensairity	667
I-beam	483
Truss	460

Table 11 For identical buckling load, the weight of the Tensairity frame is 6.2 times lower than the I-beam frame and 9.1 times lower than the truss frame

Type of frame	Buckling load (N/m)	Weight (t)
Tensairity	667	32.6
I-beam	635	202.2
Truss	646	298.4

Table 12 Geometrical dimensions of the I-profile for identical loading to the Tensairity frame

Base (mm)	Height (mm)	Base thickness (mm)	Height thickness (mm)
320	1500	60	30

than the one for the static analysis (stress and strain), as shown in Table 11. However, the buckling loads shown in Table 10 are already higher than the loading conditions, and thus it is not necessary to increase the sections of the I-beam and the truss frames (Table 12).

As already mentioned, the obtained results have not been validated experimentally. From previous research [33] we can expect deviations from the obtained results, since theoretical models do not consider geometrical imperfections or defects in the material. In addition, the numerical model assumes that the struts are tightly attached to the membrane, without allowing relative displacements. Those differences might decrease the experimental buckling load. In addition, there is no previous research on the effect of large-scale on Eq. (5), and we have not included the gravity load for the estimation of the buckling load. Consequently, the obtained results must be considered preliminary, as part of the analysis of the potential of Tensairity on airships.

7 Conclusions

In this research, a novel concept of light-weight structure for an airship has been introduced: Tensairity. Tensairity is a lightweight structure based on tensegrity: the functional separation of elements. It is made of a compression element (strut), a tension element (cable) and a membrane to give cohesion. This solution envisages fulfilling the same structural requirements, but being much lighter compared to traditional structures. Consequently, Tensairity appears as a promising solution to reduce the structural weight on airships that require a keel. In this line, a Tensairity circular frame has been researched via FEA. The weight of the Tensairity frame has been compared to the weight of a similarly sized I-profile beam to understand the potential of Tensairity. We have performed such comparison under two conditions. The first condition is a fundamental case study (frame without connections to the keel, which shows the potential of Tensairity for large span structures). The second condition represents an airship frame, and includes the self-weight load and the constraints that result from the interaction with other members of the keel. The second analysis allows understanding the potential of Tensairity frames as part of an airship keel.

The performed research results in the following personal contributions:

- Development of numerical modelling on the commercial software ABAQUS that allows attaching the reinforcing elements after inflation, which has been proven necessary at large scales.
- For the fundamental case study (large-scale frame without connections to the keel): weight 4 times lighter than conventional I-beam frames if dimensioned against stress and strain (for the considered size, geometry and loading conditions).
- For the second case study (frame connected to the keel, self-weight considered): weight 8 times lighter than conventional rigid frames, dimensioned against stress, strain and stability (for the considered size, geometry and loading conditions).

Related to the development of numerical modelling on the commercial software ABAQUS that allows attaching the reinforcing elements after inflation, the air beam deformation during its inflation can pre-stress excessively the struts. To avoid that, the struts can be attached after inflation. A numerical model to account for this has been developed in the commercial ABAQUS software and validated with the analytical equations. The numerical model has been modified to account for the effect of large deformations, minimizing the inertial forces and modifying the interaction modelling between fabrics and struts. The modification consists in modelling the stiff elements as different parts and not as stringers of the inflatable beam. The numerical model has been applied to size the large Tensairity frame that has been compared in weight to the I-beam frame.

Finally, regarding the weight reduction, we have compared the Tensairity frame to an I-beam and a truss frame. We have performed such analysis under two conditions: considering only the frame without the interaction of the keel, and considering also the connections to the keel. Both analyses dimension the frames to withstand a frontal distribution load. The first condition allows understanding the potential of Tensairity on large scale-frames. The second condition, which includes the self-weight load and the constraints that result from the interaction with other members of the keel, allows understanding the potential of Tensairity frames as part of an airship keel. The first analysis showed that, if dimensioned against stress and strain, the Tensairity frame is more than 4 and 8 times lighter than the corresponding I-beam and truss frame respectively. In addition, preliminary studies show that the buckling load of the Tensairity frame is higher than the targeted load, while the buckling load of the rigid frames are one and two factors lower than the targeted load.

If the air beam is not reinforced with struts, the applied overpressure is not sufficient to avoid the resulting compression stress overall the entire membrane. This fact leads to the failure of the air beam. Thus, if the air beam has to withstand the same loading conditions, higher overpressure is required. This requirement would lead to a challenging research on fabrics able to withstand sufficiently high breaking stress.

For the second case, we have considered the Tensairity frame dimensioned for the first case to estimate the minimum number of connections to the keel, which has resulted in 12 (i.e. 6 frames). After adding the loads of gravity and drag, the maximum stress and the maximum displacement are both below the maximum allowable values. For the I-beam frame, the maximum stress exceeds the yield stress due to its self-weight. Aiming to obtain stress and deformations below maximum values, it would require 24 points of connection, which means that the keel would require 12 frames more, which would increase the global structural weight (i.e. the double to the Tensairity frame). On the contrary, the displacements and stress of the truss frame are both below the maximum allowable values. Regarding stability, the buckling load of the 3 frames is above the targeted buckling load. As general conclusion, considering stress, strain and stability, the lightest rigid keel would be the truss structure, which is 8 times heavier than the Tensairity frame.

It is important to highlight that the considered case study is preliminary, with different simplifications as using only one air volume for the entire frame, or a constant pressure and load distribution. In addition, the buckling results are preliminary and the model must be further developed to understand the scalability of the developed model, as well as to include the self-weight load.

Taking into account the detailed findings, the developed equations are useful tools to preliminary assessment of the Tensairity frame design, as well as, to validate a numerical simulation. In addition, the obtained results suggest a large potential of Tensairity structures on airships. Future work must be focused on buckling, three dimensional loading, torsional rigidity and connection of different Tensairity beams and frames to create a Tensairity keel.

Acknowledgements The presented work in this paper was performed as part of the Multibody Advanced Airship for Transport (MAAT) project, supported by European Commission through the 7th Framework Programme, and which is gratefully acknowledged.

References

1. Liao L, Pasternak I (2009) A review of airship structural research and development. Prog Aerosp Sci 45:83–96
2. Khoury GA (2004) Airship technology. Cambridge University Press

3. Goodyear (2016) http://www.goodyearblimp.com. Retrieved 29 March, 2019
4. Aeros 2016 http://aeroscraft.com/. Retrieved 29 March, 2019
5. RosAeroSystems 2015 http://rosaerosystems.com. Retrieved 29 March,2019
6. Aerolift (2012) https://www.aero-lift.de/en/home-en/l. Retrieved 29 March, 2019
7. Lockheed Martin (2016) http://www.lockheedmartin.com. Retrieved 29 March, 2019
8. Balaskovic P (2011) Lenticular airship. US Patent US 7866601 B2, 11 01 2011
9. MAAT project (2012) MAAT project website. [Online]. Available: http://www.eumaat.info. Retrieved 25 March, 2017
10. MAAT Consortium (2011) MAAT Part B. Description of Work. Seventh Framework Programme Theme 7-Transport.
11. Dumas A, Anzillotti S, Madonia M, Trancossi M (2013) Photovoltaic production of hydrogen at stratospheric altitude. Journal of Solar Energy Engineering, 135(1)
12. Pedretti A, Steingruber P, Pedretti M (2014) The new structural concept Tensairity: FE-modeling and applications. Progress in Structural Engineering, Mechanics and Computation, Zigoni
13. Luchsinger RH, Pedretti A, Pedretti M, Steingruber P (2004) The new structural concept tensairity: basic principles. Progress in Structural Engineering, Mechanics and Computation, Zigoni
14. Buildair (2015) http://www.buildair.com. [Online]. Retrieved 29 March, 2019
15. De Laet L (2010) Deployable tensairity structures: development, design and analysis, doctoral thesis. VUB, Brussels
16. Luchsinger RH, Pedretti A, Steingruber P, Pedretti M (2004b) Light weight structures with tensairity
17. Galliot C, Luchsinger RH (2013) Structural behavior of symmetric spindle-shaped Tensairity girders with reinforced chord coupling. Eng Struct 56:407–416
18. Suñol A (2017) Development, analysis and validation of novel airship systems, Brussels: PhD thesis at Vrije Universiteit Brussel, VUB Pres
19. ASM International (1990) ASM Handbook volume 2: properties and selection: nonferrous alloys and special-purpose materials, vol. 2
20. FAA (2014) Title 14—aeronautics and space. Chapter I—Federal Aviation Administration, Department Of Transportation. Subchapter C—Aircraft. Part 25—Airworthiness Standards: transport category airplanes. Subpart C—Structure.—Flight Maneuver And Gust,"
21. Barbero E, Viacheslav V (2014) D8.3 Functional model realization report, MAAT
22. Bessert N, Frederich O (2005) Nonlinear airship aerolasticity. J Fluids Struct 21:731–742
23. A. Manual Abaqus Analysis User's (2016) Abaqus Analysis User's Manual, Abaqus 6.12, Simulia
24. Roekens J., Luchsinger RH, Mollaert M, De Laet L (2011) Experimental investigation of a deployable Tensairity arch. In: 35th annual symposium of IABSE 52nd annual symposium of IASS/6th international conference on space structures, London
25. Liao L (2013) Implementation of PATRAN/NASTRAN into the development of advanced buoyancy air vehicles. In: MSC Software Users Conference
26. Blum R, Bögner H, Némoz, G (2004) Material properties and testing. In: European design guide for tensile surface structures, vol. 9, Brussels, Tensinet
27. Kang W, Suh Y, Woo K, Lee I (2006) Mechanical property characterization of film-fabric laminate for stratospheric airship envelope 75:151–155
28. Wever TE, Plagianakos TS, Luchsinger RH, Marti P (2010) Effect of fabric webs on the static response of spindle-shaped tensairity columns 136(4)
29. Wever T, Plagianakos T, Luchsinger RH, Wageman L (2009) Experimental study of the effect of fabric webs on the static response of Tensairity columns to axial compression. In: International association for shell and spatial structures (IASS) Symposium, Valencia
30. Timoshenko SP, Gere JM (1989) Theory of elastic stability. Dover Publications, New York, p 297

31. Nguyen T, Ronel S, Massenzio M, Jacquelin E, Apedo KL, Phan-Dinh H (2013) Numerical buckling analysis of an inflatable beam made of orthotropic technical textiles. Thin-Walled Struct 72:61–75
32. NASA (1969) Buckling of thin-walled circular cylinders. In: NASA space vehicle design criteria, Vols. SP-8007, Langley, NASA
33. Wever T (2008) The effect of internal stiffeners on the buckling behaviour of an inflatable column. Master's thesis, Delft University of Technology

Realization of Subtitle Support in Hybrid Digital TV Applications

Ana Bilandžić and Zvonimir Kaprocki

Abstract In the last few years Hybrid Broadcast Broadband TV (HbbTV) become very popular because it provides some new services to the users when watching TV. One of those services are subtitles, which allows users to watch TV with subtitles in different languages. This chapter describes the application for the subtitles support in the hybrid digital TV applications with relatively simple hardware equipment. In case when the respective language is selected, the subtitle files are downloaded from server, subtitle file is parsed and time elapsed from the beginning of the show is calculated so that subtitle can be shown on the display on time. Downloading of subtitle files from the server is implemented using *thttpd* server and *cURL* library and graphical elements are displayed by using *DirectFB* library. The application is written in C programming language and consists of several modules which are main module, module for parsing of service information (SI) and program-specific (PSI) tables, module for graphics, module for handling inputs from remote controller, module for downloading of the .srt file from the server and module for parsing of the .srt file.

Keywords Subtitle · DVB · .srt file · DirectFB · thttpd · cURL

1 Introduction

Television is a telecommunication medium for sending and receiving motion picture and sound. The digitalization of television has affected the development of new services and applications. One of those services are subtitles that are of great importance in digital television application because they allow viewers to view

A. Bilandžić (✉) · Z. Kaprocki
Institute RT-RK Osijek LLC, Cara Hadrijana 10B, Osijek, Croatia
e-mail: ana.bilandzic@rt-rk.com

Z. Kaprocki
e-mail: zvonimir.kaprocki@rt-rk.com

D. Vucinic et al. (eds.), *Advances in Visualization and Optimization Techniques for Multidisciplinary Research*, Lecture Notes in Mechanical Engineering, https://doi.org/10.1007/978-981-13-9806-3_7

content in different languages with translation. For the show currently being watched, the user may have a choice between several subtitle languages available, whose display can be activated or deactivated at any time.

1.1 Digital Television and DVB Standard

The development of television is intertwined with the advances of other technologies. Due to the development of new technologies, it is possible to display higher resolution. The development of integrated chips has led to increased speed, memory, and, consequently, lower prices. New resolutions use a larger amount of bandwidth so new, better video compression algorithms with improved implementation and modulation schemes need to be defined. Digital television includes transmission of television signal from transmitters to viewers digitally, using digital modulation techniques. The signal is sent as a digital transport stream. Packages related to audio and video content, metadata, and additional content related to multiple services (programs) are transmitted [1].

Digital television uses digital compression, which allows broadcasting more programs than in analogue television, and ensures high quality of sound and picture. The variety of services has increased, such as audio and video subtitle language, electronic program guide with reminders, parental control, teletext, Internet access, etc.

DVB (Digital Video Broadcasting) is an industry-led consortium of the world leading digital TV and technology companies, such as software developers, broadcasters and regulators and manufacturers. They are working on designing technical standards for digital TV and other broadcast media. The main goal is to define standards for delivering digital audio and video content with a special focus on digital broadcasting [2].

DVB is one of the most widely used standards for digital TV broadcasting in the world. It covers a wide range of uses in digital television such as satellite, cable or terrestrial transmission of digital television, IP and Internet television, mobile TV, architecture, and more.

1.2 MPEG-2

The MPEG-2 system is part of the MPEG standard that defines the mode of multiplexing the streams of video signals and audio signals into one data stream for transfer or storage. Depending on the portable media, the MPEG-2 standard defines different multiplexing formats. Multiplexing of audio and video signals is required to enable their joint transmission, correct decoding and display [3].

If transfer media is not susceptible to transmission errors (DVD, magnetic tape, etc.), MPEG-2 elementary streams are multiplexed into the Program Stream (PS). This format enables easier data management. It is used for audio and video playback, and in some network applications. The video signal is multiplexed with audio signals (multi-channel sound) and data. All signals belong to one program.

If the portable medium is susceptible to errors (ground, cable or satellite connections), the elementary streams are multiplexed into the Transport Stream (TS). It is possible to multiplex audio signals, video signals and data belonging to different programs into a single stream of information.

The MPEG-2 standard defines a hierarchy of multiplexing that consists of three types of data streams (Fig. 1):

1. Elementary Stream, ES
2. Packetized Elementary Stream, PES
3. Program or Transport Stream, PS or TS.

Elementary stream is a stream that is generated at the output from the encoder and contains everything needed for decoding in order to reconstruct the original audio signal or image. Elementary streams are divided into PES packets whose maximum length is 8 kB. The packets have a constant or variable length [3].

Packets of Program Stream have a variable length that causes difficulties since the decoder needs to recognize the exact start and ending of the packet. Because of that, the packet header contains information about length. The header of the

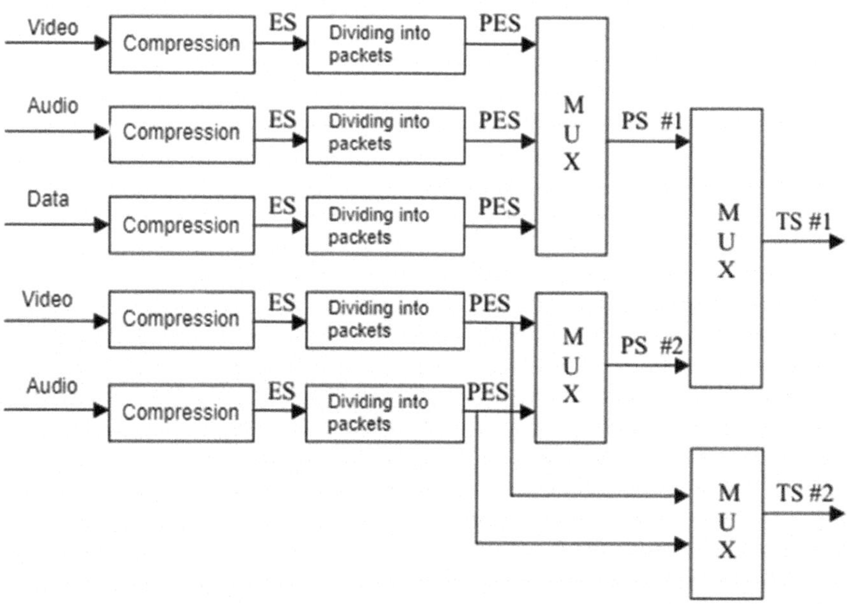

Fig. 1 Stream multiplexing in MPEG-2 standard [3]

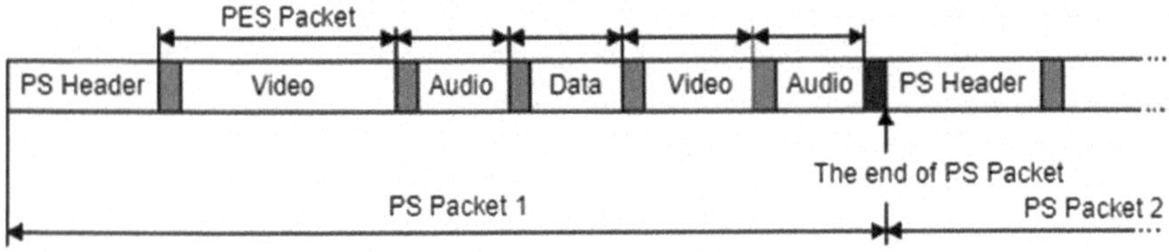

Fig. 2 Program Stream packets [3]

Fig. 3 Transport Stream packet [3]

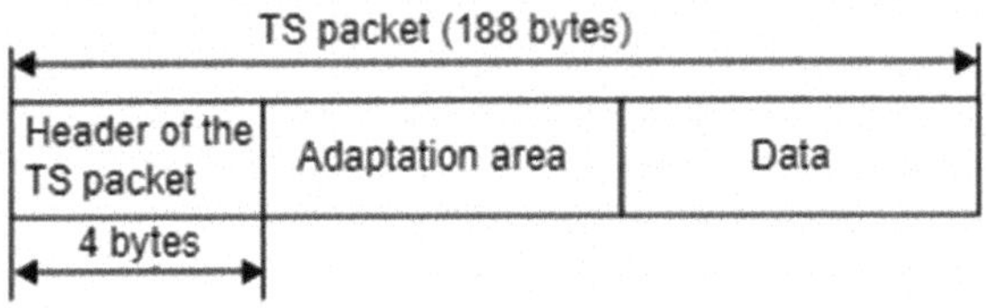

package contains start code of the package, System clock Reference (SCR), and data for multiplexing. Figure 2 shows layout of PS packet.

Transport stream is intended for transmission over the network, which means that it must be designed in such a way that is resistant to transmission errors and interference in transmission. Transport stream packet consists of header, adaptation area and data that is shown in Fig. 3. Adaptation area consists of time references, labels, and information used for data retrieval.

In the header of TS packet the most important field is Packet Identifier (PID). PID is used not only for the identification of transport stream that carries PES packets related to the same program, but also for the determination of the type of data being transmitted.

1.3 DVB Signalling Information

In order for a user to receive a specific Transport Stream, it is first necessary to determine which PID is used and then filter the packets that have this PID value. As an aid in identifying which PID corresponds to which program, special control streams were created. Those streams contain Signalling Tables, SI. Signalling tables carry data for each DTV service that is transmitted within the Transport Stream. They are transmitted as separate streams within the Transport Stream and are multiplexed with other packets.

These tables are called Program Specific Information, and they are carried in the form of a table structure. Each table is divided into one or more PSI sections. Sections have variable length. Each section is protected by a CRC (cyclic redundancy check) algorithm used for the verification of the integrity of the table.

Fig. 4 PAT table syntax [1]

syntax	bit index	# of bits
table_id	0	8
section_syntax_indicator	8	1
'0'	9	1
reserved	10	2
section_length	12	12
transport_stream_id	24	16
reserved	40	2
version_number	42	5
current_next_indicator	47	1
section_number	48	8
last_section_number	56	8
for i = 0 to N		
program_number	56 + (i * 4)	16
reserved	72 + (i * 4)	3
if program_number = 0		
network_PID	75 + (i * 4)	13
else		
program_map_pid	75 + (i * 4)	13
end if		
next		
CRC_32	88 + (i * 4)	32

MPEG-2 signalling tables

PAT (Program Association Table) contains a list of PID values of TS packets with PSI tables carrying data on DTV service broadcast on that transponder or frequency. TS packets containing PAT information always have PID 0x0000. [4] Syntax of PAT Table is shown in Fig. 4.

PMT (Program Map Table) contains a list of PID values of TS packets that contain information about specific program (audio, video, teletext). PID of PMT table is 0x0002. [4] Syntax of PMT Table is shown in Fig. 5.

NIT (Network Information Table) table contains information about network emitting TS multiplex with its belonging DTV service and information about transmission frequencies. PID of NIT table is 0x0010 [4].

CAT (Conditional Access Table) table is used for protected services. It defines the coding type and PID values of PID packets that contain information necessary for content decoding. The PID of CAT table is 0x0001 [4].

DCM-CC (Digital Storage Media Command and Control) table contain commands sent to the receiver [4].

In order to identify the appropriate PID for PES demultiplexing, first it is necessary to search the PAT table to get a list of all programs in the multiplex. Each program is associated with several PIDs (one for each PES) and they correspond to PMT table that is transmitted in separated PSI sections. There is one PMT table for each program.

Fig. 5 PMT table syntax [1]

syntax	bit index	# of bits
table_id	0	8
section_syntax_indicator	8	1
'0'	9	1
reserved	10	2
section_length	12	12
program_number	24	16
reserved	40	2
version_number	42	5
current_next_indicator	47	1
section_number	48	8
last_section_number	56	8
reserved	64	3
PCR_PID	67	13
reserved	80	4
program_info_length	84	12
descriptor()	96	varA
for i = 0 to N		
stream_type	96 + varA	8
reserved	104 + varA	3
elementary_PID	107 + varA	13
reserved	120 + varA	4
ES_info_length	124 + varA	12
descriptor()	136 + varA	varA
next		
CRC_32		32

DVB signalling tables

DVB norm expands list of signalling tables with Service Information tables—SI Tables. Those are:

- **BAT** (Bouquet Association Table)—grouping of services in logic groups (music, sport, movies…)
- **SDT** (Service Description Table)—service description
- **TDT** (Time and Date Table)—current date and time
- **RST** (Running Status Table)—service status and enabling of automatic event switching
- **EIT** (Event Information Table)—service details.

Of the above-mentioned tables, SDT, EIT, and TDT tables are parsed in this project. Syntax of SDT Table is shown in Fig. 6, syntax of EIT Table in Fig. 7 and syntax of TDT Table is shown in Fig. 8. The first column of the table shows all table fields, while the second column contains the number of bits for each table element [4].

Each table has descriptors that contain descriptive data. In [4] in Chap. 6.1. *Descriptors identification and location* it can be seen which descriptor is related to which table.

Syntax	Number of bits
service_description_section(){	
table_id	8
section_syntax_indicator	1
reserved_future_use	1
reserved	2
section_length	12
transport_stream_id	16
reserved	2
version_number	5
current_next_indicator	1
section_number	8
last_section_number	8
original_network_id	16
reserved_future_use	8
for (i=0;i<N;i++){	
service_id	16
reserved_future_use	6
EIT_schedule_flag	1
EIT_present_following_flag	1
running_status	3
free_CA_mode	1
descriptors_loop_length	12
for (j=0;j<N;j++){	
descriptor()	
}	
}	
CRC_32	32
}	

Fig. 6 SDT table syntax [4]

For this project, a descriptor with short event description (short event descriptor) belonging to EIT Table and a descriptor with service description (service descriptor) belonging to SDT Table are parsed.

The EIT table has a loop for descriptors for each event in it. The short event descriptor is used for the transmission of name and short description of events. In addition, language code is transmitted. This code provides information about the language in which the name and descriptor are written [5]. Figure 9 shows syntax of short event descriptor [4].

The SDT table contains a loop for descriptors for each event described in it. The Service descriptor contains basic information, such as the name of the service and the name of the service provider [5]. Figure 10 shows syntax of service descriptor [4].

Syntax	Number of bits
event_information_section(){	
table_id	8
section_syntax_indicator	1
reserved_future_use	1
reserved	2
section_length	12
service_id	16
reserved	2
version_number	5
current_next_indicator	1
section_number	8
last_section_number	8
transport_stream_id	16
original_network_id	16
segment_last_section_number	8
last_table_id	8
for(i=0;i<N;i++){	
event_id	16
start_time	40
duration	24
running_status	3
free_CA_mode	1
descriptors_loop_length	12
for(i=0;i<N;i++){	
descriptor()	
}	
}	
CRC_32	32
}	

Fig. 7 EIT table syntax [4]

Syntax	Number of bits
time_date_section(){	
table_id	8
section_syntax_indicator	1
reserved_future_use	1
reserved	2
section_length	12
UTC_time	40
}	

Fig. 8 TDT table syntax [4]

1.4 SRT Format

The .srt format is one of the most commonly used formats for subtitles display. Depending on the video content, one .srt file consist of several subtitle sets. One subtitle set consists of the ordinal number of the set, the time consisting of two

Syntax	Number of bits
short_event_descriptor(){	
descriptor_tag	8
descriptor_length	8
ISO_639_language_code	24
event_name_length	8
for (i=0;i<event_name_length;i++){	
event_name_char	8
}	
text_length	8
for (i=0;i<text_length;i++){	
text_char	8
}	
}	

Fig. 9 Short event descriptor syntax [4]

Syntax	Number of bits
service_descriptor(){	
descriptor_tag	8
descriptor_length	8
service_type	8
service_provider_name_length	8
for (i=0;i<N;I++){	
char	8
}	
service_name_length	8
for (i=0;i<N;I++){	
Char	8
}	
}	

Fig. 10 Service descriptor syntax [4]

components—start and end time of subtitle set in hours, minutes, seconds and milliseconds –, the subtitle text, and of the empty row [6]. The number of characters in one line is not strictly defined but usually varies between 25 and 40. The subtitle text can be shown in one or two rows. Table 1 shows example of .srt format (part of cartoon „Mascotas Papanatas").

2 Used Libraries, Hardware and Software

To implement this application, it is necessary to set up hardware and software, install external libraries and a server on which subtitle files are located.

Table 1 Example of .srt format

Ordinal number of subtitle set in file	4
Start time → End time	00:14:13.725 → 00:14:14.826
Subtitle text	What's going on here?
Empty row	
Ordinal number of subtitle set in file	5
Start time → End time	00:14:13.725 → 00:14:14.826
Subtitle text	– They are stained so they left

2.1 Hardware and Software

The application was created in virtual machine for Linux operating system.

It is necessary to start TS from PC. For the purpose of this application, application TS Player was used. Figure 11 shows the program interface and parameters set to match used TS.

A set-top box is a device that enables a television set to become a user's interface to the Internet and enables a television set to receive and decode digital television broadcasts. A set-top box is necessary for television viewers who wish to use their analogue television sets to receive digital broadcasts.

Figure 12 shows set-top box used in the project. Figure 13 shows which connectors on set-top box are used in the project. An antenna cable connected to the DVB-T signal modulator via amplifier and signal divider must be connected to the antenna connector (1). The HDMI cable (2) is connected to the monitor, and the USB port (4) is connected to the speakers which are connected to the monitor.

A remote control is required to operate the application. Figure 14 shows remote control with indicated buttons that have functionality in the application.

Those are:

- Button for shutdown (0)
- Buttons for program control:
  - Channel select (1–9)
  - Channel up (12)
  - Channel down (13)
- Buttons for volume control:
  - Volume up (10)
  - Volume down (11)
  - Mute (14)
- Button for showing information about current event (18)
- Button for removing graphics from screen (20)
- Button for subtitle language selection menu (22)
- Button for stopping of the subtitles (21)

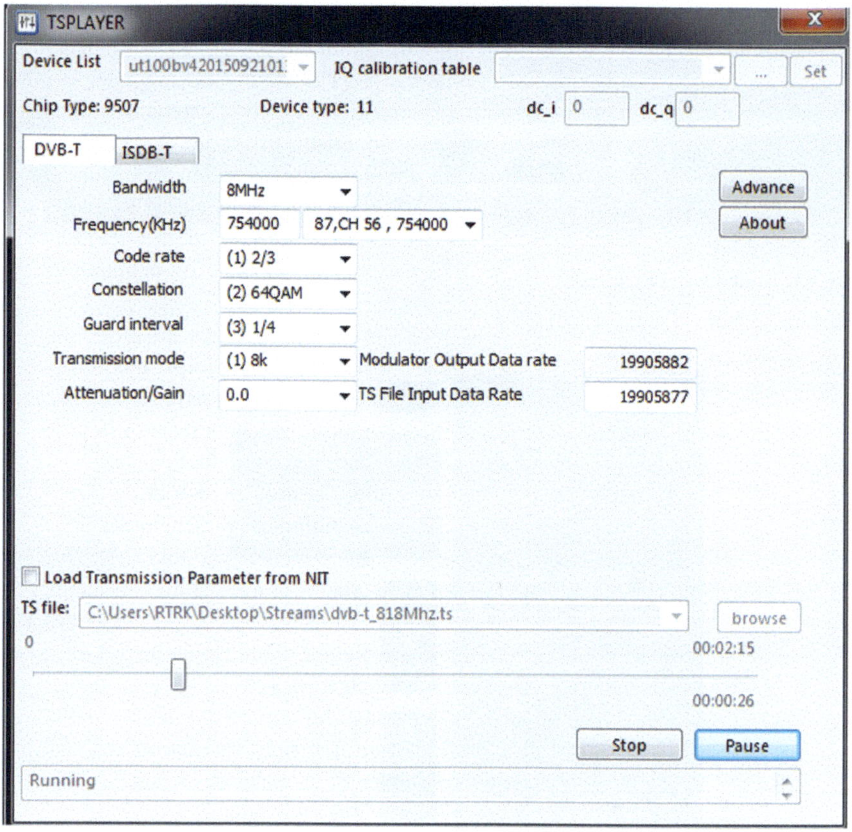

Fig. 11 TS Player interface

Fig. 12 Set-top box—front side

Fig. 13 Set-top box—back side

Fig. 14 Remote control

- Buttons for position manipulation in subtitle language selection menu:
  - Left (15)
  - Right (16)
- Button for selection in subtitle language selection menu (17)

Every time when the application is started it is necessary to connect PC and set-top box via TELNET protocol. More details on this topic can be found in [7].

2.2 *Transport Stream*

In the used Transport Stream, the total number of channels is seven. Four of those channels are video and three audio channels. All video channels have EIT table. The subtitles are realized for the event "Doña Barbara (Eps. 151 Al 190)" on the first channel La 1, and for the cartoon "Mascotas Papanatas" on the fourth channel Clan.

2.3 *Used Libraries and Programming Language*

DirectFB (Direct Frame Buffer) graphics API is used to show data on the display. DirectFB is an open-source software requiring little memory usage. It can display several geometric elements; in addition, it supports working with fonts, pictures, etc. [8]. The used server is *thttpd* [9] and the library used for download of files is cURL [10]. The PC running the application is connected to the set-top box with TELNET protocol. More details on how to install server and cURL library can be found in [7].

The application is written in C programming language because of bit manipulation, memory management and speed that C provides.

3 Solution Concept

The solution concept of the application is shown in Fig. 15. The TS file is stored on the PC from which it is started. Then, the DVB-T modulator sends a signal to the set-top box via signal amplifier and splitter. The application itself is stored on STB. In case that the internet connection is available, .srt file will be downloaded from the server. Otherwise, subtitles will not be available for usage. Finally, video and data from TS as well as subtitles downloaded from the server are shown on display via application.

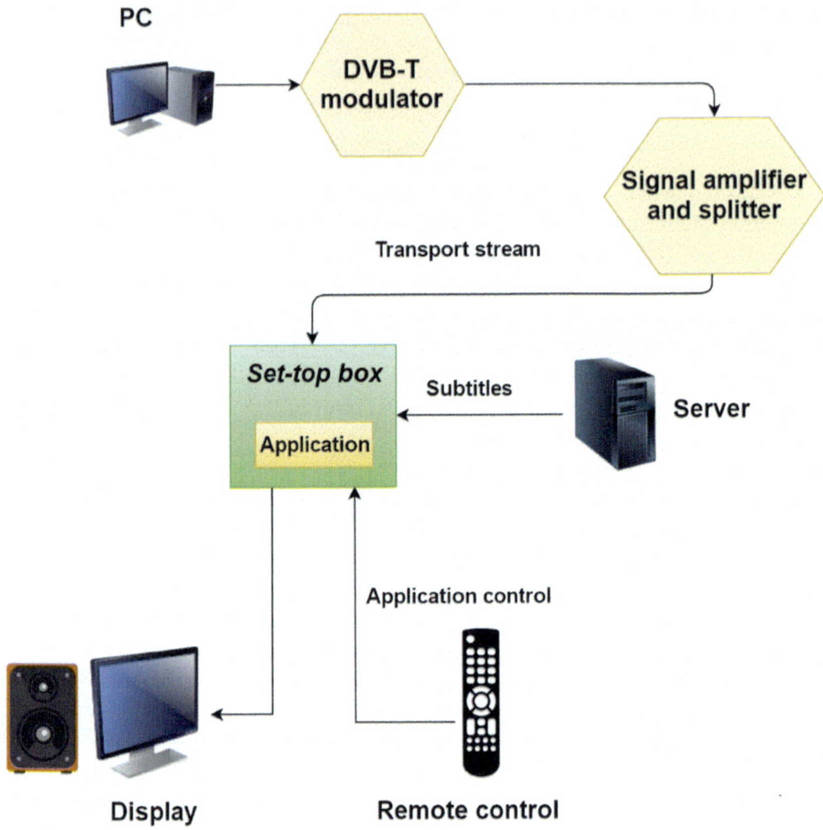

Fig. 15 Solution concept

3.1 *Modules Description*

The application is realized in several modules and several layers, which are shown in Fig. 16. Each module has its functionality required for the application to work. The arrows show the interdependence of the modules.

The module *main* is the central module of the application. It contains functions for opening and parsing of configuration file. The configuration file contains information about Transport Stream and initial channel used for the project. In addition, this module calls initialization functions.

The module *player* is the backbone of the whole application. This module integrates remote control, parsing of configuration file and Signalling Tables, handling of .srt file, graphics and the rest of the functionalities. Those functionalities are implemented in several functions.

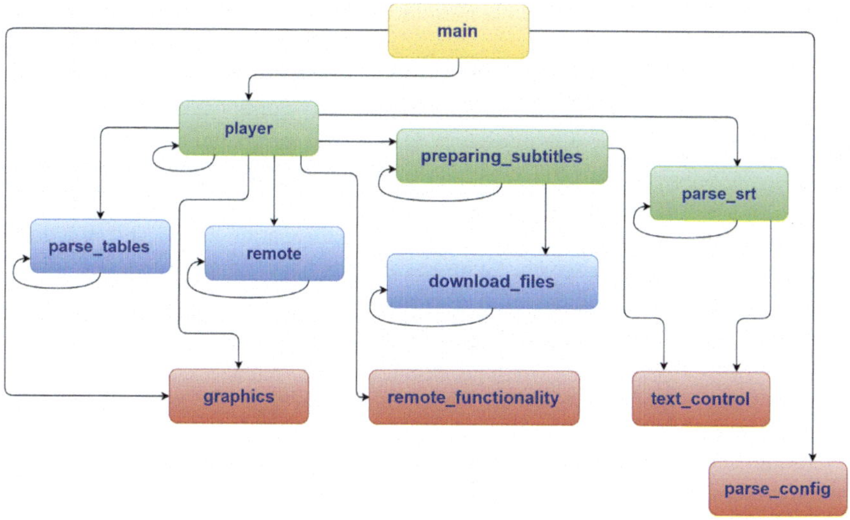

Fig. 16 Module dependency

In the initialization function, tuner, player and source of the player are initialized. When tuner is locked to frequency, the application can continue with execution. PAT, PMT and SDT tables are parsed and memory for EIT and TDT tables is allocated. During shutdown in deinitialization function, video and audio are removed, and source of the player, player, tuner, graphics and allocated memory are deinitialized.

The function *initialConfiguration* creates initial video and audio based on content parsed from the configuration file.

The function *eitLoop* is executed on a separate thread and is used to parse EIT tables. The EIT parsing works in a way that firstly the parsing filter is set to EIT table, then the current time is obtained, and after that application waits for one second for EIT table to be parsed. If parsing is successfully finished, a variable that counts the number of parsed EIT tables is increased by one. When the number of parsed tables is equal to number of EIT tables in the stream, parsing is finished and timer is set to 15 s. When expired, the parsing of EIT table starts again.

The function *tdtParsing* is a function that is executed on a separate thread and is used to parse TDT tables. It operates on a similar principle as *eitLoop*. When TDT table is found in the stream, the received time is saved in system time structure and timer is set to 10 s. After expiration, the parsing of TDT table is started again.

The function *handleRemoteKeyPress* is a function that is executed on a separate thread and processes events from the remote control.

The function *findEventForParsing* is scanning EIT table of current program and it calls the function for parsing of .srt file of a currently displayed event. If parsing of .srt file is successfully executed, the thread for searching through parsed .srt file is created. This thread is searching for the correct subtitle to be shown on screen.

The function *findAndDrawInfo* is used to find data in parsed data to fill out info dialogue.

The function *findAndDrawSubtitle* is a function that is executed on a separate thread and is used to search for parsed .srt file. The current time calculated in a separate thread is used for subtitle search. That time is compared to the start time of a current event parsed from EIT table. Based on those times, the time elapsed from start of the event is calculated.

The function *drawingElements* is executed on a separate thread and is used to draw graphic elements on screen. Graphic elements have a control flag and timers. When the flag is set, the element is drawn on screen. When the timer expires, a respective graphic element will be removed from display.

The function *countTime* is executed on a separate thread and is used to calculate time. It waits for a predefined time; later, when that time expires, it calculates the real time passed so as to get the current system time before and after the timer expiration, and calculates their difference in milliseconds. After that, the calculated milliseconds are forwarded to function *addTime* in which current time saved in struct *sysTime* is updated.

The module *preparing_subtitles* is used to prepare subtitles for download from the server. Figure 17 shows flowchart of subtitle download process.

For subtitle download, the application needs to be notified about the status of Internet connection, server and subtitles. The status is monitored with control flags. When a function for preparing subtitles for usage is called, *prepareSubtitles*, the first step is to reset values of control flags. The next step is to check if the Internet connection is available. The connection is checked by connecting to a chosen web page. The Internet connection is available if the connection to the page is established, so the next step, the check of sever status, can be performed. The server status is checked with features from library *libcurl* so that the test HTML file is downloaded from the server. If this check is conducted successfully, the next step is to generate names of .srt files that will be searched on the server. Names are created in a way that EIT table of current channel is searched; then, when the name of a current event is found, that name will be copied in variables that hold values of file names. Afterwards on event name language and file type are added—"-lang.srt". Having created names, a search can be performed if the subtitles are available on the server. The subtitles are created for two shows—„Dona Barbara (Eps. 151 Al 190)“ on the program La1 and „Mascotas Papanatas“ on the program Clan. If the name of the shows with subtitles is to the same as the name of the show currently displayed, the function for downloading files from server is called. After the download process is completed, a check if the subtitles are available in working directory is performed. If those files exist, the flags that subtitles are downloaded and available are set and users can use subtitles.

The module *parse_srt* is used to parse .srt files and find appropriate subtitle that should be shown on the display.

A function that is used to parse .srt files, *parseSrtFile*, receives language of subtitles picked by user and event name for which .srt file needs to be parsed as input parameters. The name of event will be aligned with standard ASCII code by

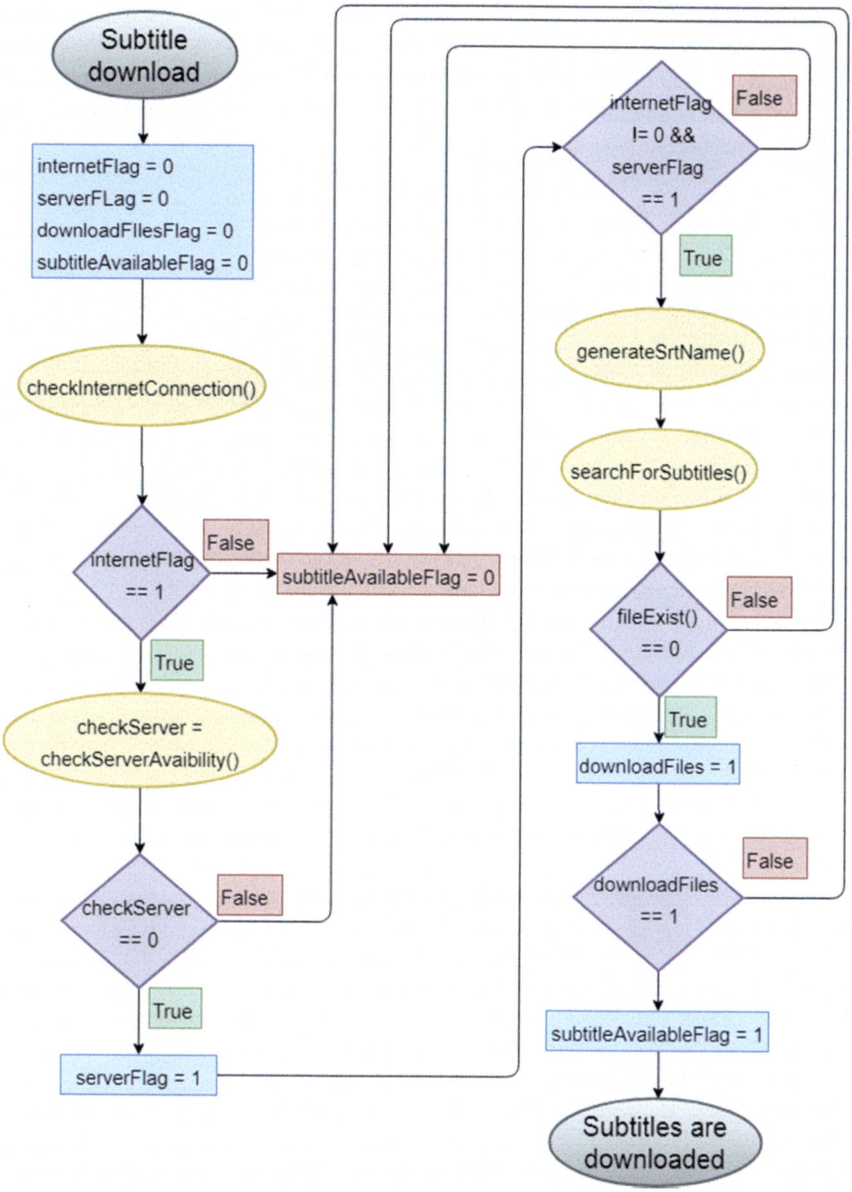

Fig. 17 Flowchart of subtitle download

calling function *checkASCII* from module *text_control,* followed by adding name extension "-lang.srt". The language extension is added for the language forwarded to function. The subtitle file is parsed by iterating the file line by line. Each subtitle is saved in one node that contains subtitle data and the pointer to the next subtitle node that contains next subtitle data. Upon parsing completion, the file is closed.

Search function is used to find appropriate subtitle that has to be shown on the display. Time in hours, minutes, seconds and milliseconds elapsed since the event started, and a pointer to the first element in the list is passed to the function. The search is executed in milliseconds because of a larger time precision. Inside linked list, two conditions are checked. The first condition checks if the current event time is within the time of the subtitle. If it is, the subtitle is shown. The next condition checks if the current event time is in-between the time frame of the two subtitles. No subtitles will be shown then or they will be removed from the display. Figure 18 shows the conditions mentioned above.

The module *parse_tables* is used to parse Signalling Tables (PAT, PMT, SDT, EIT and TDT). This module comprises function to parse and display content of tables on screen. In the parsing function of aforementioned tables the data stored in forwarded parameter are processed. The data is stored in the dynamically allocated memory. If parsing is successfully done, a mark for successful parsing is returned. Otherwise, a mark for unsuccessful parsing will be returned.

The module *remote* is used to initialize remote control and read user input from it. The events are received via remote control file located on STB. In the initialization function the file is opened, memory for events is dynamically allocated, and

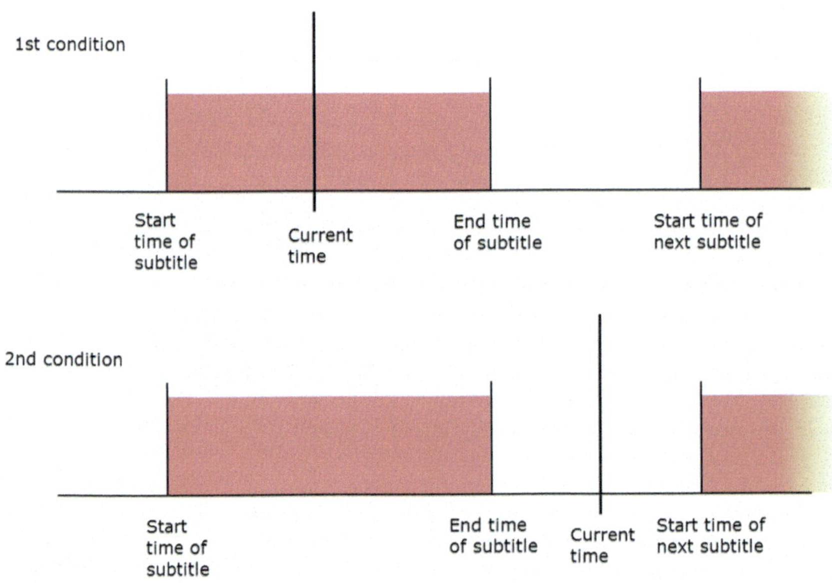

Fig. 18 Conditions to check subtitles

thread for registering callback for remote control is created. The key that is pressed on remote control is retrieved using the local function. These parameters are forwarded to the callback. When application is shut down, remote control will be deinitialized, thread will be joined and device will be closed.

The module *download_files* is used to download files from server and to check if files are downloaded successfully. Files are downloaded with the help of a library *libcurl*. The *libcurl* needs to be initialized each time the application is started, and *libcurl* session should be started. The download of files is implemented in a function that receives URL to server location, and file name under which it will be saved as input parameters. Files are downloaded to working directory on set-top box.

The module *graphics* is used to initialize graphics and draw graphics elements on the display with the help of *DirectFB* library. Graphics elements that are drawn on the display are the following: loading screen, radio program layout, info dialogue, volume dialogue, menu for subtitle language selection, and subtitles. Each graphical element has its own flag. Depending on the flag value, the element will either be drawn or not.

A function for graphic initialization, *initGraphics*, is called on every time the application is started. In this function, DirectFB is initialized, parameters for graphics are set, and main area on which elements will be drawn is created. The area for drawing encompasses the whole display. Double buffering is enabled which prevents flickering on the display when graphic elements are shown. It works so that one buffer graphics is drawn and the other buffer is shown on display. When a function to switch buffers, *flipScreen*, is called, buffers are switched. Figure 19 shows how this principle works.

In the initialization function, the interfaces for fonts are created. Figure 20 shows an example of creating of font for radio channel.

On application shutdown, a function for deinitialization of DirecFB, *deinitGraphics*, and for graphics remove, *clearScreen*, is called.

Loading screen is shown while data is being loaded on application start. Once the data is loaded, loading screen is removed and initial video and audio content are shown. The screen consists of black background and text that notifies user that content is being loaded. Figure 21 shows loading screen.

Due to API limitation when channel is switched from video to radio channel, video from previous channel stays on display. In order to enhance the user experience and the look of the graphical interface, radio display is implemented as black

Fig. 19 Double buffer principle

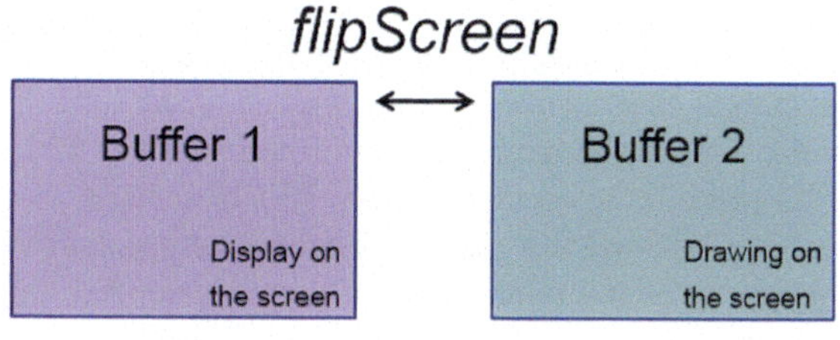

```
fontDesc.flags = DFDESC_HEIGHT;
fontDesc.height = screenHeight/FONT_HEIGHT_RADIO;
DFBCHECK(dfbInterface->CreateFont(dfbInterface,
"/home/galois/fonts/DejaVuSerif.ttf", &fontDesc, &fontInterfaceRadio));
```

Fig. 20 Code example of creating font for radio

Fig. 21 Loading screen

Fig. 22 Radio channel layout

background with info that the current channel is a radio channel, as well as with the number and name of the channel. Figure 22 shows radio channel layout.

The rest of the graphic elements will be described in the chapter "Flow chart diagrams remote control buttons".

The module *remote_functionality* is used for processing of asynchronous events from remote control, creating of audio and video content, changing volume and removing graphics from the display.

The function used for creating of audio and video stream, *playVideoAndAudio*, is called on every program change. When channel is changed, audio and video content of current channel needs to be removed; in addition, it is necessary to find audio and video PID of the next channel in the respective PMT table. In the case that video PID exists, a new video content will be created. If video PID does not exist, the selected channel is a radio channel so the flag for radio channel will be set to one. Subtitle availability and look of graphics depend on the status of the radio flag.

A function used to change volume, *setSound*, is called when the button for volume change or mute is pressed. The code of the pressed button is forwarded to the function which returns volume value.

Graphics elements are removed on user input on the remote and during shutdown. Elements that are removed are the info dialogue, volume dialogue and menu for subtitle language selection.

The module *text_control* is used to align strings used in application with standard ASCII code. If a sign is a part of extended ASCII table, it will be switched with equivalent sign from a standard part of the ASCII table.

The module *parse_config* is used for parsing of configuration file. The configuration file consists of the following parameters: frequency, bandwidth, module, and data about the starting program. These data include audio PID, video PID, audio type and video type.

The function *parseConfigFile* checks if the configuration file is successfully opened. If not, application will not be started. If the file is successfully opened, it can be parsed. The file is iterated line by line. When a line is read it is saved in the buffer, and the NULL pointer is returned to mark the end of file. The line in the buffer is searched for keywords: „frequency=", „bandwidth=", „module=", „apid=", „vpid=", „atype=" and „vtype=". After the keyword is found, related data will be saved in a specified field in the struct used to save data from the configuration file. Data in the struct is used later on to start stream on the program specified in the configuration file.

3.2 Flow Chart Diagrams for Remote Control Buttons

In the application, several remote control functionalities are implemented. The following chapter will describe what happens in the application when buttons on the remote control are pressed.

Button for application shutdown

Pressing the button for application shutdown (on remote control number 0) shuts down the application.

Figure 23 shows what happens in the application when the button is pressed.

If the thread for searching through linked list with subtitles is started (*workSrtParse*) it will be stopped, graphics will be removed from screen (*removeGraphics*) and all remaining threads will be stopped (*threadsRunning*). If a pointer to the first node in the linked list with subtitles is not NULL, memory will be freed and message queue for inputs from remote will be cleared. After that, the application shuts down.

Button for select program number

By pressing one of program selection buttons (numbers 1–9 on the remote control), the channel number is changed to the selected channel if the chosen number is different than the current channel. Figure 24 shows what happens in the application when a button is pressed.

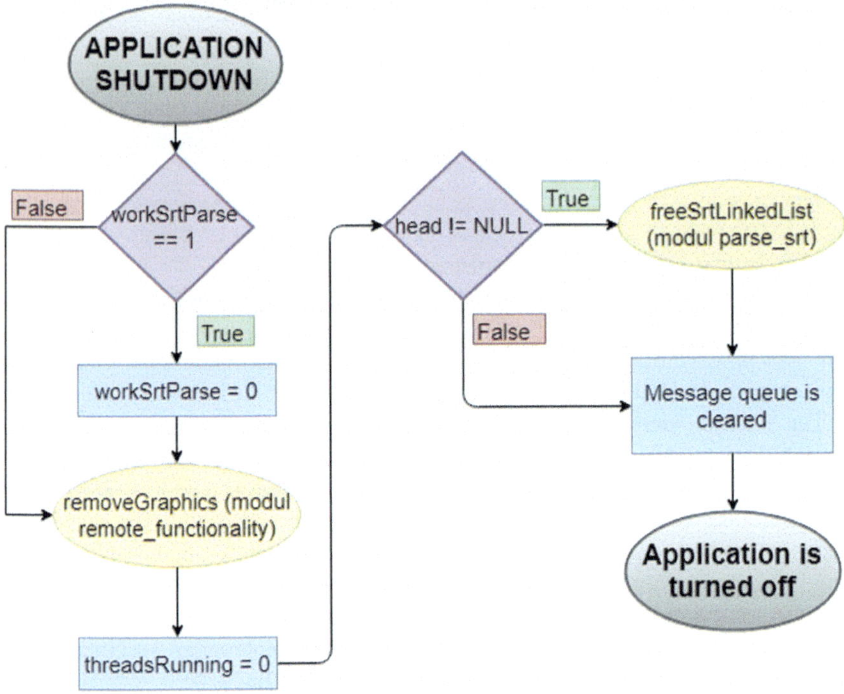

Fig. 23 Flowchart for the application shutdown button

In simple words, the changes will go as follows:

Currently watched channel → change to channel equal or less than maximum channel number → switch to wanted channel

Currently watched channel → change to channel larger than maximum channel number or to currently watched channel → there is no change

If the thread for searching of the linked list with subtitles is started, it will be stopped. The value of the previous channel is saved to *beforePrevious* variable which is used to detect if the selected number is different from the current program. The current channel is saved to *previousChannel* variable.

If the selected channel, *chanNumber*, is larger than the number of channels, the value saved in *previousChannel* will be saved in *chanNumber* indicating that the channel will not be changed.

The next step is to check if the selected channel is different from the current channel. If yes, it is checked if the pointer to the first node of the linked list that contains subtitles is different than NULL. If it is, memory allocated for the linked list will be freed; if the subtitles are shown they will be stopped and new audio and video content will be created. In continuation, it is checked if the radio flag is set. If not, the subtitles are prepared for the user. The flag for the info dialogue is shown and the info dialogue is shown for 8 s. When this time expires, the info dialogue is removed from display.

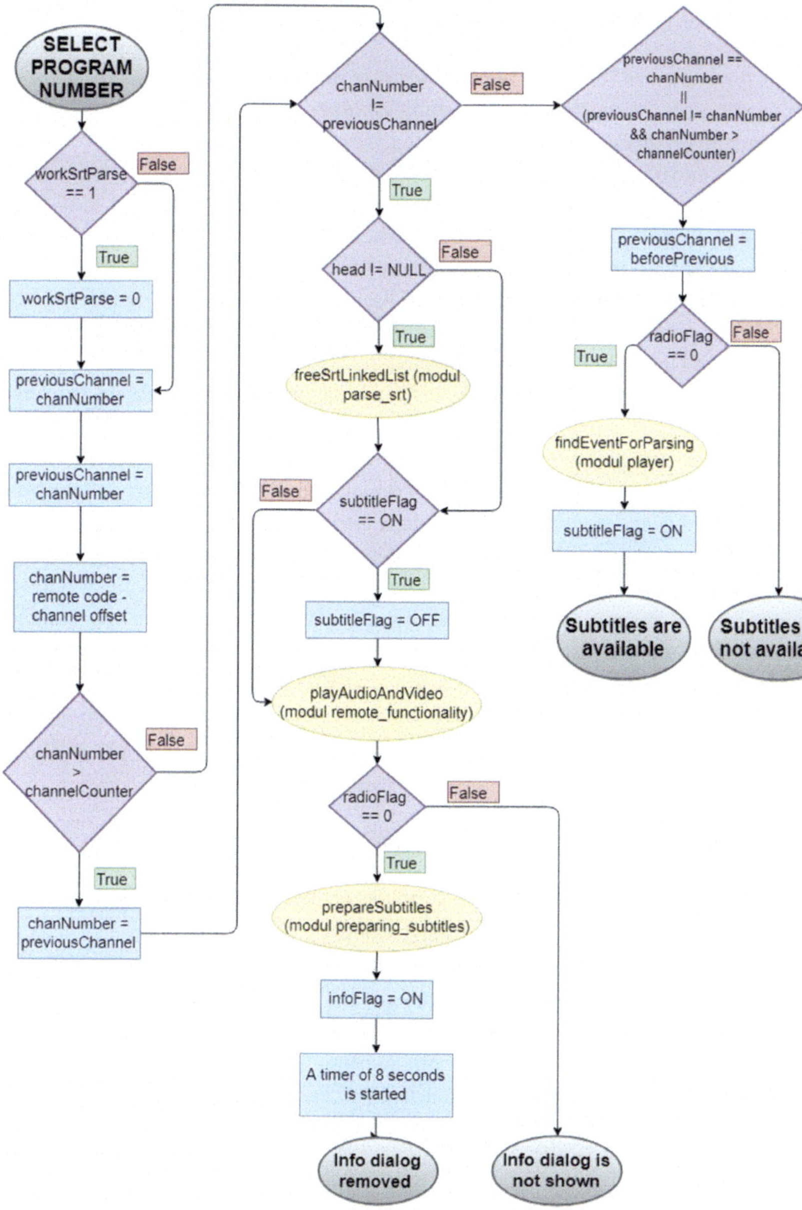

Fig. 24 Flowchart for select channel buttons

If the selected channel is not different from the current channel, the channel is not changed.

Button for switching channel number for one number up or down

Implementation for buttons for switching channel number for a one number up or down is similar. Pressing button for switching channel number for a number up (on remote control number 12) or down (on remote control number 13) will increase or decrease channel number for one. Figure 25 shows the flowchart depicting the working of the application when one number up is pressed and Fig. 26 shows what happens when the button for a number down is pressed.

If the thread for searching of the linked list with subtitles is started, it will be stopped. If the pointer to the first node of the linked list that contains subtitles is different than NULL, the memory allocated for linked list will be freed, and if subtitles are shown they will be stopped. After that, graphics is removed from the display.

The next step is the channel switch. The next channel number depends if the pressed button is a button for a channel up or down. Pressing the button to change the channel number for one number up will change current channel to one channel up. If the current channel is the last channel in the channel list, the channel will be switched to channel number one. Pressing button to change channel number for one down will change the current channel to one channel down or if current channel is one, channel will be switched to the last channel in channel list. Then, new audio and video content is created. Next step is to check if video PID is different than zero which means that the current channel is a video channel. If it is, subtitles are prepared so that user can use them. the flag for the info dialogue is set and info dialogue is shown for 8 s. After the timer expires, the info dialogue is removed from display.

Button for switching to previously watched program

Pressing the button for switching to previously watched channel (number 19 on the remote control), the channel will be changed to previously watched channel. Figure 27 shows what happens in the application when the button is pressed.

If the thread for searching of the linked list with subtitles is started, it will be stopped. If the pointer to the first node of the linked list that contains subtitles is different than NULL, the allocated memory for the linked list will be freed, and if the subtitles are shown they will be stopped. The next step is the algorithm for replacing current and previous channel with auxiliary *temp* variable after which new video and audio content is created. What is checked next is whether video PID is different from zero, which means that the current channel is a video channel. If it is, subtitles are prepared so that user can use them. The flag for the info dialogue is set and the info dialogue is shown for 8 s. When the time expires, the info dialogue is removed from display.

Buttons for mute, volume up and volume down

When the button for mute or volume change (numbers 10, 11 and 14 on the remote control) is pressed, the volume is changed or muted. Figure 28 shows what happens in the application when one of buttons is pressed.

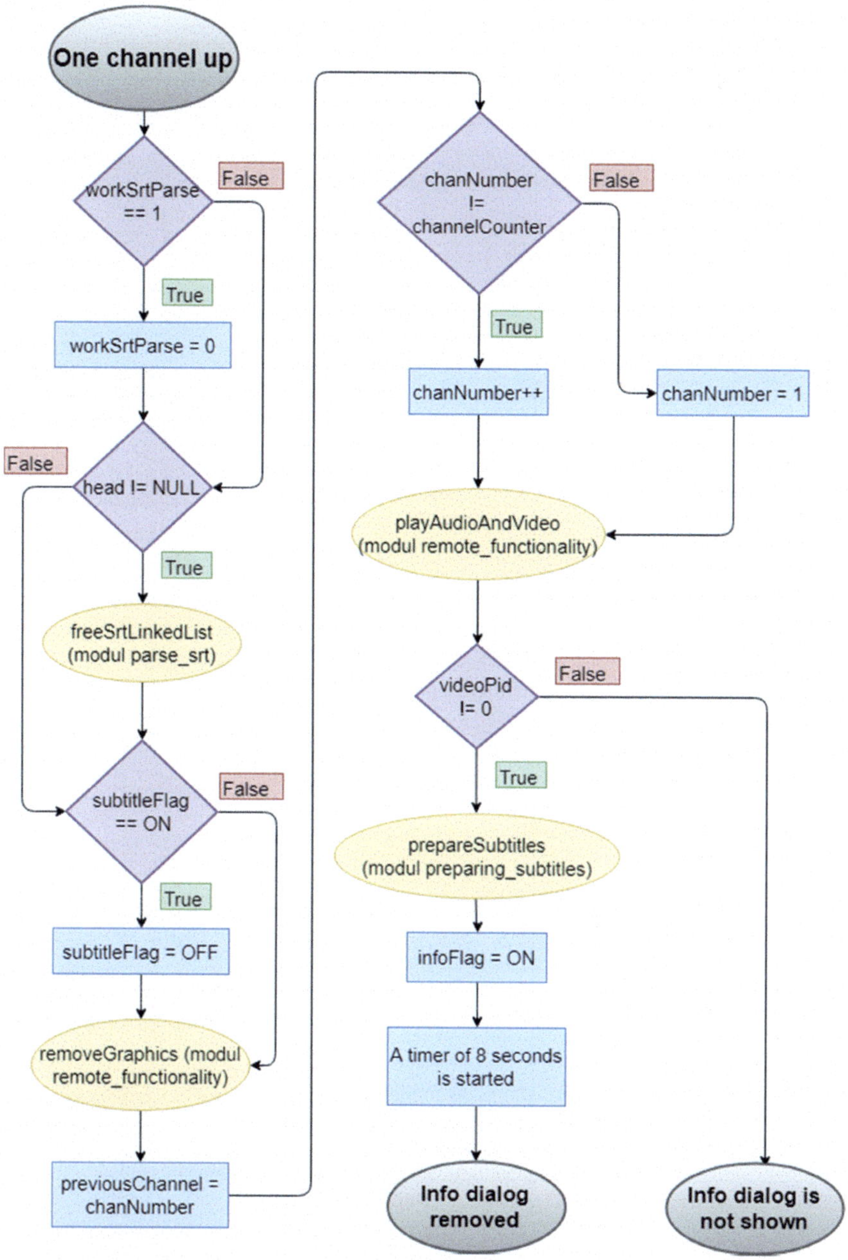

Fig. 25 Flowchart for one channel up

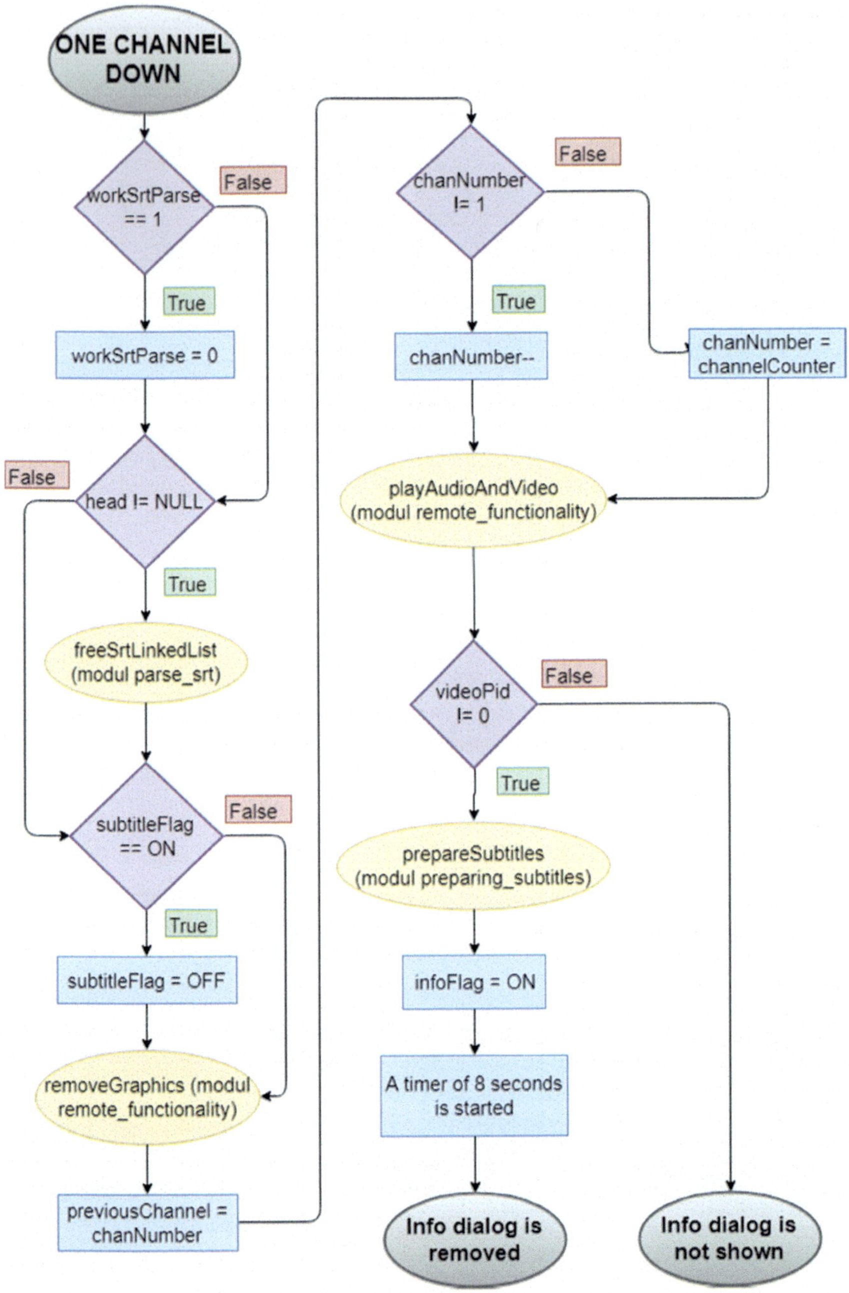

Fig. 26 Flowchart for one channel down

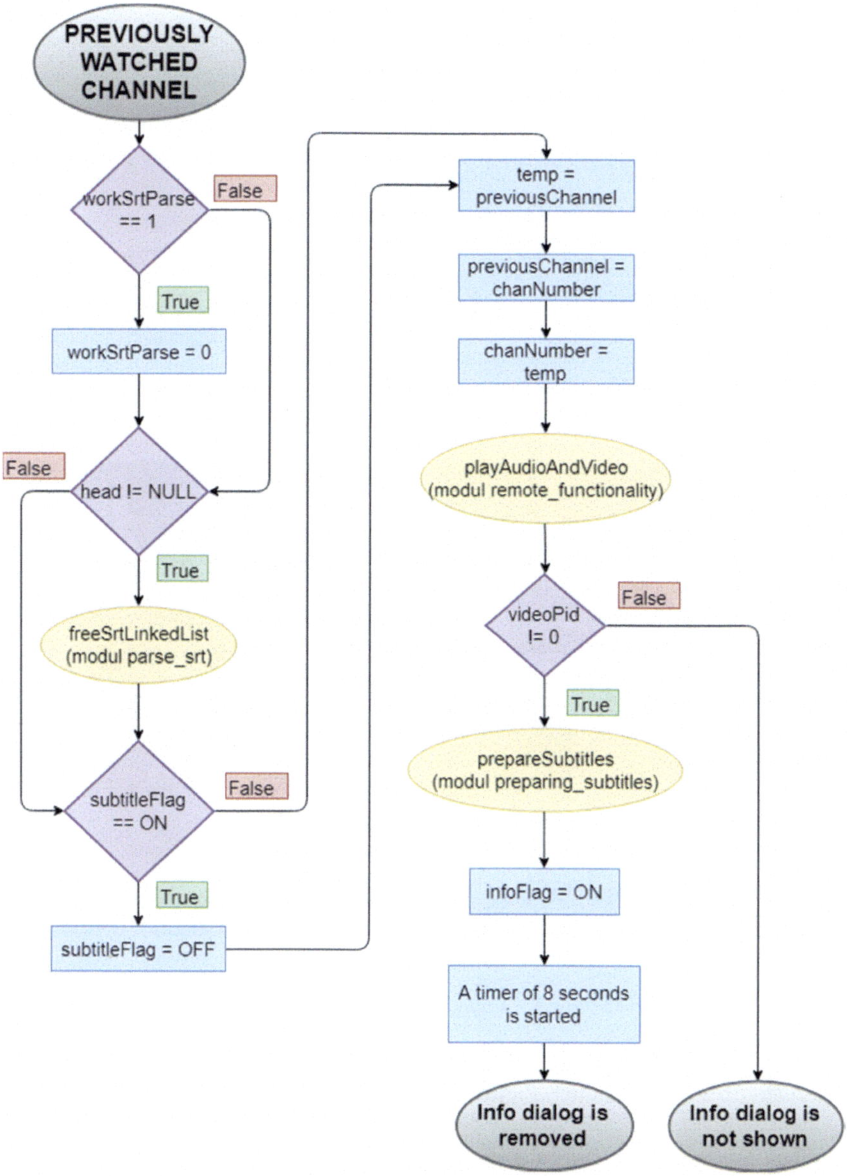

Fig. 27 Flowchart for previously watched channel

After user input, a function for calculating new volume, *setSound*, will be called and it will return new volume value. That value will be forwarded to API function *Player_Volume_Set* to set new volume in the stream. The volume flag will be set indicating that graphic for volume will be shown. When the timer expires, graphics is removed from display.

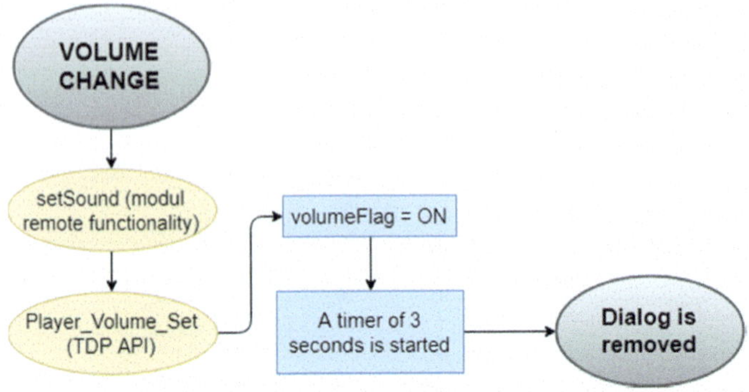

Fig. 28 Flowchart for volume change

Fig. 29 Volume bar

The look of the volume bar is adapted depending on whether the current channel is a video or a radio channel. Figure 29 shows volume bar.

Button for info dialogue display

When the button for information display is pressed (number 18 on the remote control), the info dialogue with information about the current channel and event is shown. Figure 30 shows what happens in the application when the button is pressed.

The first step is to check if the video PID is different from zero. If video PID is zero, the current channel is a radio channel and the info dialogue will not be available. If video PID is different from zero, a semaphore for parsing of EIT table is signalled to refresh data, a flag for info dialogue is set so graphics is displayed, and timer is started. Once the timer expires, graphics is removed from display.

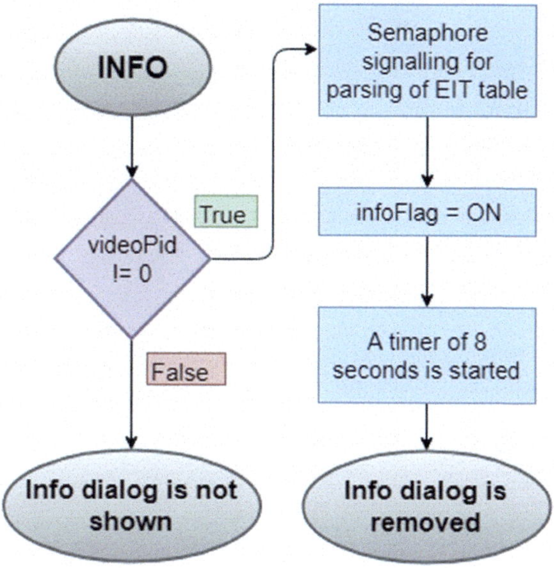

Fig. 30 Flowchart for info dialogue

<table>
<tr><td rowspan="2">Channel
number</td><td>Channel name</td><td rowspan="2">Start and end
time of event</td></tr>
<tr><td>Event name</td></tr>
<tr><td colspan="2">Short description of event</td><td>Current time</td></tr>
</table>

Fig. 31 Info dialogue layout

The type of displayed data can be seen on Fig. 31, the info dialogue with event description on Fig. 32 and the event without description on Fig. 33.

Button for subtitle language selection menu

By pressing the button for language selection menu (number 22 on the remote control), a menu for subtitle language selection will be shown. Figure 34 shows what happens in the application when the button is pressed.

The first step is to check if the info dialogue is shown. If it is, the dialogue will be removed. The reason for this is to keep the display with as little overlaying graphics as possible. The next step is to check if the current channel is a radio channel. If not, a flag for subtitle language selection menu is set and graphics is shown for 8 s. After the timer expires, the menu will be removed from display. If the current channel is a radio channel, the subtitles are not available.

The menu will be adapted depending on the subtitles availability. If subtitles are not available, an info message will be shown as can be seen on Fig. 35. If subtitles are available, a menu will be shown as can be seen on Fig. 36. Users can navigate through the menu which colours are adapted depending on the current position.

Fig. 32 Info dialogue with event description

Fig. 33 Info dialogue without event description

Buttons for shifting position in subtitle language selection menu

A position in the menu for subtitle language selection can be changed by pressing the button for shifting right (number 16 on the remote control) or left (number 15 on the remote control). Figure 37 shows happens in the application when the button for right is pressed and Fig. 38 shows what is happening when the button for left is pressed.

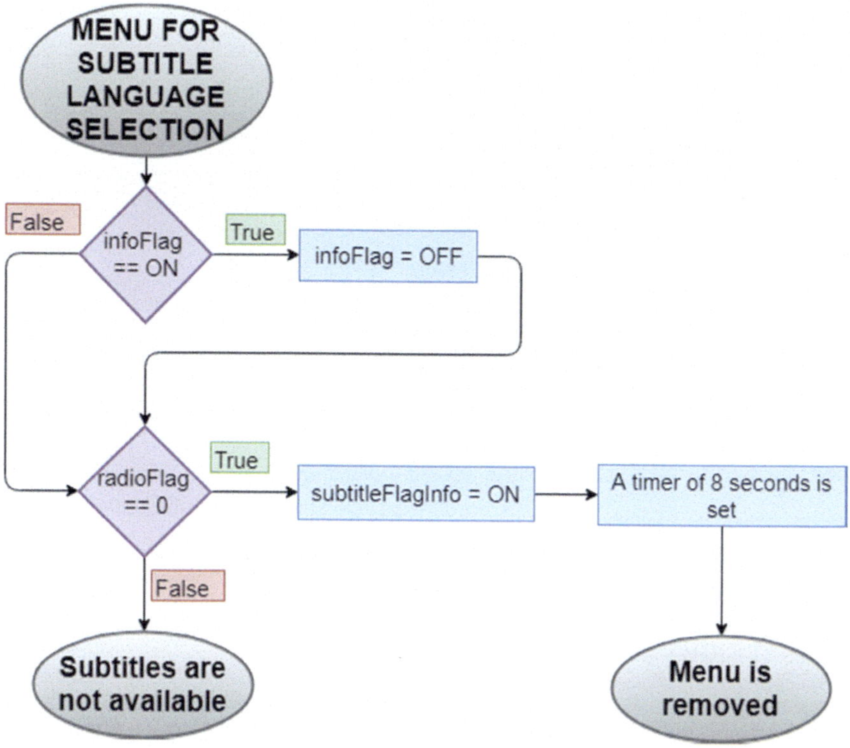

Fig. 34 Flowchart for the menu for subtitle language selection

Fig. 35 Message when subtitles are not available

Fig. 36 Menu when subtitles are available

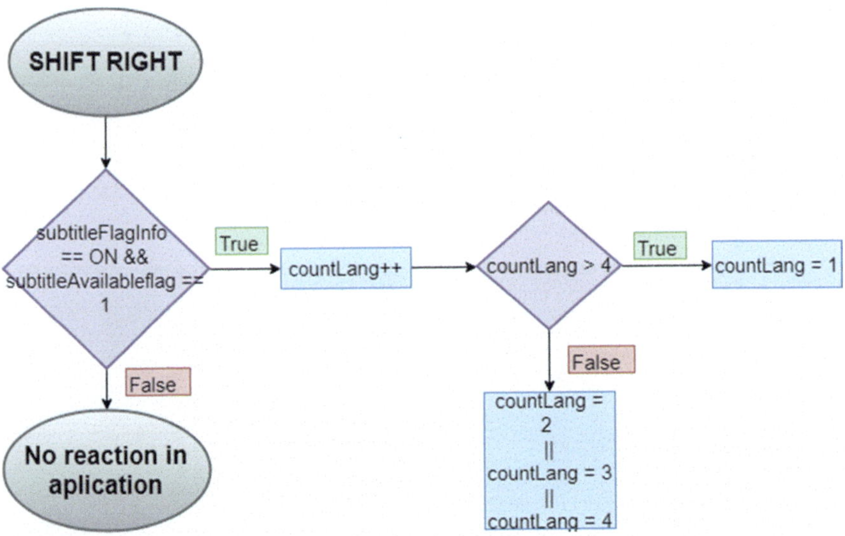

Fig. 37 Flowchart for the button shift right

The first step is to check if the subtitles are available and if the menu for subtitle language selection is displayed. If conditions are not met, the button press will be registered but nothing will happen in the application.

If the pressed button is the button for left, value will be decreased for one and if the pressed button is the button for right, value will be increased for one.

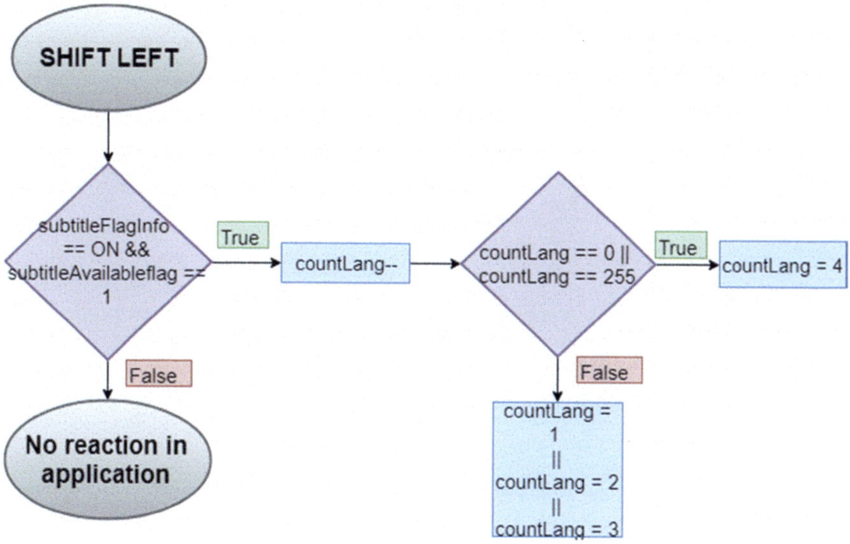

Fig. 38 Flowchart for the button shift left

The variable that handles the current value is of the type *uint8* and therefore it is necessary to make conversions in both cases. In the case of "button left", if after subtraction the value is zero or 255, the value four will be assigned to variable; contrarily, in the case of "button right", if after addition value is higher than 4, value one will be assigned to variable.

Button for language selection

By pressing the select button (number 17 on the remote control), the subtitle language is selected if the menu for subtitle language selection is shown. Figure 39 shows what happens in the application when the button is pressed.

The first check is whether the menu for subtitle language selection is shown. If not, the button press will be registered but nothing will happen in the application.

If it is shown, the next step is to check if the thread for searching of the linked list with subtitles is started. If it is, it will be stopped, and if the pointer to the first node of the linked list that contains subtitles is different from NULL, the memory allocated for the linked list will be freed.

Additionally, it is checked if the current channel is a radio channel but this step is considered redundant because menu will not be available if the current channel is a radio channel. After that, the menu is removed from the display, the current event is found in EIT table, respective .srt file is parsed and subtitles are shown on the display.

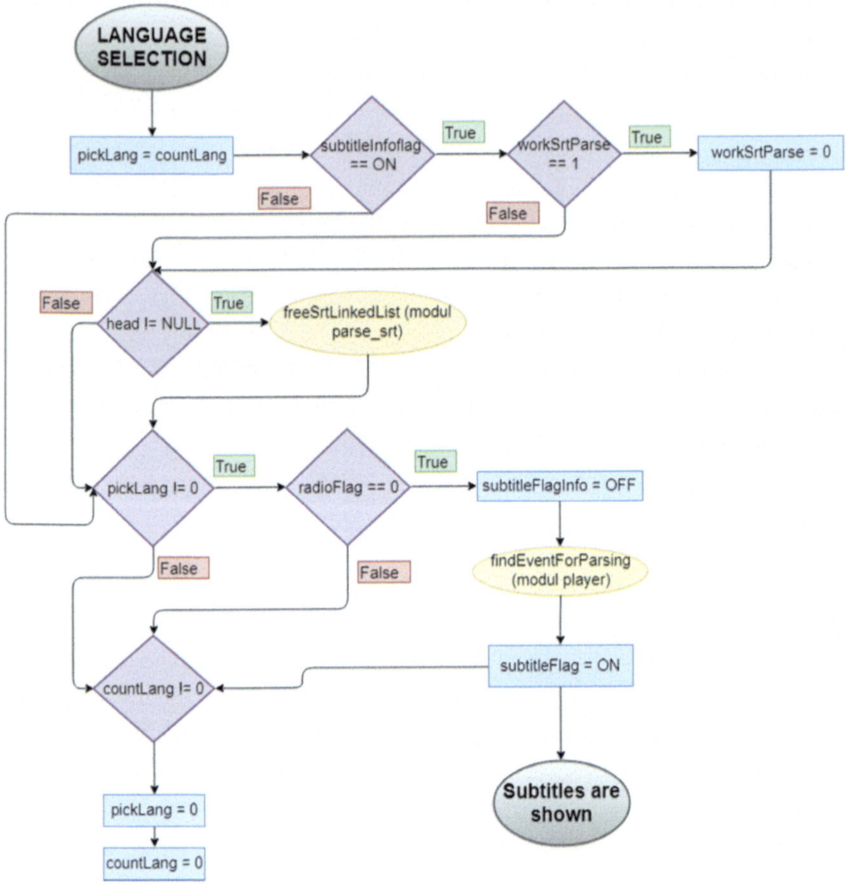

Fig. 39 Flowchart for the language selection button

If subtitles exist, the first and the second line of subtitles are forwarded to the function used for drawing subtitles on the display, where they will be shown. If the second line of subtitles exists, the position of subtitles is adapted accordingly. Figure 40 shows subtitles with one row and Fig. 41 shows subtitles with two rows.

Button for stopping subtitles display

Pressing the STOP button on the remote control will stop subtitles if they are shown on display. Figure 42 shows what happens in the application when the button is pressed.

If the thread for searching of linked list with subtitles is started, it will be stopped. If the pointer to the first node of the linked list that contains subtitles is different than NULL, the allocated memory for the linked list will be freed, and if subtitles are shown, they will be stopped.

Fig. 40 Subtitles with one row

Fig. 41 Subtitles with two rows

Button for removing graphics from the display

By pressing the button for removing graphics (number 20 on the remote control), all graphics currently being displayed will be removed (info dialogue, volume bar or menu for subtitle language selection). Figure 43 shows what happens in the application when the button is pressed.

Graphics removal is implemented very simply by calling function *removeGraphics* from module *remote_functionality*.

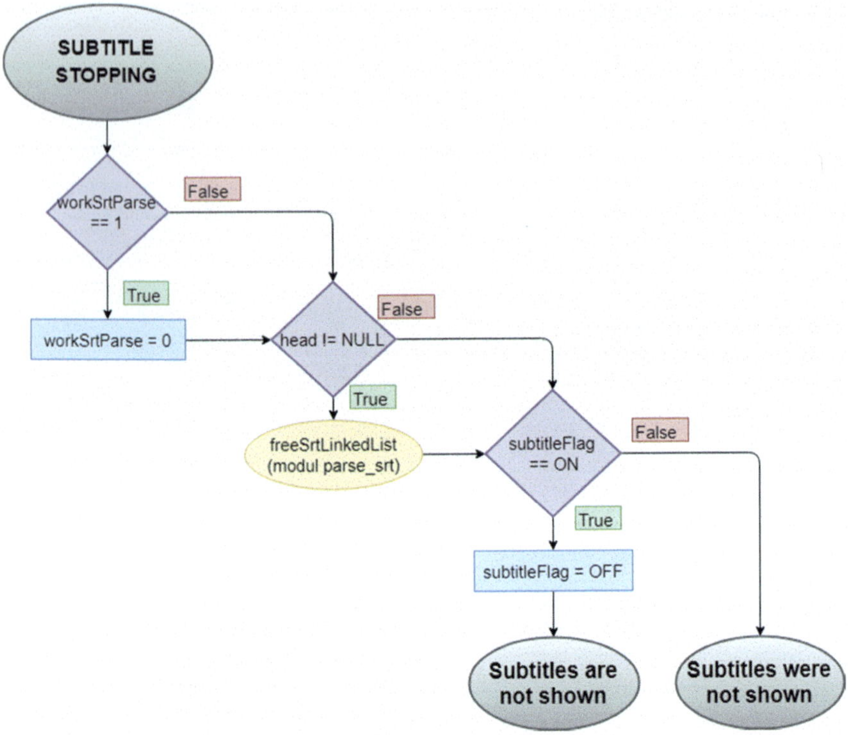

Fig. 42 Flowchart for the button for subtitle stopping

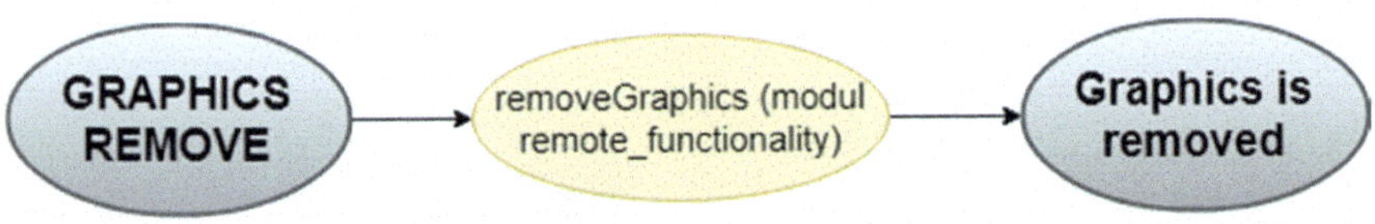

Fig. 43 Flowchart for the graphics remove button

4 Application Testing

The application was tested in order to check whether the functionality, stability and robustness meet the quality expectations. Tests containing checks of remote control keys that hold functionalities, and tests of displaying graphical elements in different use cases were conducted. The main part of the testing was the integration of subtitles and it was tested in several use cases. All the tests gave a passing result. More details on conducted tests can be found in [7]. Although all test passed, the application has some imperfections such as assumptions that only video channels have EIT table and consequently there are no available subtitles for radio channels.

5 Conclusion

Subtitles are a very important feature for TV service. They are used for translating events on television. In this chapter, the emphasis is on subtitle realization to languages different from the language in the Transport Stream. This solution does not require expensive hardware. It is enough to have the set-top box with Internet access, DVB-T modulator, display, remote control and audio speakers.

The application consists of several layered modules. Each modules has its own functionality. The application is written in C programming language because of bit manipulation, memory management and speed that C provides.

Acknowledgements We would like to thank Faculty of Electrical Engineering, Computer Science and Information Technology and Institute RT-RK Osijek LLC and in particular Mr. Mario Vranješ, PhD, Associate Professor at Faculty of Electrical Engineering, Computer Science and Information Technology (FERIT), Josip Juraj Strossmayer University of Osijek, Croatia for all support and advice.

References

1. Bjelica M, Teslić N, Mihić V (2016) Softver u televiziji i obradi slike 1. Faculty of Techincal Sciences, Novi Sad, Serbia
2. About DVB. https://www.dvb.org/about. February 2016
3. Prof. dr. sc Branka Zovko-Cihlar, Utjecaj digitalnih TV kanala (DVB-T) na postojeću analognu mrežu u Republici Hrvatskoj, Faculty of Electrical Engineering and Computing, Zagreb, Croatia, February 2004
4. European Telecommunication Standard Institute. Digital Video Broadcasting (DVB); Specification for Service Information (SI) in DVB systems, v1.11.1 (2010-04)
5. European Telecommunication Standard Institute. Digital Video Broadcasting (DVB); Guidelines on implementation and usage of Service Information (SI), v1.9.1 (2009-06)
6. SRT Subtitles. https://matroska.org/technical/specs/subtitles/srt.html. February 2016
7. Bilandžić A (2016) Realizacija podrške za prijevode u hibridnim digitalnim TV aplikacijama. Master Thesis, Faculty of Electrical Engineering, Computer Science and Information Technology Osijek, Osijek, Croatia
8. DirectFB. https://elinux.org/DirectFB. February 2016
9. thttpd—tiny/turbo/throttling HTTP server. https://acme.com/software/thttpd/. February 2016
10. CURL. https://curl.haxx.se/. February 2016

Human Heart Blood Flow Numerical Modelling and Simulations

Andrey Aksenov, Sergey Zhluktov, Wojciech Zietak, Ross Cotton and Dean Vučinić

Abstract In order to make a correct choice of the artificial heart valve for curing patients with cardio vascular disorder it is very important to know in detail the blood flow in the heart of the specific patient. In this book chapter, two separate approaches are presented on how to model the blood flow within the real human heart. The first approach is based on Magnetic Resonance Imaging/Tomography to get shape ("geometry") of internal heart volumes in time (tens or hundreds of frames for one heartbeat), and applies a Computational Fluid Dynamics (CFD) tool for simulating and visualizing a blood flow. The second approach is based on the SIMULIA Abaqus LHHM model, combining the structural heart Finite Element Model (FEM) with the CFD simulation to solve the fluid structure interaction problem. The blood flow circulation through the heart is numerically simulated with a modern FlowVision code that has a fully automatic mesh generation using arbitrary cell shapes for dynamic mesh refinement and taking into account the motion of the heart surface, applying the Euler coordinates. The second approach solves the bidirectional fluid-structure interaction problem applying the general purpose CFD code *FlowVision* and the SIMULIA *Living Heart Human Model*

A. Aksenov (✉) · S. Zhluktov
Joint Institute for High Temperatures, Russian Academy of Sciences,
Izhorskaya str. 13, Moscow 125412, Russia
e-mail: andrey@tesis.com.ru

S. Zhluktov
e-mail: sz@flowvision.ru

W. Zietak
Capvidia, Technologielaan 3, 3001 Louvain, Belgium
e-mail: wz@capvidia.com

R. Cotton
Simpleware, Synopsys Northern Europe Ltd, Bradninch Hall, Castle Street,
Exeter EX4 3PL, UK
e-mail: r.cotton@simpleware.com

D. Vučinić
Vesalius College (VeCo), Vrije Universiteit Brussel (VUB), Pleinlaan 2,
1050 Brussels, Belgium
e-mail: dean.vucinic@vub.be

D. Vucinic et al. (eds.), *Advances in Visualization and Optimization Techniques for Multidisciplinary Research*, Lecture Notes in Mechanical Engineering, https://doi.org/10.1007/978-981-13-9806-3_8

(LHHM), a dynamic, anatomically realistic, 4-chamber heart model with mechanical valves that considers the interplay and coupling of electrical and mechanical fields, which are acting simultaneously to regulate the heart filling, ejection, and overall like pump functions. LHHM natively includes a 1D fluid network model capable of representing dynamic pressure/volume changes in the intra- and extra-cardiac circulation. In the current work, this network model is first replaced with a full 3D blood model (solved in FlowVision) that provides detailed spatial and temporal resolution of cardiac hemodynamic driven by motions of the beating heart and constrained with appropriate time-varying boundary conditions derived from the literature. After validating this approach, the bidirectional coupling between the blood flow CFD model and the LHHM electromechanical model is activated by using the SIMULIA co-simulation engine and in conclusion the modelling details and results of interest are discussed. Using the real heart MRI/ MRT has tested the described approaches of simulating the blood flow and SIMULIA Abaqus LHHM and the simulation results are presented.

Keywords CFD simulation · Human heart · Internal flow · Time-variant shape

1 Introduction

Today, CVDs (Cardio Vascular Disorder) are the first cause of death. By 2030, CVD-related deaths will reach 23.3 million each year [1]. The lack of realistic human models limits ability to predict device behaviour under realistic real life conditions. Ultimately, 95% of all new devices are not tested in a human environment before approval. Growth in recalls suggests room for improvements. The numerical methods to model and simulate the human heart flow makes feasible to design or test an artificial medical device more efficiently and at low-costs. Moreover, the heart function under different, even abnormal conditions, can be in depth studied, especially interesting becomes to visualize the internal heart blood flow under such conditions.

One heart cycle contains 2 phases: (1) diastole (ventricle filling) and (2) systole (ventricle contraction). Atrium is filled with fresh blood. During diastole, the atria and ventricles of the heart relax and begin to fill with blood. The atria then begin to relax. In the next phase the heart's ventricles contract (ventricular systole) and pump blood out of the heart. At the end of diastole, the heart's atria contract (atrial systole) and pump blood into the ventricles. A heartbeat may seem like a simple, repeated event. However, it's a complex series of very precise and coordinated events. These events take place inside and around the heart.

A typical heartbeat cycle can be illustrated in Fig. 1, it can be noticed that the shape and volume always keep changing during the process. The time dependent geometry change brings new challenges to CFD simulations, which are difficult to solve with traditional numerical method, and they are (1) the time-variant shape of geometry having big deformations requiring high-quality mesh deformation

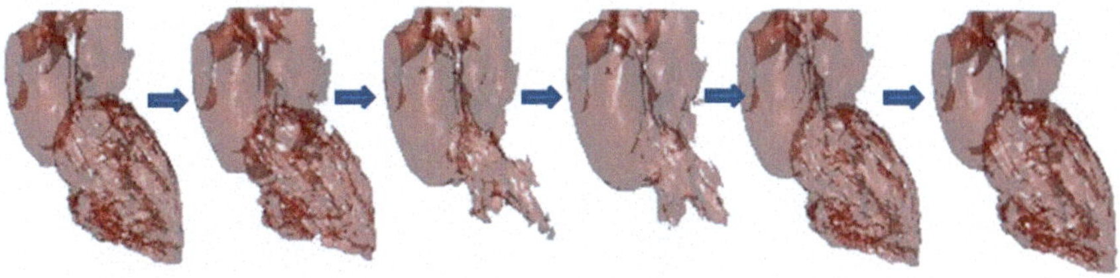

Fig. 1 The heart shape change within the heartbeat cycle

detection for its re-meshing; (2) the boundary conditions are changing in time, and from Wiggers Diagram (a standard diagram used in cardiac physiology) they are extracted for setting inlet or outlet BC; (3) the complex heart geometry has plenty of tiny features, which capturing needs fine mesh creation in order to guarantee the robustness for the applied numerical scheme.

In recent years, the Computational Fluid Dynamics (CFD) simulation of the heart flow has made great progress and has been playing an important role in investigating the heart function. The most active CFD approaches to simulate the heart flow might be roughly classified in 3 types: (1) geometry-prescribed CFD methods, (2) fictitious Fluid-Structure Interaction (FSI) methods and (3) realistic FSI methods [2]. The geometry-prescribed CFD method is based on the prescribed moving meshes or boundaries constructed, which are mostly defined from Computerized Tomography (CT) or Magnetic Resonance Imaging/Tomography (MRI/MRT) data [3–6]. However, almost all of them made simplifications when modelling the time dependent heart geometries within the heartbeat process. In this chapter, a dynamically adaptive grid method based of the moving body in is described, and combined with the automatic geometry replacement; enabling the dynamic simulation of the heartbeat process and modelling the internal flow mechanism.

2 Computerized Tomography (CT)

The applied numerical schemes are based on the Cartesian grid and are solving the URANS equations, applying the FlowVision CFD software. The important feature is that the dynamic mesh adaptation is generated automatically, and reconstructs the mesh dynamically in real-time, which is especially interesting to model the dynamic time dependent geometry of the heartbeat, and fully supported in FlowVision, which solver automatically replaces the geometry each time the heart shape changes.

2.1 *Heart Frames Reconstruction*

The heart model is a dynamic high-fidelity model of a normal (healthy), 4-chamber adult male human heart. It includes well-defined anatomic details of the heart as well as proximal vasculature, such as the aortic arch, pulmonary artery, and superior

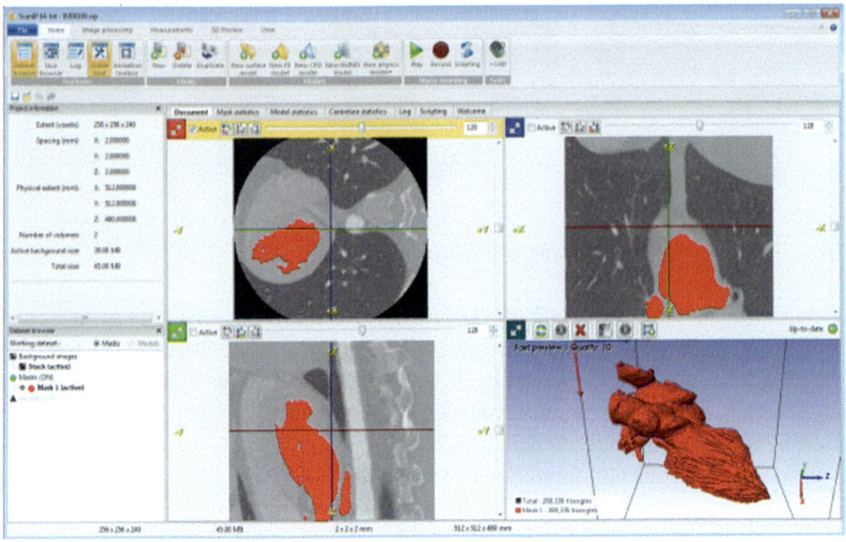

Fig. 2 Heart frames generation from DICOM files with ScanIP software

vena cava (SVC). Assuming the duration of a heartbeat cycle is 1 s, which corresponds to a heart rate of 60 beats per minute, a typical adult (resting) heart rate. Heart geometry frames originated from DICOM file, which are obtained from CT scan. By inserting the set of DICOM files in Simpleware ScanIP software (480 slices for 1 frame), the 100 heart frames (geometries) can be generated, as shown in Fig. 2.

The scripting language available in Simpleware ScanIP software makes possible to automatize the process of generating 100 heart geometries. Since the shape of heart is very complex and includes many tiny features such as: gaps, overlaps, narrow regions, etc. They could possibly cause some problems in mesh generation and it is essential to check and fix such created geometries before CFD modelling takes place. By using the described workflow the heartbeat frames (geometries) representing the heartbeat cycle can be automatically extracted, discretized and simulated in the time dependent way.

3 Heart Blood Flow Governing Equations

The heart blood flow can be modelled and considered to be incompressible and isothermal. The numerical flow simulation solves the full Navier-Stokes equations for the incompressible fluid. The following equations are integrated.

The continuity equation:

$$\nabla \cdot \boldsymbol{V} = 0 \tag{1}$$

where $\boldsymbol{V}$ is the flow velocity.

The momentum equation:

$$\frac{\partial \boldsymbol{V}}{\partial t} + (\boldsymbol{V} \cdot \nabla)\boldsymbol{V} = -\frac{1}{\rho}\nabla p + \frac{1}{\rho}\nabla \cdot \hat{\boldsymbol{\tau}}, \tag{2}$$

$$\hat{\boldsymbol{\tau}} = (\mu + \mu_t)2\hat{\boldsymbol{S}}, \quad S_{ij} = \frac{1}{2}\left(\frac{\partial V_i}{\partial x_j} + \frac{\partial V_j}{\partial x_i}\right)$$

where ρ is fluid density, t is time, p is static pressure, $\hat{\boldsymbol{\tau}}$ is shear stress tensor, $\hat{\boldsymbol{S}}$ is deformation rate tensor, μ is dynamic coefficient of blood viscosity.

The red blood cells have the unique mechanical behaviour, which can be defined in terms of erythrocyte deformability and erythrocyte aggregation [7], and because of that fact, the blood behaves like a non-Newtonian fluid. However, it can be modelled as a Newtonian fluid, when the arteries diameter is larger than 1 mm [8]. The diameter of the heart ventricle is much bigger than 1 mm. Therefore it can be modelled as Newtonian fluid to simulate the blood flow inside the heart. The considered blood viscosity was 0.0035 kg/(m s) (i.e. 3.5 mPa s) and used in the performed calculations.

The main blood vessels are relatively wide, the blood velocity in such vessels is high enough and results in the high Reynolds number value. It means that the blood flow is turbulent. For this reason, the turbulent eddy viscosity μ_t is introduced into the momentum equation. This approach is traditionally called URANS (Unsteady Reynolds-Averaged Navier-Stokes Equations)—see, e.g. [9]. Different turbulence models follow different ways to compute the eddy shear stress tensor—see [9, 10]. The 'standard' k-ε turbulence model [10] is used in the current study. This model is traditionally used to simulate the blood macro-circulation [11]. Within this model, the turbulent eddy viscosity is computed as follows:

$$\mu_t = C_\mu \rho \frac{k^2}{\varepsilon} \tag{3}$$

where k is turbulent energy, ε is the dissipation rate of turbulent energy.

To obtain solution for these two quantities, the following convection-diffusion equations are integrated in space and time:

$$\frac{\partial k}{d} + \nabla \cdot (\boldsymbol{V}k) = \frac{1}{\rho}\nabla \cdot \left(\left(\mu + \frac{\mu_t}{\sigma_k}\right)\nabla k\right) + \frac{\mu_t}{\rho}S - \varepsilon, \tag{4}$$

$$\frac{\partial \varepsilon}{\partial t} + \nabla \cdot (\boldsymbol{V}\varepsilon) = \frac{1}{\rho}\nabla \cdot \left(\left(\mu + \frac{\mu_t}{\sigma_\varepsilon}\right)\nabla \varepsilon\right) + \frac{\varepsilon}{k}\left(C_{s1}\frac{\mu_t}{\rho}S - C_{s2}\varepsilon\right), \tag{5}$$

$$S = 2\sum_{i,j} S_{ij}S_{ij} = \sum_{i,j}\left(\frac{\partial V_i}{\partial x_j} + \frac{\partial V_j}{\partial x_i}\right)\frac{\partial V_j}{\partial x_i},$$

the model constants being as follows:

$$C_\mu = 0.09, \sigma_k = 1.0, \sigma_\varepsilon = 1.3, C_{\varepsilon 1} = 1.44, C_{\varepsilon 2} = 1.92.$$

The deformation of the flexible hart walls is simulated in Abaqus/Explicit. The governing equation of the deformed structure within the finite-element approach (FE) is as follows:

$$\boldsymbol{M}\frac{d^2\boldsymbol{u}}{dt^2} = \boldsymbol{P} + \boldsymbol{P}_{hd} - \boldsymbol{I} \tag{6}$$

where $\boldsymbol{M}$ is the mass matrix attributed to a finite-element node, $\boldsymbol{u}$ is the vector of the node displacement. $\boldsymbol{P}$ is non-hydrodynamic force acting on the node, $\boldsymbol{P}_{hd}$ is the hydrodynamic force exerted onto the node, $\boldsymbol{I}$ is the internal force due to the structure resistance.

The nodal hydrodynamic force is computed using the values of the forces exerted onto the facets surrounding the node:

$$\boldsymbol{P}_{f,wall} = p_{wall}\boldsymbol{n} - \widehat{\boldsymbol{\tau}}_{wall} \cdot \boldsymbol{n}$$

4 Dynamic Mesh Adaptation

The internal surface of a living heart has extremely complex shape. This makes the generation of ordinary curvilinear meshes impossible and the application of the Lagrange approach is necessary in order to allow taking into account the motion of the heart walls. IN order to solve it FlowVision is applied as it has a fully automated grid generator based on the Sub-Grid Geometry Resolution method (SGGR) [12]. In a sense, this method can be regarded as an advanced analogue of the well-known cut-cell method. The algorithm can be briefly described as follows. First, the FlowVision grid generator builds a Cartesian initial mesh in the computational domain. In doing so, irregular distributions of coordinate planes are specified in the X, Y, and Z directions. Then, the hexahedral cells are cut by the complex triangulated boundaries of the domain—see Fig. 3.

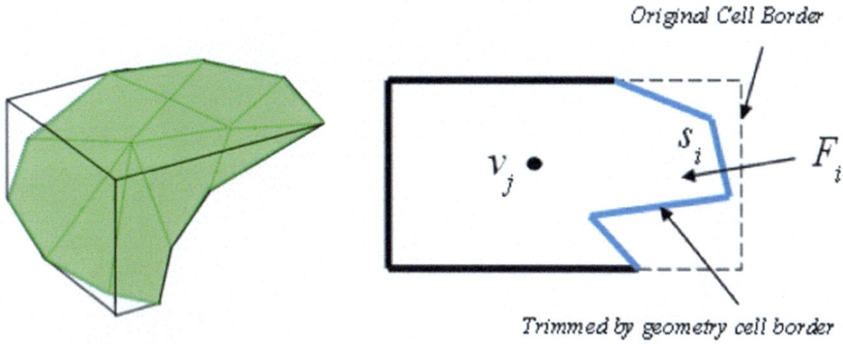

Fig. 3 Sub-Grid Geometry Resolution (SGGR) method

The boundary cells truncation is performed in a natural way, i.e. the cells adjacent to the boundaries become polyhedrons of the arbitrary shape. The governing equations are approximated by a high-order numerical scheme in all these cells. In the regions of high gradients of the sought-for variables (blood velocity and pressure) and around fine elements of the heart surface, the mesh is automatically adapted (refined) in order to reach the required accuracy required for accurate numerical results. In FlowVision, adaptation by one level means division of a hexahedral cell (parent cell) in 8 equal cells. The child cells can be split further forming an "octree" of cells, which are dynamically stored in the computer memory during the computation. The cell-adaptation parameter has to be provided as input through the FlowVision interface. A user can specify this adaptation parameter in several ways: (a) to a surface (for which a particular boundary condition is specified), (b) inside an object (a FlowVision geometric primitive like box, cone, cylinder, sphere …), as well as (c) adaptation to solution (to the value of a selected variable or to the gradient of the variable). When the adaptation to a surface is required, several layers of cells with different sizes can be specified near the surface in order to fulfil the adaptation criteria. The adaptation parameter to drive the solution convergence is dynamic: the mesh is periodically rebuilt in the predefined regions by following the evolution of the flow field. A user defines the maximum number of cells, which are generated, when cells are split and merged during the process of dynamic mesh adaptation.

The motion of the heart walls is modelled with the Euler approach. This approach assumes that the walls move over a stationary mesh. The computational cells are created, deleted and are able to change shape in order to adapt to the near wall in a best possible way. The approximation of the governing equations allows that these cells change their volume. Since the blood flow is unsteady, the capability of the FlowVision grid generator to rebuild the computational mesh automatically by following the motion of the boundaries and taking into account the variations of pressure and velocity becomes extremely valuable.

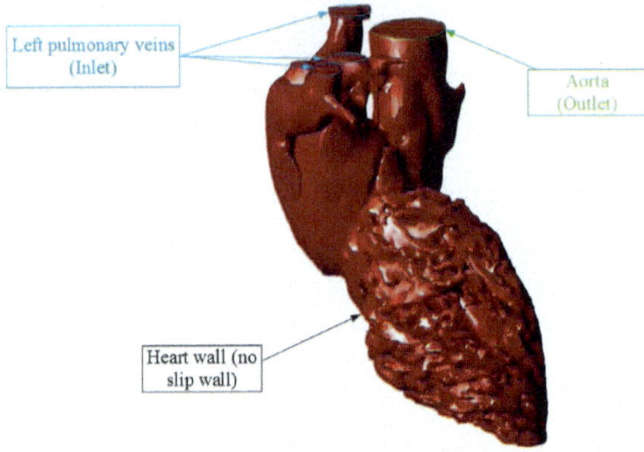

Fig. 4 Boundaries of the heart computational domain

The values of specific mass flow rate are stored on the cell faces. The values of the other sought-for variables are stored in the cell canters.

The FlowVision dynamic mesh adaptation capability uses the mesh convergence test. The simulation of the blood flow during heart beating is performed taking into account several mesh adaptation criteria. In this work the dependence of the flow characteristics on the mesh refinement level is investigated.

The initial number of cells in X, Y and Z direction has been set respectively to 90, 83, and 92. Two planes are created as inlet and outlet boundaries, as shown in Fig. 4, combined with the heart volume, and then a closed simply connected domain forms the volume of control. The volume adaption level number 1 is used to refine the computational domain mesh, and the surface adaption level 3 with 2 cells for the heart surface. Since this is an internal flow problem and the irregular heart shape, a cuboid domain (Fig. 5) is introduced at the start to create the initial mesh.

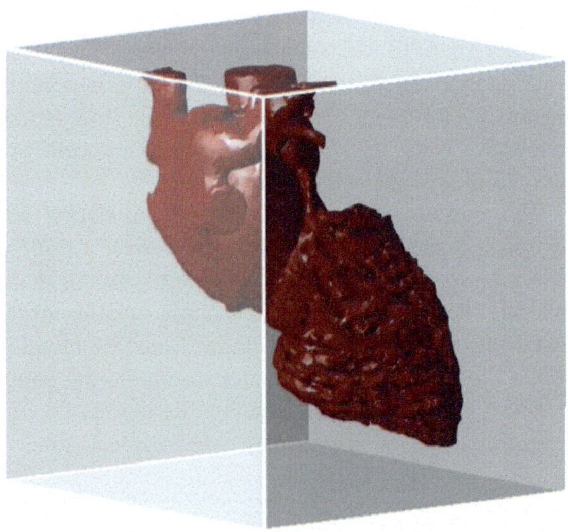

Fig. 5 Cuboid domain to initialised the mesh generation

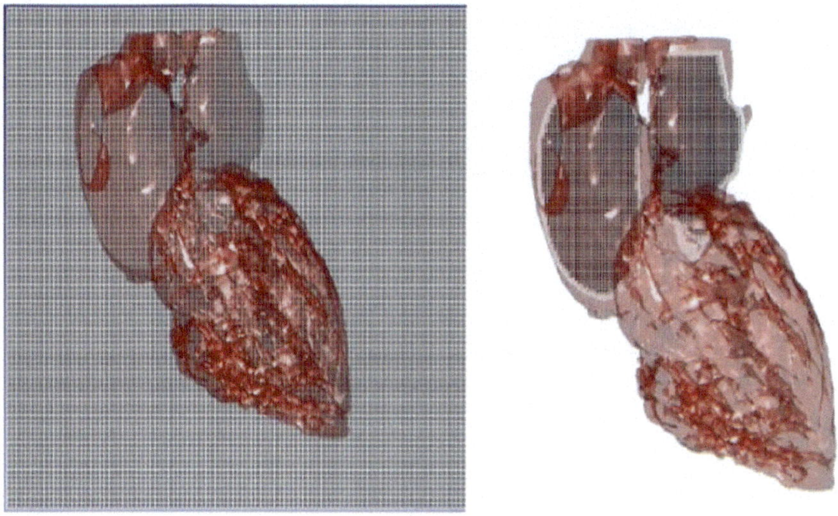

Fig. 6 Mesh applied for heartbeat simulation

The heart shape is changing during the heartbeat process, requiring the reconstruction of the mesh, which is done dynamically in real-time. The total number of cells is also dynamically changing in time and in the presented example it is ranging from 2.0 to 3.2 million cells, as shown in Fig. 6.

5 Numerical Modelling

It is well known that for the internal flow of heartbeat cycle the blood properties have a big influence on the obtained results. The blood viscosity is determined by plasma viscosity, haematocrit (volume fraction of red blood cell, which constitutes 99.9% of the cellular elements) and mechanical properties of red blood cells. The red blood cells have a unique mechanical behaviour, which can be discussed under the terms erythrocyte deformability and erythrocyte aggregation [7]. Because of that, blood behaves as a non-Newtonian fluid. However, it is reasonable to regard it as a Newtonian fluid, when modelling arteries, which diameter is larger than 1 mm [8]. By considering the size of the heart ventricle, which is much bigger than 1 mm, it is a good approximation to apply Newtonian fluid to model the blood flow inside the heart. The assumed blood viscosity is 0.0035 kg/(m s) (i.e. 3.5 mPa s), as shown in Fig. 7, the blood can meet the Newtonian fluid requirement in rheology of human blood. The main parameters applied for modelling the blood are summarized in Table 1.

The numerical setup, listed in Table 2, is used for solving the URANS equations in FlowVision, for simulating the heartbeat process. The blood flow is considered to

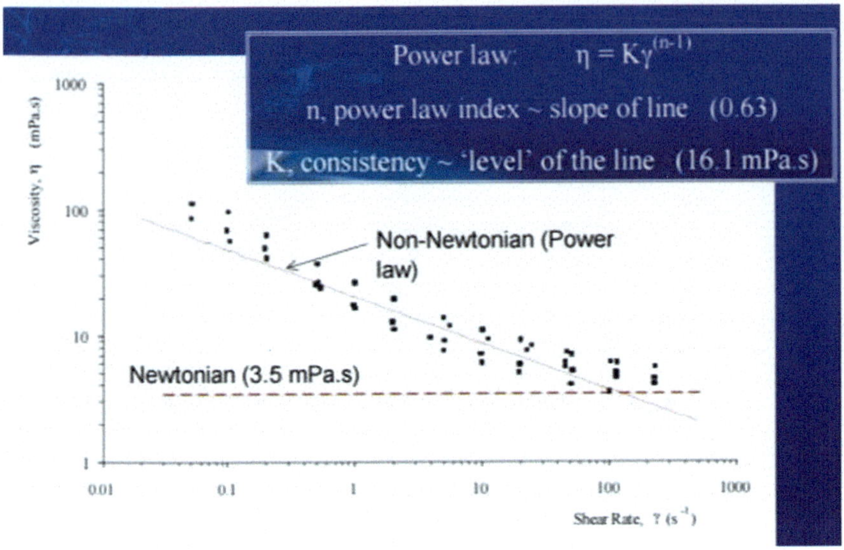

Fig. 7 Rheology of human blood [7]

Table 1 Parameters applied for blood

Symbol	Physical quantity	Value
μ	Viscosity	0.0035 kg/(m s)
ρ	Density	1035 kg/m^3
Cp	Specific heat	3470 J/(kg K)

Table 2 Numerical setup applied in FlowVision

Numerical setup	Quantity	Value
Inlet	Type normal mass velocity	Inlet time-variant, kg/(m^2 s)
Outlet	Type pressure	Outlet Time-variant, Pa
Initial flow field	Velocity pressure	0 m/s 10666 Pa
Physical processes	Heat transfer turbulence model	None KES (standard k-ε model)
Numerical scheme	Implicit/Explicit time step	Implicit new 0.001 s

be at low speed, thus for such internal flow it is reasonable to assume the flow model to be incompressible and without heat transfer, while the applied turbulence model is standard k-ε turbulence model (KES). In the macro-circulation, the blood vessels are relatively large and the fluid velocity is relatively high. Therefore, the standard k-ε turbulence model is commonly used to simulate the flow field under such condition [11].

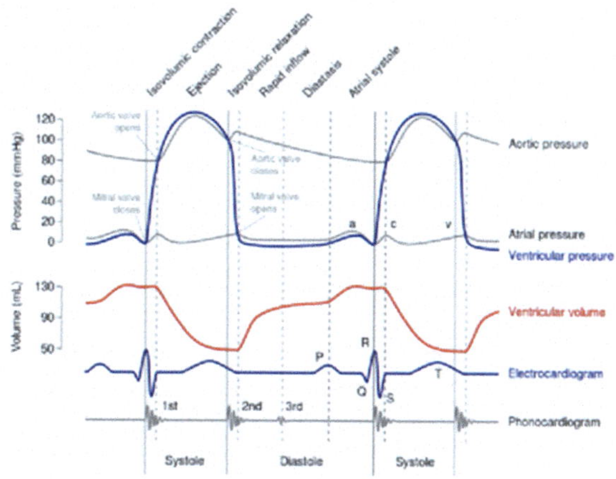

Fig. 8 Wiggers diagram

The time-dependent boundary conditions are extracted from Wiggers Diagram (Fig. 8), which is a very standard diagram used for the cardiac physiology. The cardiac cycle is the sequence of events that occurs during the heartbeats. There are 2 phases in the cardiac cycle. The first one is the diastole phase, when the heart ventricles are relaxed and the heart fills with blood. In the second phase, the systole phase, the ventricles contract and pump blood to the arteries. One cardiac cycle is completed when the heart, which has been filled with blood, pumps out of this blood to the arteries.

The specific mass flow rate (normal mass velocity) at the inlet is shown in Fig. 9a. The aortic pressure is shown in Fig. 9b, which can be specified as the static pressure at the outlet. It becomes clear that the time-dependent boundary conditions have to be specified at the inlet and at the outlet. For this purpose, graphical pressure data shown in Fig. 8 are converted into the continuous piecewise-linear functions (with a small time step) and assigned to the surfaces of the inlet and outlet respectively.

Time step independency has been pursed for the performed unsteady simulations. Consequently, different time steps (0.01, 0.005, 0.002, 0.001 and 0.0005 s) have been tested to ensure the time independent results. When the time step is

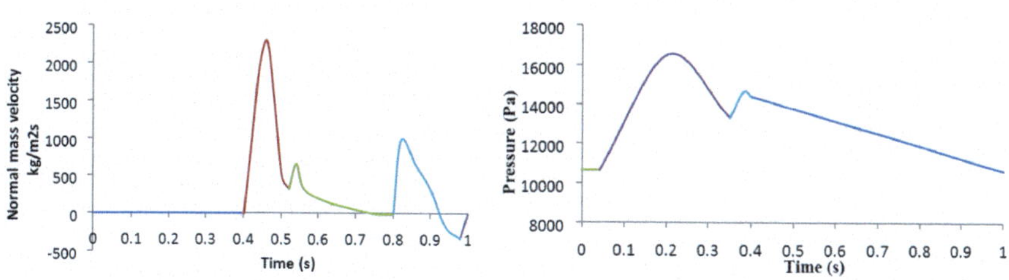

(a) Normal mass velocity at the inlet

(b) Static pressure at outlet

Fig. 9 Flow characteristics used for setting the boundary conditions

bigger than 0.001 s, the simulations are unstable and diverge approximately after 0.4 s (as the inlet mass velocity becomes too high). When time step is equal to 0.001 and 0.0005 s, the calculations are stable. The results obtained with these time steps produce quite close results. The time step 0.001 s was finally selected, since it provides reliable results and cut-down the calculation time.

6 Dynamic Simulation with Geometry Automatic Replacement

The heartbeat process was simulated with the time-variant heart shape applying 100 heart frames from CT scan images to generate different stage of heartbeat circulation, taking into account even the smallest geometric features of a real individual human heart. The transient heartbeat process is simulated by solving URANS Eqs. (1)–(5) in which the heart-shape frames are replaced, one after another, in accordance with the pre-defined time step.

It is obvious that too many heart frames geometries are inappropriate to be replaced manually. Thus, the command lines script was implemented for batch mode processing (see Fig. 10a) to enable the automatic geometry replacement. Such approach makes possible to simulate 100 heart frames, one after another in chronological order, and ensures the heartbeat simulation is performed smoothly, and in such way increases the robustness of calculation, which might be effected by

(a)

```
SS_PROJECTLOAD</Heartbeat/Heartbeat.fvproj><><><>

SS_APPLYNEWGEOMETRY<OBJECT[66]></Heartbeat_Geom/frame1.wrl>
SS_SOLVESTART_EX<0><RI><10>
SS_DO_NOT_DISTURB
        .
        .
        .
SS_APPLYNEWGEOMETRY<OBJECT[66]></Heartbeat_Geom/frame100.wrl>
SS_SOLVESTART_EX<7><RI><10>
SS_DO_NOT_DISTURB

SS_PRJDATASAVE
SS_PROJECTUNLOAD
SS_SHUTDOWNSOLVER
```

(b)

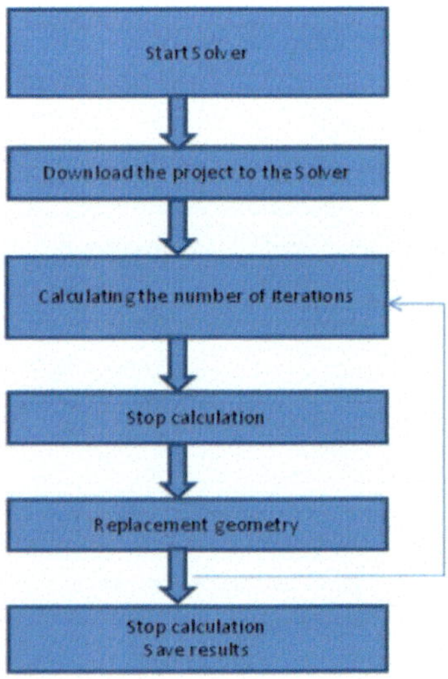

Fig. 10 Flow chart of the unsteady simulation (**a**) and automatic replacement in batch mode (**b**)

geometric data. Figure 10b illustrates the mechanism of unsteady simulation for which the automatic geometry replacement is done. There are 10 time steps performed between the intervals in which the 2 adjacent frames are replaced. It is important to note that before each frame replacement is performed, the computed results are automatically saved.

6.1 Simulations Result Analysis

The achieved dynamic simulation of the heartbeat process is taking into account one complete heartbeat. Figure 11 shows the convergence history of residuals, which are fluctuating in time. Each curve peak corresponds to a heart frame replacement, and it is visible that it soon re-converges in 10 time steps, followed with a new frame replacement cycle.

The dynamic simulation results reveals the main characteristics of the real heart internal flow, Figs. 12, 13 and 14 illustrate the internal flow behaviour at different time intervals. The general view shows that the flow is unsteady and has significant vortices, which makes more complex the flow mechanism analysis.

In Fig. 12, 6 sections are defined to show the pressure distribution within the flow field. There can be noticed that the pressure considerably over time, and in addition, even at a specific time interval; there are significant pressure gradients in-between different sections. Moreover, the pressure variation nearby the outlet region is in agreement with the tendency shown in Fig. 9b.

During the atrial systole period, the inside pressure is much bigger than during the diastole period, especially around the aorta area, the pressure is relatively small due to the high flow velocity. Similar to the pressure, the blood flow speed reaches

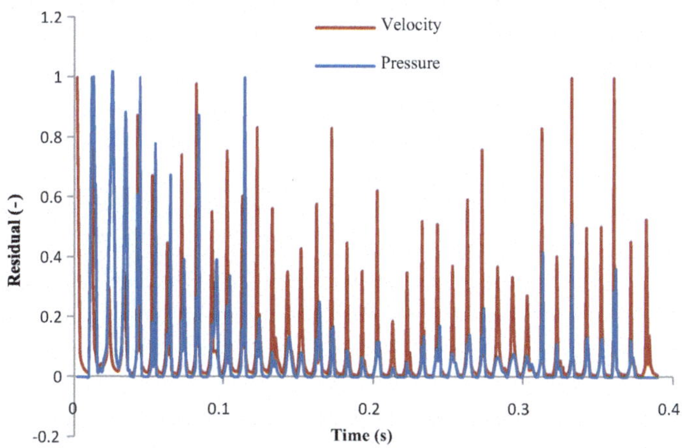

Fig. 11 Convergence history

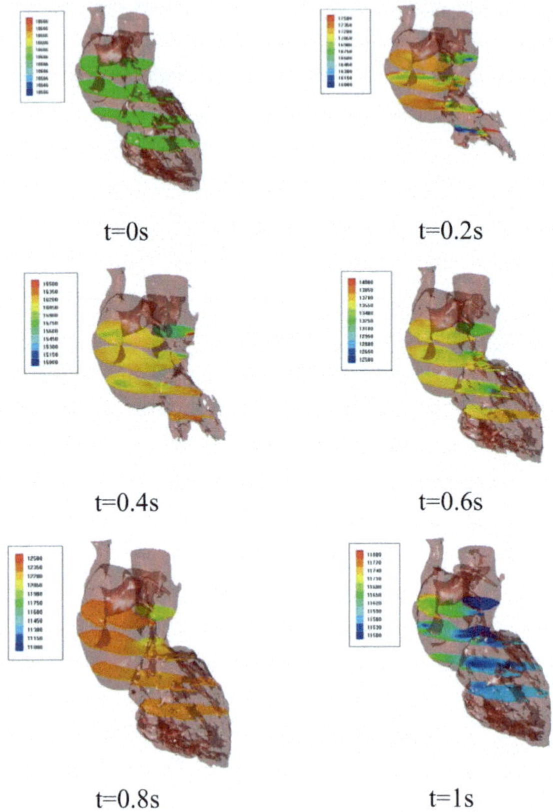

Fig. 12 The pressure distribution on section planes at different time intervals

the maximum approximately at t = 0.2 s, which corresponds to the peak pressure in Fig. 9b. In a cardiac cycle, the blood flow speed is extremely slow (see Figs. 12 and 13), which magnitude is always ranging from 0 to 1 m/s. The average velocity during systole is also bigger than diastole period, and the peak velocity at different stages is in agreement with the results found in the reference literature [13]. Generally, in the human body, the blood flow is laminar, as shown in Fig. 13 (a) t = 0 s (b) t = 0.2 s However, under the conditions of the high velocity flow, particularly in the region of the ascending aorta, the laminar flow can be disrupted and becomes turbulent, as shown in Fig. 14. In a cardiac cycle, the turbulence viscosity in the aorta is always much bigger than the viscosity of the laminar flow. Even, in most of the other heart parts, the blood flow is turbulent. Thus, the internal flow within the human heart can be regarded to be fully turbulent.

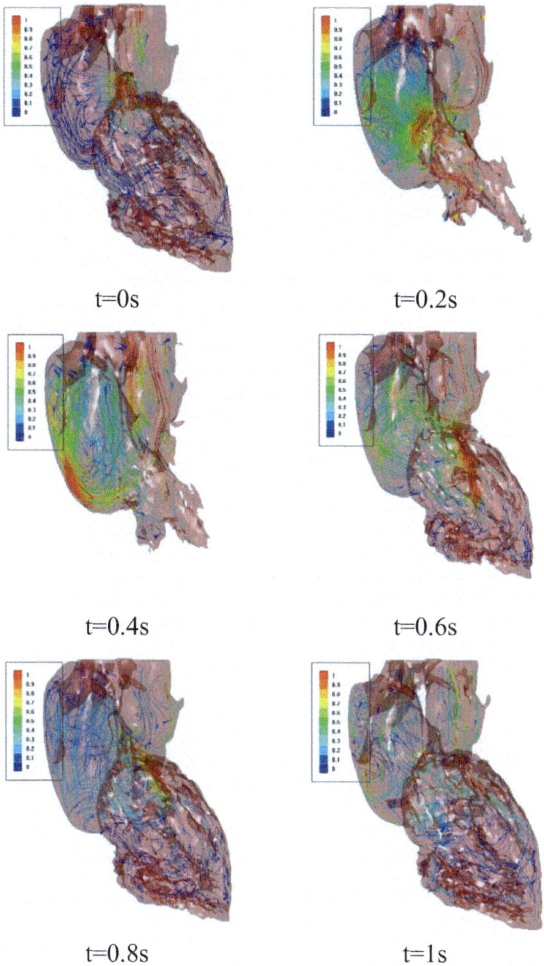

Fig. 13 3D streamlines inside the heart

7 SIMULIA Abaqus Living Human Heart Model

The Finite Element (FE) Living Human Heart Model (LHHM) is a high fidelity representation of a normal (healthy) 4-chamber adult male human heart. It includes the well-defined anatomic details of the atria, ventricles, heart valves and proximal vasculature, such as the aortic arch, pulmonary artery, and superior vena cava (Fig. 15). The dynamic response of this heart model is governed by a realistic representation of the electrical and mechanical properties. An initial implicit analysis computes the electrical potential distribution in space and time, and a subsequent explicit analysis computes the mechanical displacement by using the electric

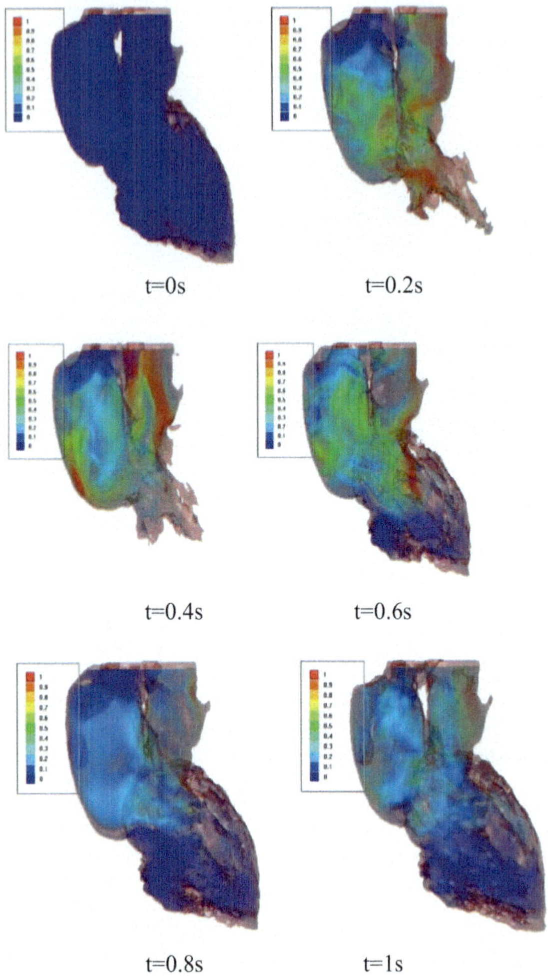

Fig. 14 Volume visualization of turbulence viscosity

potentials, as the excitation sources. In the basic (native) model, the blood flow is modelled by a system of fluid chambers and fluid links (representing the heart valves and extra-cardiac circulation), and driven by the volume changes computed from the mechanical analysis. This lumped parameter approach is computationally efficient and able to capture the average dynamic quantities, such as chamber pressures and volumes; as such, it has been successfully applied in several applications, where in vivo mechanical effects are dominant, such as extending the fatigue life of pacemaker leads and coronary stents. However, in some applications, it is also necessary to understand in vivo localized hemodynamic quantities, such as flow velocity and wall shear stress. To examine these effects, the network blood flow model in SIMULIA Abaqus LHHM is replaced with a CFD 3D blood flow

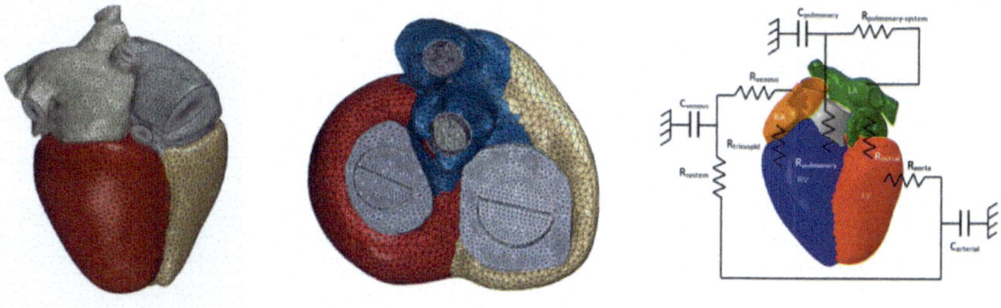

Fig. 15 LHHM—external geometry (left), mechanical valves (middle), schematic representation of the blood flow model (right)

model with coupling Abaqus with the FlowVision CFD solver, to enable the fluid structure interaction simulations.

The heart movement is defined from the electrical analysis and the resulting electric potentials are used as the excitation source for a subsequent mechanical analysis. It is important to mention that the electrical and mechanical analysis is using the same mesh topology. The electrical analysis begins at 70% ventricular diastole within the cardiac cycle. The electrical analysis is performed by using the Heart Transfer procedure in Abaqus/Standard, where the electrical analysis time step is 500 ms. The mechanical analysis contains multiple steps of Dynamic, Explicit types, which are: Pre-load, Beat1 and Recovery1. Once the Pre-load step is achieved, the approximated pre-stressed state of the heart is linearly ramping up the pressure in the chambers, where the time step is 0.3 s. For Beat1, which is atrial and ventricular contraction phase of the cardiac cycle, during which the voltages resulting from the electrical analysis are used, with the time step of 0.5 s for the Recovery1 step, which is the cardiac relaxation and for the ventricle filling phase, the time step is 0.5 s.

One heart cycle contains 2 phases: (1) diastole (ventricle filling) and (2) systole (ventricle contraction). Atrium is filled with the fresh blood during diastole. The atria and ventricles of the heart relax and begin to be filled up with blood. In the follow up the atria relaxes, and further on, the heart's ventricles contract (ventricular systole) and pump the blood out of the heart. At the end of diastole, the heart's atria contract (atrial systole) and pump the blood into the ventricles. The LHHM does not contain the blood volume entity.

8 CFD Surface Boundaries

In order to simulate the heart bit cycle with the blood flow the wetted surface needs to be extracted from the SIMULIA/Abaqus LHHM, shown in Fig. 16. The extracted blood flow volume is automatically inspected by checking its consistency. Examples are overlaps, intersections and gaps, which are eliminated in order to generate a watertight blood volume.

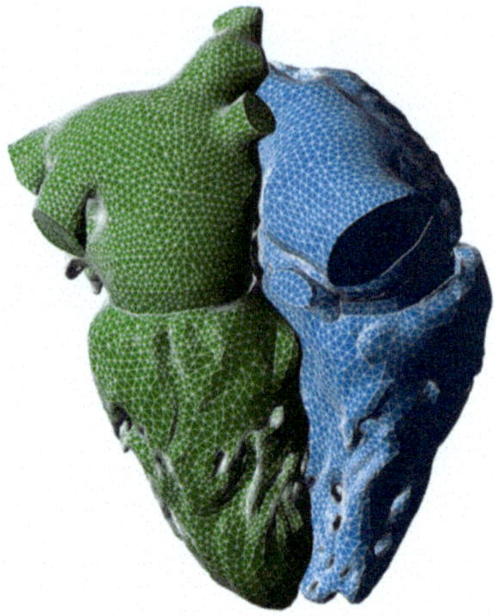

Fig. 16 Watertight blood volume

The wetted surface is bounding the watertight blood volume, representing the outer boundary interface used in the FSI simulation.

9 Mechanical Valves Modelling

The Abaqus LHHM consists of 4 mechanical valves as shown in Fig. 17, which is adding more complexity to the simulation. A unique capability of the FlowVision CFD software allows adding valves as separate bodies and to simulate their movement in CFD domain according to their space displacement as computed in FEA.

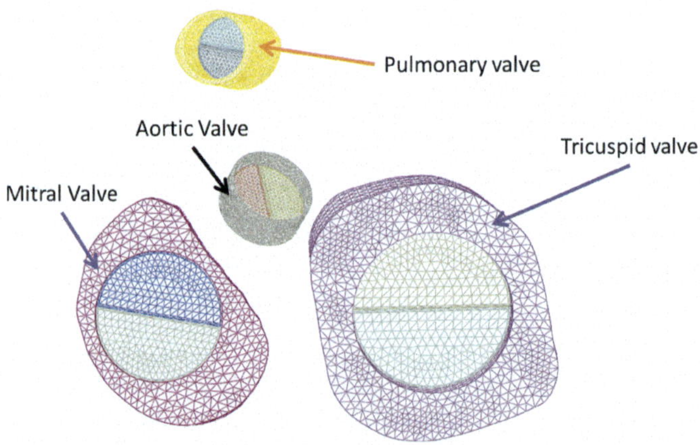

Fig. 17 Abaqus LHH-FEM—Mechanical valves

Fig. 18 Valves kinematic schema

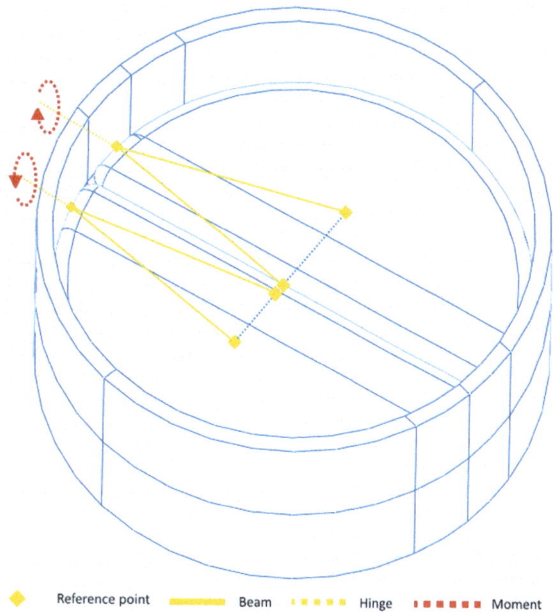

The valve kinematic is based on a simple schema, as shown in Fig. 18, where the moment is applied to the beam and the hinge helps to open the valve leaf.

For each valve, their behaviour depends on the specific heart simulation stage, such as systole or diastole, where the adequate movement of the valves leafs needs to be considered. The valves movement kinematics is setup in Abaqus, as shown in Fig. 19.

By using the FlowVision unique capabilities, the mechanical valves are placed into the CFD surface as the individual object called moving body. This approach enables to transfer the valve movement into CFD calculation domain during the simulation and calculate the fluid interaction between those moving objects. The CFD surface with the mechanical valves is shown in Fig. 20.

10 CFD Grid Generation

The initial number of cells in X, Y and Z direction are respectively X = 100, Y = 100, and Z = 100. The planar surfaces are created as follows: the inlet and outlet boundaries, which are combined with the CFD surface enabling the formation of the control volume. The volume adaption level 1 is used to refine the mesh in the computational domain.

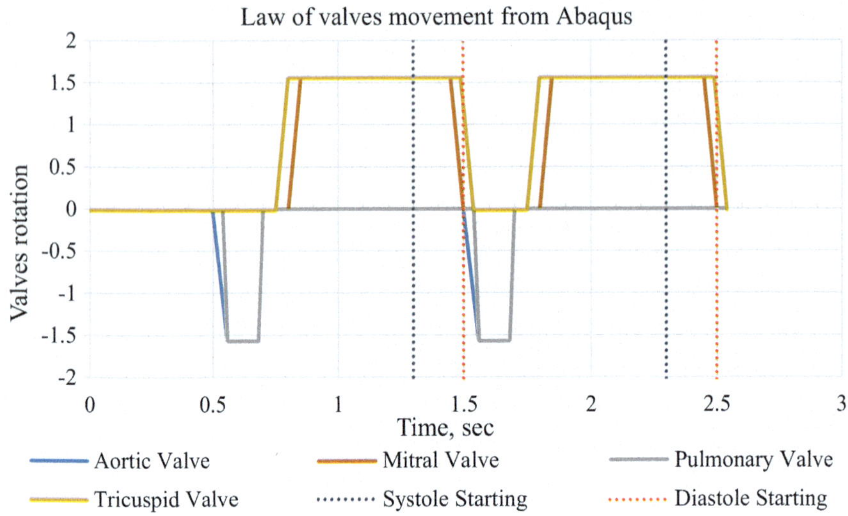

Fig. 19 Valves movement settings in Abaqus

Fig. 20 CFD surface with mechanical valves setup

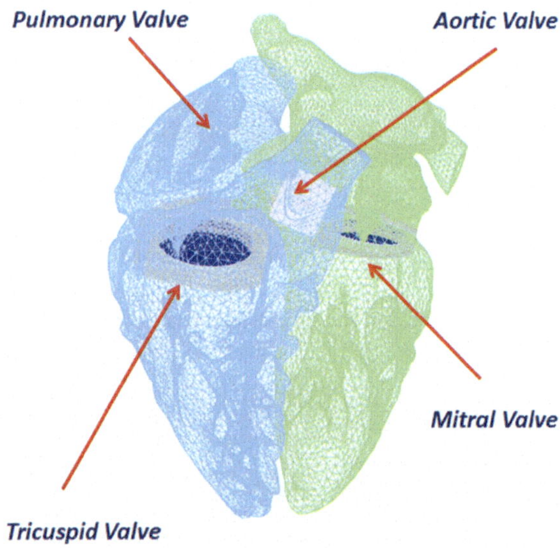

This results in a relatively coarse mesh. Since the heart shape changes during the heartbeat process, the mesh is dynamically rebuilt in real time. In this case, 177500 mesh cells were created in the computational domain, for which the initial mesh is shown in Fig. 21.

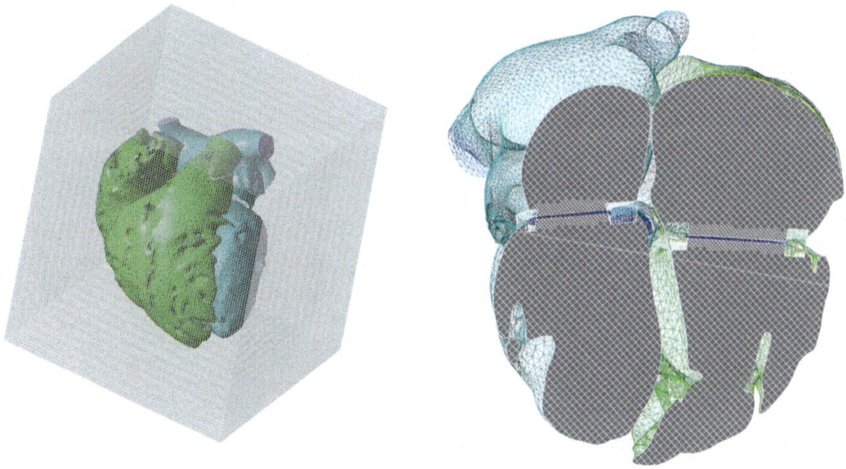

Fig. 21 Initial mesh

11 Fluid Structure Interaction

The FSI simulation takes advantage of the Abaqus co-simulation engine, by activating the by-pass option. This enables the mapping of the CFD surface to the FEM geometry by using the FlowVision mapper, as shown in Fig. 22. The cells of the fluid domain are intersected by the surface mesh, which encloses the heart internal blood volume. The resulting cells are not simplified and they are stored in the form of complex polyhedrons. The cell sides, which are part of the surface mesh, have links to the original faces of the finite-element mesh. By solving the fluid dynamics equations in the presence of such a grid, a natural data transfer mechanism allows the transfer of data, from the fluid to structural computational domain, without a need for any intermediate geometry interpolation algorithm.

The link with ABAQUS is made applying the user DLOAD function (VDLOAD for ABAQUS explicit) in order to transfer the loads calculated in FlowVision. The node positions are obtained by FlowVision via the ODB utilities.

FlowVision and Abaqus exchange information at predefined synchronization time instances T_n. The time interval between two such instances is called the *FSI time step*:

$$\theta^{n+1} = T^{n+1} - T^{n} \tag{7}$$

Both the codes can perform their own time steps (one or several) during the one FSI time step. An explicit split algorithm applied for the FSI calculations is discussed. The algorithm requires that the FSI time steps are small. However, it is efficient, since it excludes the iterative solution of the combined system resulting from the algebraic equations generated by the two codes.

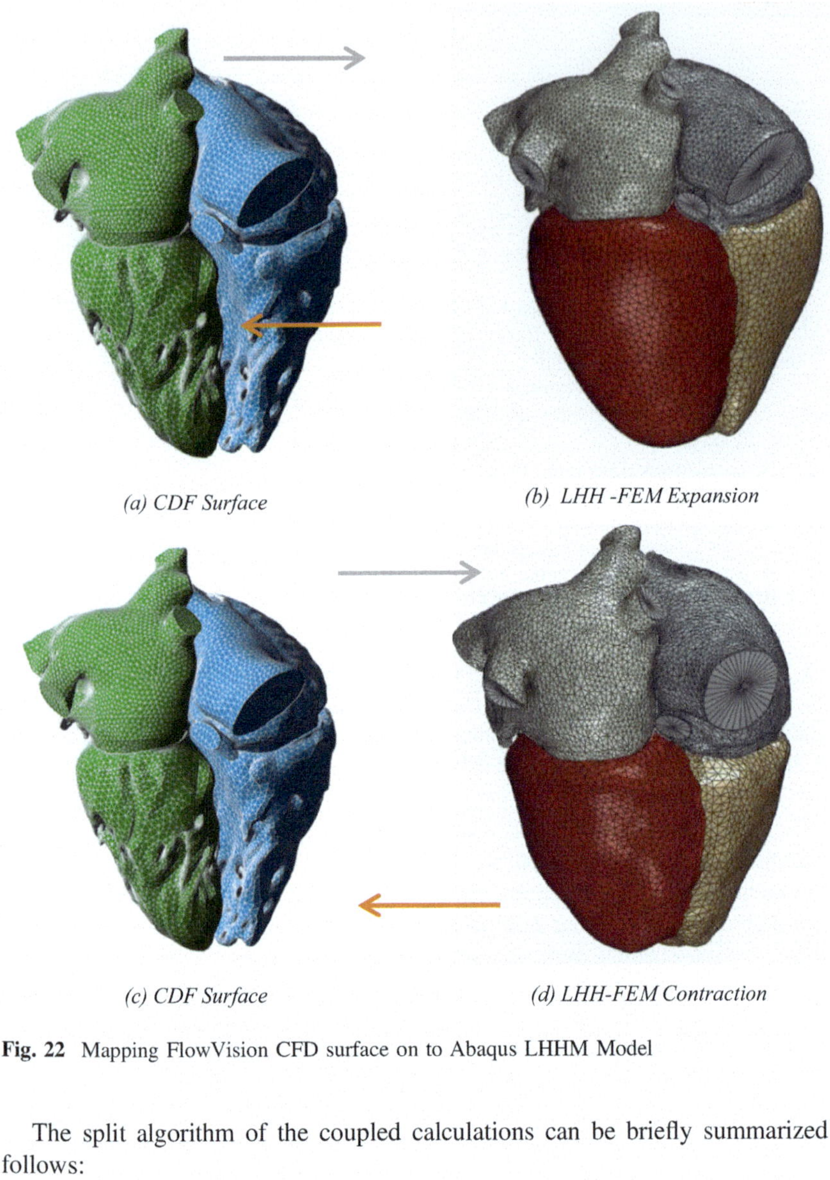

(a) CDF Surface (b) LHH -FEM Expansion

(c) CDF Surface (d) LHH-FEM Contraction

Fig. 22 Mapping FlowVision CFD surface on to Abaqus LHHM Model

The split algorithm of the coupled calculations can be briefly summarized as follows:

1. FlowVision calculates concentrated forces exerted onto the nodes of the FE mesh surface (loadings onto the ABAQUS nodes).
2. FlowVision calculates FSI time step θ^{n+1} on the basis of its own time step.
3. FlowVision transmits the concentrated forces at θ^{n+1} to ABAQUS.

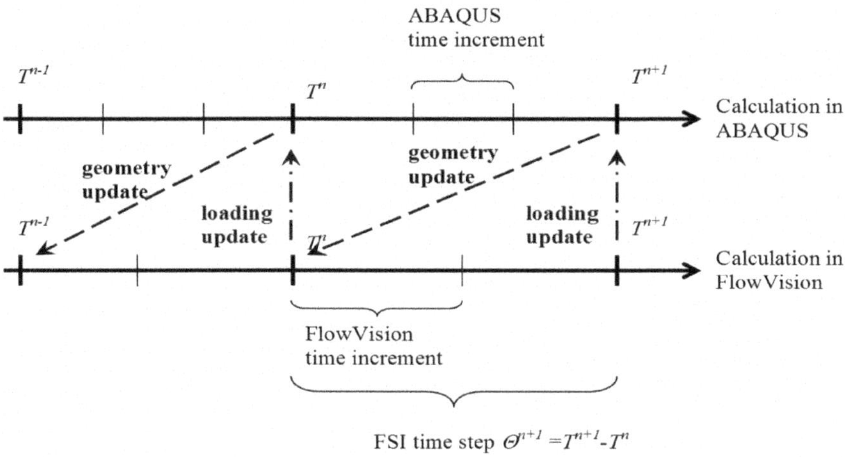

Fig. 23 Explicit coupling procedure of FSI calculations in FlowVision and Abacus

4. ABAQUS performs the FE calculations during θ^{n+1} by using the loadings obtained from FlowVision. The new locations of the surface mesh nodes are resulting from these calculations.
5. ABAQUS transmits the new surface mesh nodes to FlowVision.
6. FlowVision calculates the new concentrated forces and the new exchange time step (the cycle is repeated).

The explicit coupling procedure of FSI calculations in FlowVision and Abacus is shown in Fig. 23.

The momentum equation, as integrated in FlowVision, is modified for the cells adjacent to the structure walls. The momentum equation in a semi-discrete form for a near-wall cell i, is defined as follows:

$$\frac{\boldsymbol{V}_i^{n+1} - \boldsymbol{V}_i^n}{\Delta t} + \frac{1}{\Omega} \sum_{f=\text{faces}} \boldsymbol{V}_f^{n+1} \left(\boldsymbol{V}_f^n \cdot \boldsymbol{n} \right) \Delta S_f = -\frac{1}{\rho} \nabla p_i^n + \frac{1}{\Omega} \frac{1}{\rho} \sum_{f=faces}^{n} \boldsymbol{\tau}_{n,f}^n \cdot \Delta S_f. \quad (8)$$

The face f of the cell is a part of the corresponding wall facet, and the normal velocity of the face is modified as follows:

$$\left(\boldsymbol{V}_f^n \cdot \boldsymbol{n} \right)_{wall} \rightarrow \left(\boldsymbol{V}_f^n \cdot \boldsymbol{n} \right)_{wall} + \left(C + dt^2 M \right) \frac{dp}{\partial t}, \quad (9)$$

$$C = \frac{ds}{dp}, \quad (10)$$

$$M = \frac{\Delta S_f}{m_f}. \quad (11)$$

where *ds* is the displacement of the face centre, m_f is the mass associated to the face. The quantity *C* is called *flexibility*, which is characterized as the response of the structure to the pressure load onto the structure surface. The quantity *M* is called *mobility, which* is a local inertia property of the structure. The both quantities take into account the motion of the structure surface during the FSI time step. All these quantities are set upfront through the FlowVision interface. For the explained technology it is found to augment the stability of the calculations by supporting the split algorithm implementation, as described in this section.

12 Boundary Conditions

The parameters of the time-dependent boundary conditions are extracted from the Wiggers Diagram – see section D, where Fig. 24 shows the surfaces at which these boundary conditions have been specified.

13 Simulations Results Analysis

The cardiac hemodynamic is driven both (1) by the motion of the myocardium as well as (2) by the motions of the 4 heart valves. To simplify the problem, the initial FSI simulations are considered only for the myocardial motion, while neglecting the

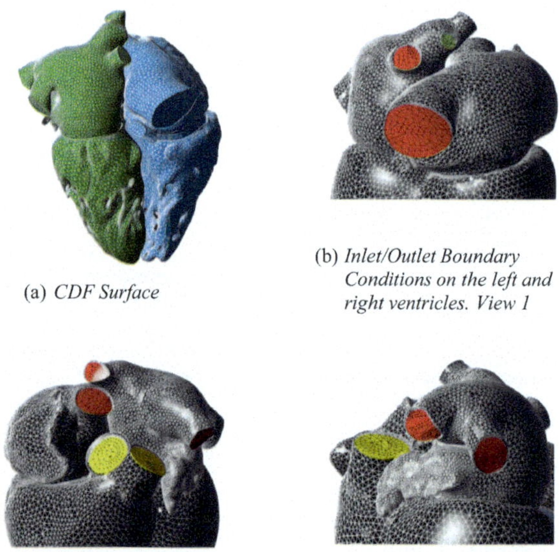

(a) *CDF Surface*

(b) *Inlet/Outlet Boundary Conditions on the left and right ventricles. View 1*

(c) *Inlet/Outlet Boundary Conditions on the left and right ventricles. View 2*

(d) *Inlet/Outlet Boundary Conditions on the left and right ventricles. View 3*

Fig. 24 Inlet and outlet boundary conditions

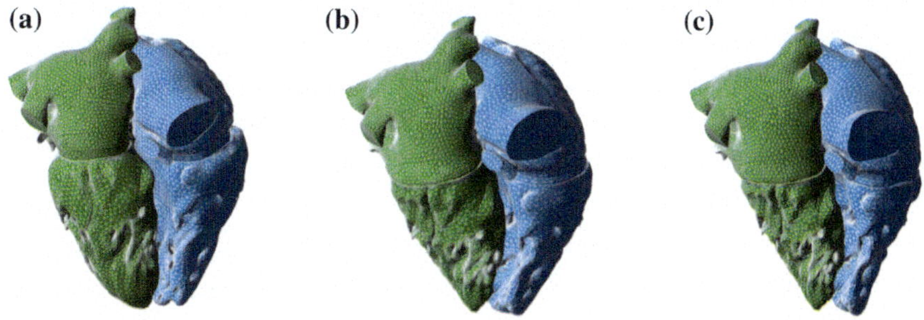

Fig. 25 Highest and lowest volume of the CFD mesh deformations at different points in time of the FSI simulation **a** t = 0.2 s, **b** t = 0.6 s, **c** t = 0.8 s

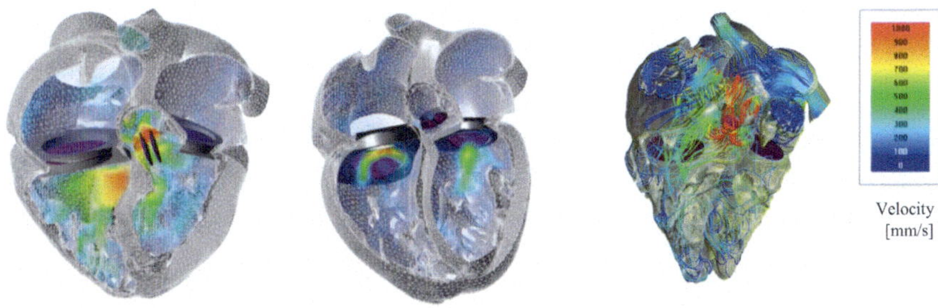

Fig. 26 Blood flow velocity in the FSI simulation

internal valves. The resulting CFD mesh deformations, for the predefined time instances of the cardiac cycle are shown in Fig. 25. Figure 26 shows the internal flow patterns at 3 time instances. In general, the flow field is unsteady and shows significant presence of vortices. The blood flow velocity reaches the maximum approximately at t = 0.2 s, which corresponds to the peak pressure. In a cardiac cycle, the blood flow velocity is extremely slow, ranging from 0 to 1 m/s. During systole, the average velocity is greater than in diastole period. The peak velocity at different stages matches the results published in the reference literature [11, 13].

It's also possible to get information about the flow rate through the aortic valve, as shown in Fig. 27.

14 Conclusion

The novel approach for simulation of the blood flow using the data from the MRI/ MRT scanners or Finite Element Living Heart Human Model together with the FlowVision CFD software is presented in the given article. The FlowVision capability to simulate the motion of the boundaries of the computational domain

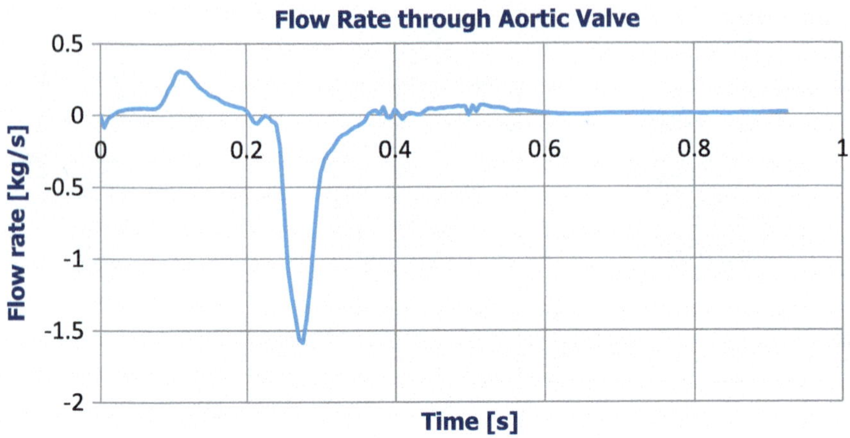

Fig. 27 Flow rate through the aortic valve

within the Euler approach enables one to account for the motion of the walls of a living heart. The automated Sub-Grid Geometry Resolution method implemented in FlowVision allows resolution of the flow patterns in boundary layers, in vortices, and around small elements of the considered geometrical model. FlowVision provides numerous tools for visualization of the distributions of the pressure, temperature and velocity of blood inside the heart. For this purpose, both planar and volume rendering techniques can be used. The presented technology is especially appealing for simulation of valves. The influence of mechanical cardiac valves on the internal blood flow has been investigated within the performed research. The article describes how the technical challenges have been resolved. The outcomes of the work prove that the combination of Abaqus FEM and FlowVision CFD enables solving an entire fluid- structure interaction problem. Applied together, these software represents a promising tool for the Cardio Vascular Disorders diagnosis. The developed technology will help cardiologists to find healing solutions for customized valves and individual hearts. Besides that, the technology can be used for improvement of the design of existing medical heart devices.

References

1. Kumar A (2014) Changing trends of cardiovascular risk factors among Indians: a review of emerging risks. Asian Pac J Trop Biomed 4(12):1001–1008
2. Cheng Y, Oertel H, Schenkel T (2005) Fluid-structure coupled CFD simulation of the left ventricular flow during filling phase. Ann Biomed Eng 33(5):567–576
3. Baccani B, Domenichini F, Pedrizzetti G (2003) Model and influence of mitral valve opening during the left ventricular filling. J Biomech 36:355–361
4. Keber R (2003) Computational fluid dynamics simulation of human left ventricular flow. PhD Dissertation, University of Karlsruhe, Karlsruhe

5. Long Q, Merrifield R, Yang GZ, Xu XY, Kilner PJ, Firman DN (2003) The influence of inflow boundary conditions on intra left ventricle flow predictions. J Biomech Eng 125: 922–927
6. Nakamura M, Wada S, Mikami T, Kitabatake A, Karino T (2002) A computational fluid mechanical study on the effects of opening and closing of the mitral orifice on a transmitral flow velocity profile and an early diastolic intraventricular flow. JSME Int J Ser C—Mech Syst Mach Elem Manufact 45:913–922
7. Baskurt OK, Meiselman HJ (2003) Blood rheology and hemodynamics. Seminars Thromb Haemost 29:435–450
8. Noutchie SCO (2005) Flow of a Newtonian fluid: the case of blood in large arteries. Master of Science, University of South Africa
9. Rodi W (1980) Turbulence models and their application in hydraulics, a state of the art review. Report. University of Karlsruhe, Karlsruhe, Federal Republic of Germany, 104p
10. Wilcox DC (1994) Turbulence modeling for CFD, DCWIndustries, Inc., 460p
11. Rubenstein D, Yin W, Mary D (2015) Frame, biofluid mechanics: an introduction to fluid mechanics, 2nd edn. Elsevier, London, p 466
12. Aksenov A, Dyadkin A, Pokhilko V (1998) Overcoming of barrier between CAD and CFD by modified finite volume method. In: Proceedings of 1998 ASME pressure vessels and piping division conference, San Diego, ASME PVP, vol 377-2, pp 79–86
13. Wexler L, Bergel DH, Gabe IT (1968) Velocity of blood flow in normal human venae cavae. Circ Res 23(3):349–359

A Review on Computer Vision Applied to Mechanical Tests in Search for Better Accuracy

Felipe de Oliveira Baldner, Pedro Bastos Costa, Juliana Freitas Santos Gomes and Fabiana Rodrigues Leta

Abstract Mechanical tests are used to determine a material's behaviour under certain conditions of stress. Computer Vision Systems have been used since the 1980s in mechanical tests, with techniques evolving in order to accommodate modern needs. This paper has the purpose of reviewing research that combines Computer Vision Systems and Mechanical Tests, analysing the main steps for better results from such application.

Keywords Computer vision · Mechanical tests · Image acquisition · Lighting conditions · Computer vision algorithms

1 Introduction

Mechanical tests have the purpose of determining the property of a material and how it behaves under certain conditions. Having knowledge about such behaviour is essential for the conformity of a manufactured piece, which will, thus, determine

---

F. de Oliveira Baldner
Catholic University of Petrópolis—UCP, R. Barão do Amazonas, 124—Centro, Petrópolis, Rio de Janeiro, Brazil
e-mail: felipe.baldner@ucp.br

P. B. Costa
Minas Gerais Federal University—UFMG, Av. Antônio Carlos 6627, Belo Horizonte, Minas Gerais, Brazil
e-mail: pedro@demec.ufmg.br

J. F. S. Gomes
National Institute of Metrology, Quality and Technology—Inmetro, Av. Nossa Senhora das Graças, 50, Duque de Caxias, Rio de Janeiro, Brazil
e-mail: jfgomes@inmetro.gov.br

F. R. Leta (✉)
Fluminense Federal University, R. Passo da Pátria, 156, Niterói, Rio de Janeiro, Brazil
e-mail: fabianaleta@id.uff.br

D. Vucinic et al. (eds.), *Advances in Visualization and Optimization Techniques for Multidisciplinary Research*, Lecture Notes in Mechanical Engineering, https://doi.org/10.1007/978-981-13-9806-3_9

the quality of the end product in which this piece is a component. The mechanical tests are classified by how they affect a material sample, either by stressing it until failure (destructive test), or by evaluating a material sample, without damaging it (non-destructive test).

From the 1980s, Computer Vision Systems (CVS) have been employed in mechanical tests in order to improve the results previously obtained by regular methods. Since then, computers have become increasingly cheaper and more powerful, enabling CVS' to be used in many more applications. This is a technique which replaces human response for a more efficient and unbiased result [1].

This chapter reviews publications from over 30 years ago in order to determine a timeline and evolution of the CVS applications. The mechanical tests characteristics reviewed in this chapter are: hardness, impact, residual stress, tensile, compression and torsion. After presenting their applications, the analysis is performed to highlight the most important aspects, which CVS covers.

2 Computer Vision for Hardness Testing

Hardness is a characteristic of solids regarding its resistance to deformation under compressive force. Some of the scales in which hardness is measures are the Brinell, Vickers and Rockwell scales. In the Rockwell scale, the hardness is a function of the depth of the indentation while, in the Brinell and Vickers scales, it's a function of the dimension of the indentations (which are, respectively, a sphere diameter and a pyramid with a square base).

Conventional equipment for measuring hardness relies on the application of force on a sharp object with a given shape. This will imprint on the material shape, which is then projected on a screen for measurement. However, in order to have more precision, instead of being measured on a projection with the regular dimensional measurement tools, those imprints are now viewed through microscopes with digital cameras, which imprint is digitalized into an image and enables measurement using computer vision techniques.

In order to extract the indentation shapes from the images, algorithms of segmentation and edge detections are commonly used [2–4]. Any image noise is removed by using the proper filtering and morphological operations of dilation and erosion. By comparing different methods for measuring those shapes, enables us to determine the most accurate one. Simple tests, involving geometric parameters such as perimeter, area, difference of coordinates are shown to be quite effective [2, 3]. More advanced techniques such as Edge detection algorithms [4] and the Hugh Transform [5] are also shown to improve on these results, in comparison with results from regular hardness measurements.

Departing from the Computer Vision algorithms, another point worth studying is the image focusing, in which several auto-focus techniques are analysed for proper image acquisition of the Vickers indentation [6].

Micro and Nano hardness techniques propose a challenge due to the diminished dimension of the imprint, and are also analysed by using several techniques, such as edge detection, vertex detection and shape matching [7].

3 Computer Vision for Residual Stress Testing

Residual stress is the stress present in a material when it is devoid of any external load or force, usually generated by fabrication processes and material treatment techniques. The methods for measuring residual stress are varied, some being destructive and others non-destructive. One of the destructive tests is the hole drilling technique, which tracks deformation before and after a hole is drilled on the material using the strain gauges.

Digital Image Correlation (DIC) and its variations [8–11] is the most established Computer Vision technique, usually paired with the hole drilling technique for residual stress measurement. Thus, instead of installing strain gauges on the material, image processing is used to detect the deformation by tracking a visual pattern on the surface of the material. In addition, it represents the robust DIC technique, being mostly insensitive to lighting variations and vibrations.

As an alternative, the colour analysis is proposed, by using a photo elastic coating, before the hole has been drilled and thus showing how this coat behaves after each drilling iteration [12].

4 Computer Vision for Impact Testing

Impact testing determines the energy absorbed by a material during fracture. The most common methods of impact testing are the Charpy and Izod tests. They both consist of the pendulum with known mass that is dropped from a known height and strikes a specimen at a given point. By analysing the conservation of energy, the impact energy can be inferred.

Computer Vision algorithms are mainly used for the analysis in the fracture area of a specimen after the Charpy impact test has been performed, providing the information on brittle, ductile and cleavage regions. Brightness [13, 14] and fractal analysis [15] are methods used to partition the fracture surface. Segmenting an image into several regions is based on the brightness difference [13] which is based on the idea that brittle fractures are brighter than ductile fractures. However, brightness is highly dependent on the lighting conditions, when the image is being acquired, and the resulting samples are considered harder to analyse under the same conditions and measuring parameters. Texture analysis can be used to overcome the problems with brightness analysis, since the texture is dependent on several image parameters that will all change along with different lighting settings [14]. By analysing the fractal dimensions of the impact surfaces, it is possible to determine the absorbed energy [15]. Shape detection and counting [16] is also used to determine what kind of inclusions are present in a material when analysing its fracture surface.

The measurement of the dynamic response of bending by means of a linear impact test can be achieved by using the ultra-high speed cameras along with DIC [17]. With an impact velocity of 28 m/s on the specimen, a total of 128 images of the bent specimen are acquired in 128 μs measuring the impact at the frame rate of 1,000,000 fps.

5 Computer Vision for Tensile, Compression and Torsion Testing

The tensile, compression and torsion tests all based on the same principle: a force (or a torque, in case of the torsion test) is applied to a specimen and its response is measured, usually to determine the maximum force (or maximum torsion) the material specimen can withstand. The standard measurements are done by load cells, extensometers or strain gauges.

Just like described for Residual Stress testing, Digital Image Correlation and its variations, which are already established as the standard Computer Vision technique for deformation measurement, we need going back to the early 1980s [18–21]. There are several advantages of the DIC technique over the standard methods. One of them measures multiple cracks, which are common in compression test of concrete blocks [22–24] and also on Pre-stressed concrete structures [25]. The lack of a strain gauge makes the DIC technique much more general, enabling them to measure displacements over the entire specimen [26], using as much as 4 cameras to be able to measure the specimen whole-field deformation [27]. As previously stated, in the residual stress testing, DIC is a very robust technique insensitive to vibrations generated by the test machines [28, 29].

The application of DIC has been expanded for the measurement of microscopic strains [30] and also for 3-dimensional strains [24, 25, 27, 31–37], greatly expanding the technique's usability. While most 3D measurements use 2 or more cameras [22, 24, 25, 27, 31, 33, 34, 37], this kind of measurement can also be achieved by rotating a single camera around the specimen, and acquiring images at 14 different positions [35]. To study the tensile response under high temperatures, a different kind of lighting and special camera filters are used as the black-body radiation affects the DIC's regular speckle pattern [36].

Even though DIC is widely used in this field, some authors have used different approaches. An observation of the light level on images enable the prediction of formation and size of the plastic zone in tensile specimens [38]. After a noise reduction step, the standard deviation of each pixel is calculated, leading to a determination of which pixels compose the plastic zone.

By drawing patterns on a geo-synthetic fibre and acquiring images following a fixed rate, it is possible to determine the displacement of those fibres while under strain [39]. Using the same concept of pattern tracking, a tensile specimen made of rubber also had its displacement determined under tensile load [40], as well as carbon and glass composites [41].

Braided composites are also analysed by both applying tensile force and torsion, comparing the pre-existing analytical models [34], with the results from the 3D DIC analysis.

Upon deformation, the higher concentration of strain in a small region of the specimen is called necking, which is observed in several applications [42–44], along with the displacement resulting from the testing.

6 Overall Analysis

While most of the cited papers do not have a metrological nature, the approach of an Image Metrology System (IMS) [1] is well suited to analyse how Computer Vision is applied to the aforementioned tests. A total of 58 papers were analysed in order to underpin this analysis.

An IMS is composed of the lighting conditions, the material being analysed, the image acquisition hardware and the image processing software, as can be seen in the block diagram of Fig. 1.

6.1 *Analysis by Mechanical Test Type*

From the 58 analysed papers, Computer Vision is the most established technique in Tensile tests, with Hardness and Residual Stress following. Charpy impact testing mostly uses Computer Vision after the test in order to determine material characteristics. The division of tests on the analysed papers is shown in Fig. 2.

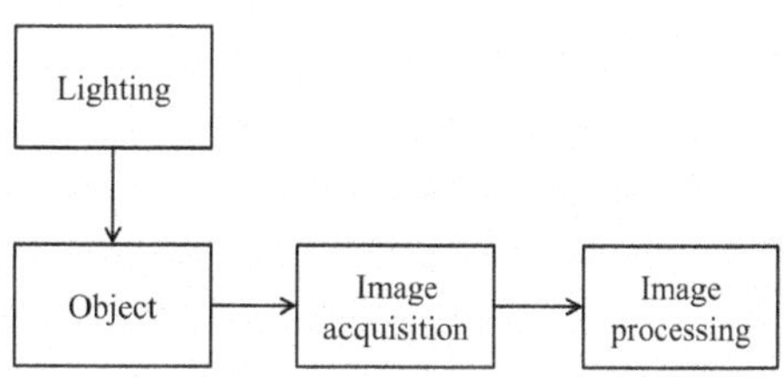

Fig. 1 Block diagram of an image metrology system (IMS)

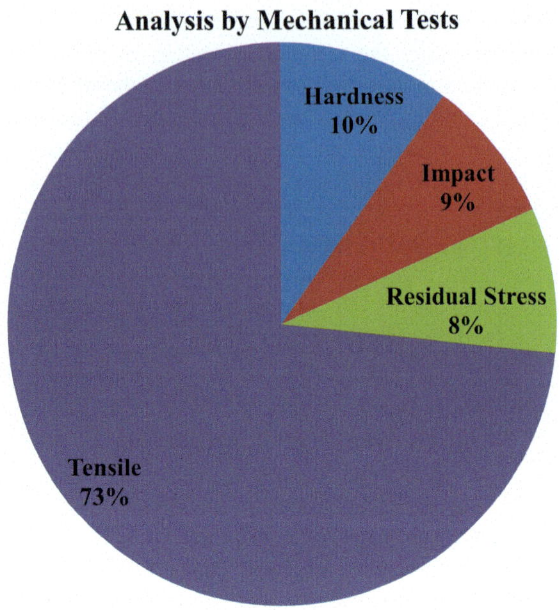

Fig. 2 Distribution of mechanical tests over reviewed papers

6.2 Analysis by Computer Vision Technique

The Computer Vision techniques used in the reviewed papers are very particular to the type of measurement done, which is completely related to the type of test. For instance, in Hardness testing, since the need is for a shape's dimension, algorithms for segmentation, edge detection and shape matching are the ones mostly used. Just as well as for the Tensile tests reviewed, in which most papers use the Digital Image Correlation (DIC) technique as a technique for linear measurement throughout several consecutive images. Figure 3 shows the percentage distribution of techniques applied in the reviewed papers.

6.3 Analysis by Lighting Conditions

In any kind of Computer Vision system, the lighting conditions greatly affect how a material is seen by the camera [45]. One of the biggest problems it may generate is regarding the dimension of the object [46], which in turn hinder the measurement.

While most papers (60%) have not described the lighting conditions or even use the lighting data of the measurement environment, from those who have described their lighting setups, fibre optics [31, 39] are addressing an interesting option due to the possibility of lighting it directionality and the ability to change a light source without intervening with the system. This choice was found in 21% of the papers.

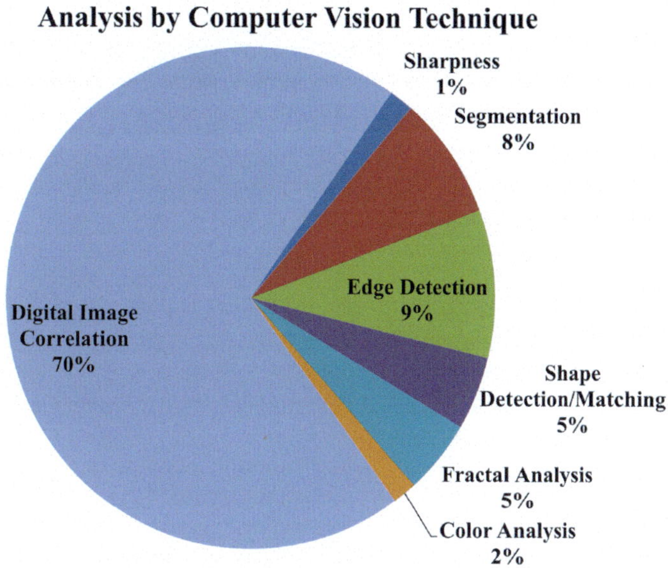

Fig. 3 Distribution of computer vision techniques over reviewed papers

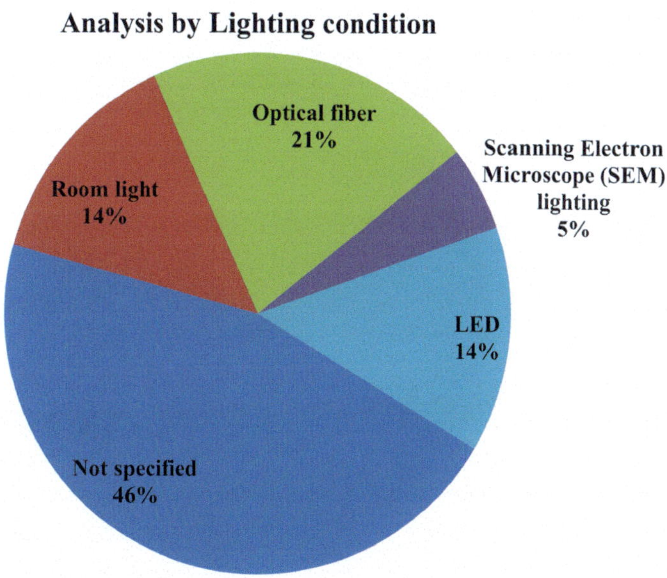

Fig. 4 Distribution of lighting conditions over reviewed papers

Those systems that employ a more specific device for the acquisition, such as the Scanning Electron Microscope, due to the measurement scale needed [13, 30, 47], end up relying on the microscope's own lighting system. The lighting conditions statistics for all the papers are shown in Fig. 4.

6.4 *Analysis by Acquisition*

In 63% of the analysed papers, the image acquisition is done with CCD cameras, while some use regular digital or DSLR cameras [12, 14]. In these cases it must be noted that many parameters, such as Exposure Value, Aperture, Exposure Time, among others [1, 46], may affect the acquired image quality impacting the desired result, motivating further exploration in Computer Vision systems and IMS'.

Some applications use high-speed cameras, which acquire images in a much faster frame rate than regular CCD cameras. As it was the case of the lighting conditions, Scanning Electron Microscopes may be used for such analysis with a much smaller scale. Figure 5 shows the statistics for devices used in the acquisition step.

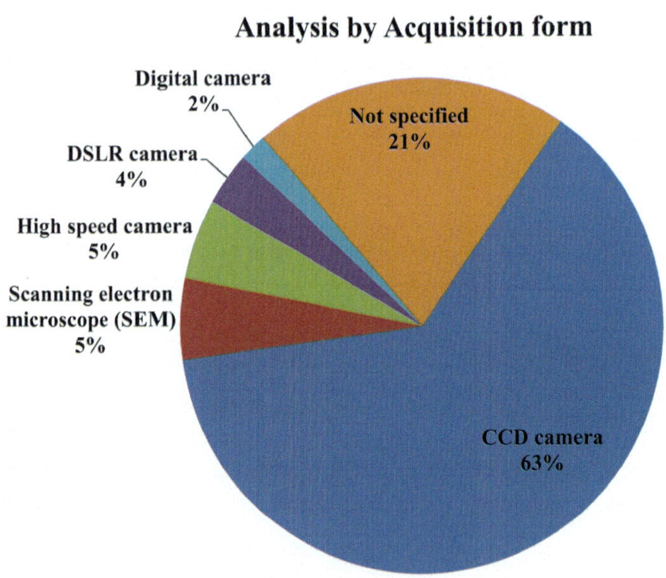

Fig. 5 Distribution of acquisition form over reviewed papers

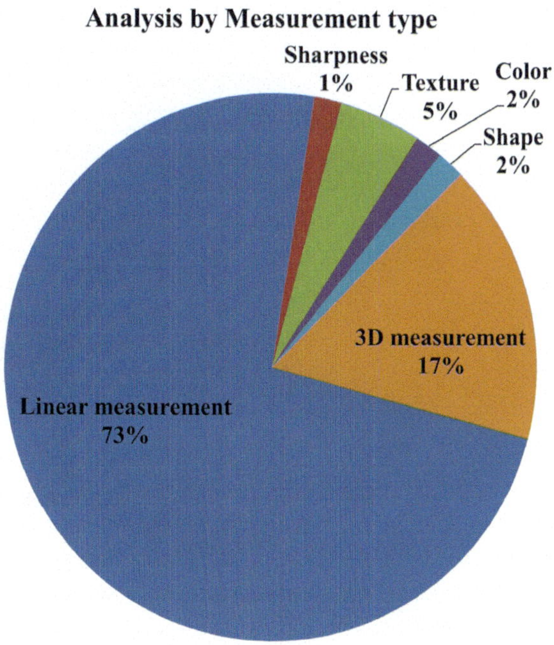

Fig. 6 Distribution of measurement type over reviewed papers

6.5 *Analysis by Measurement Type*

The majority of measurements in the reviewed papers are looking to the linear displacement of the tested material, due to the nature of their respective mechanical tests, with an increasing number of 3D measurements. Some tests, however extract other characteristics, such as texture, colour and even shape. This statistic is shown in Fig. 6.

6.6 *Analysis by Materials*

While the type of materials tested do not directly affect the Computer Vision system, the lighting conditions are indeed affected by the materials, specially due to its reflectivity [45, 46]. The statistic for materials used in the analysed papers is shown in Fig. 7.

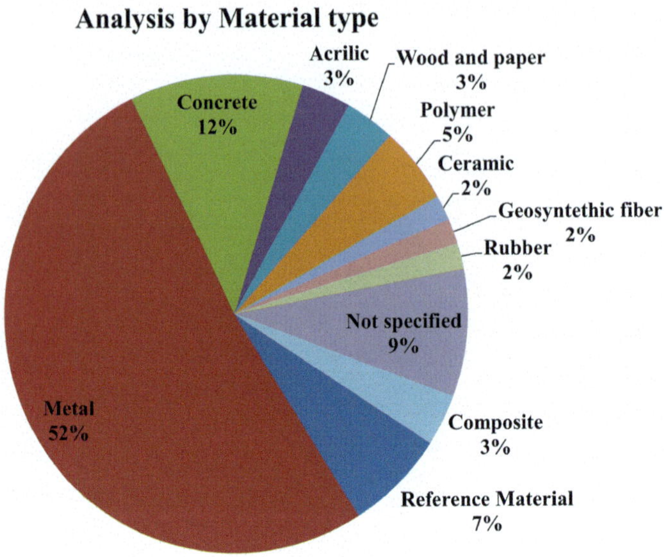

Fig. 7 Distribution of measurement type over reviewed papers

7 Summary Table

The 58 referenced papers have been analysed by generating the following tables, divided by mechanical test type, with Table 1 regarding the Hardness tests, Table 2 the Residual Stress tests, Table 3 the Impact tests and Table 4 the tensile, compression or torsion tests, highlighting the analysed parameters, as described in the previous section.

Table 1 Summary table of papers about Hardness test and their analysed parameters

References	Computer vision technique	Lighting conditions	Acquisition form	Measurement	Material
[2–4]	Segmentation, filtering, morphological operations, edge detection			Linear measurement	Reference material
[5]	Hough transform (shape detection)			Linear measurement	Reference material
[6]	Autofocus			Image sharpness	
[7]	Segmentation, edge detection, vertex detection and shape matching			Linear measurement	

Table 2 Summary table of papers about residual stress test and their analysed parameters

References	Computer vision technique	Lighting conditions	Acquisition form	Measurement	Material
[8–11]	Digital image correlation		CCD camera	Linear measurement	Metals
[12]	Colour analysis	Photo-elasticity reflector	DSLR camera	Colour	Concrete

Table 3 Summary table of papers about impact test and their analysed parameters

References	Computer vision technique	Lighting conditions	Acquisition form	Measurement	Material
[13]	Fractal analysis		CCD camera	Texture	Steel
[14]	Fractal analysis	Desk lamp	Digital camera	Texture	Steel
[15]	Fractal analysis		Scanning electron microscope (SEM)	Texture	Steel
[16]	Bernard method (edge detection)			Shape	Metal matrix composite
[17]	Digital image correlation	LED	High speed camera	Linear measurement	Metal

Table 4 Summary table of papers about tensile, compression and torsion test and their analysed parameters

References	Computer vision technique	Lighting conditions	Acquisition form	Measurement	Material
[19, 28, 49]	Digital image correlation	Optical fibre	CCD camera	Linear measurement	Acrylic
[50]	Digital image correlation		CCD camera	Linear measurement	Ceramic
[34, 41]	Digital image correlation	LED	CCD camera	3D measurement	Composite
[24, 25]	Digital image correlation	LED	CCD camera	3D measurement	Concrete
[51]	Digital image correlation		CCD camera	Linear measurement	Concrete
[22, 23, 52]	Digital image correlation	Room light	CCD camera	Linear measurement	Concrete
[39]	Shape matching	Optical fibre	CCD camera	Linear measurement	Geo-synthetic fibre
[49]	Digital image correlation	Optical fibre	CCD camera	Linear measurement	Metal

(continued)

Table 4 (continued)

References	Computer vision technique	Lighting conditions	Acquisition form	Measurement	Material
[36]	Digital image correlation	Blue LED	FLIR stereo camera	3D measurement	Metal
[27]	Digital image correlation	LED	CCD camera	3D measurement	Metal
[33]	Digital image correlation		CCD camera	3D measurement	Metal
[38, 53–55]	Segmentation, edge detection		CCD camera	Linear measurement	Metal
[35]	Digital image correlation		DSLR camera	3D measurement	Metal
[18]	Digital image correlation			Linear measurement	Metal
[31]	Digital image correlation	Optical fibre	CCD camera	3D measurement	Metal
[21, 43, 56–59]	Digital image correlation	Optical fibre	CCD camera	Linear measurement	Metal
[30, 60]	Digital image correlation	Room light	CCD camera	Linear measurement	Metal
[42]	Digital image correlation	Room light	High speed camera	Linear measurement	Metal
[44, 61]	Digital image correlation	Scanning electron microscope (SEM) lighting	Scanning electron microscope (SEM)	Linear measurement	Metal
[20]	Digital image correlation	Optical fibre	CCD camera	Linear measurement	Cardboard
[62, 63]	Digital image correlation		CCD camera	Linear measurement	Polymer
[64]	Digital image correlation	Room light	CCD camera	Linear measurement	Polymer
[65]	Shape matching	Room light	High speed camera	Linear measurement	Rubber
[37]	Digital image correlation	LED	CCD camera	3D measurement	Wood
[47]	Digital image correlation		CCD camera	Linear measurement	Wood and paper

8 Conclusions

This study has the purpose of exploring Computer Vision Systems (CVS) applied in many types of mechanical tests. While there are far more references when regarding specific tests [26, 32], this research not only shows CVS' application historically, but also shows the evolution of techniques and how these systems are established in each of the addressed area.

Even though Computer Vision is widely used for Mechanical Testing, both fields share a wide range of applications. Ranging for different materials (steel, wood, concrete), applying different techniques (Digital Image Correlation, Hough Transform, Segmentation) and perhaps it is not possible to standardize their Computer Vision application domain.

However, even if that seems impossible, there are many parameters in each and every system that may have room for improvements and, consequently, to achieve better quality measurements producing more reliable results. As stated in the analysis section, 54% of papers do not describe the lighting conditions in which their image acquisition occurs. Thus, in conjunction with an improper setup of the acquisition equipment (CCD cameras or DSLR cameras, for instance) this can lead to differences in the measurement, which are bound to affect the end result and its measurement uncertainty [46, 48].

Acknowledgements This work was supported by the National Council for the Improvement of Higher Education (CAPES) and Foundation for Research Support of the State of Rio de Janeiro (FAPERJ).

References

1. Leta FR, Gomes JFS, Costa PB, de Oliveira Baldner F (2015) Metrology by image: discussing the accuracy of the results. In: Öchsner A, Altenbach H (eds) Mechanical and Materials Engineering of Modern Structure and Component Design, 1st ed. Springer, New York, pp 413–432. https://doi.org/10.1007/978-3-319-19443-1_34
2. Mendes VB, Leta FR (2003) Automatic measurement of Brinell and Vickers hardness using computer vision techniques. In: Proceedings of the XVII IMEKO world congress, Dubrovnik, pp 992–995
3. Leta FR, Mendes VB, De Mello JCS (2004) Medição de Identações de Dureza com Algoritmos de Visão Computacional e Técnicas de Decisão com Incertezas. ENGEVISTA 6:15–35
4. Costa PB, Muniz BCO, Machado RR, Koch CA, Barros WS (2008) Comparação Entre Métodos para Medição de Impressões de Dureza Brinell e Vickers. In: 1 Congr Bras Metrol Mecânica, Rio de Janeiro
5. Macedo M, Mendes VB, Conci A, Leta FR (2006) Using hough transform as an auxiliary technique for Vickers hardness measurement. In: Proceedings of the 13th international conference on systems, signals and image processing, Budapest, pp 1–4
6. Maier A (2012) Efficient focus assessment for a computer vision-based Vickers hardness measurement system. J Electron Imaging 21:21114. https://doi.org/10.1117/1.JEI.21.2.021114

7. Gadermayr M, Maier A, Uhl A (2012) Robust algorithm for automated microindentation measurement in Vickers hardness testing. J Electron Imaging 21:21109. https://doi.org/10.1117/1.JEI.21.2.021109
8. Lord JD, Penn D, Whitehead P (2008) The application of digital image correlation for measuring residual stress by incremental hole drilling. Appl Mech Mater 13–14:65–73. https://doi.org/10.4028/www.scientific.net/AMM.13-14.65
9. Korsunsky AM, Sebastiani M, Bemporad E (2010) Residual stress evaluation at the micrometer scale: analysis of thin coatings by FIB milling and digital image correlation. Surf Coat Technol 205:2393–2403. https://doi.org/10.1016/j.surfcoat.2010.09.033
10. Sebastiani M, Eberl C, Bemporad E, Pharr GM (2011) Depth-resolved residual stress analysis of thin coatings by a new FIB–DIC method. Mater Sci Eng A 528:7901–7908. https://doi.org/10.1016/j.msea.2011.07.001
11. Baldi A (2013) Residual stress measurement using hole drilling and integrated digital image correlation techniques. Exp Mech 54:379–391. https://doi.org/10.1007/s11340-013-9814-6
12. Chang C-W, Chen PH, Lien HS (2009) Evaluation of residual stress in pre-stressed concrete material by digital image processing photoelastic coating and hole drilling method. Measurement 42:552–558. https://doi.org/10.1016/j.measurement.2008.10.004
13. Shterenlikht A, Howard IC (2004) Partition of Charpy fracture surface with digital image processing. Int J Fract 129:39–50. https://doi.org/10.1023/B:FRAC.0000038898.56538.d6
14. Camargo A, Tickle AJ, Smith JS (2010) Algorithm to measure automatically the ductile/brittle fracture of Charpy test specimens. J Electron Imaging 19:23013. https://doi.org/10.1117/1.3429721
15. Tanaka M, Tagami M, Kato R, Akiyama H, Oyama N, Ono J (2007) Quantitative evaluation of the impact fracture surfaces of SS400 steel by the three-dimensional geometrical analysis. ISIJ Int 47:178–186. https://doi.org/10.2355/isijinternational.47.178
16. Ayar F, Ayari F, Bayraktar E, Amar CB (2012) Image processing and finite element modelling for analysis of a metal matrix composite. Int J Comput Sci 9(2012):448–457
17. Koohbor B, Kidane A, Sutton MA, Zhao X (2017) Analysis of dynamic bending using DIC and virtual fields method. In: Advancement of optical methods in experimental mechanics, pp 143–150. https://doi.org/10.1007/978-3-319-41600-7_17
18. Peters WH, Ranson WF (1982) Digital imaging techniques in experimental stress analysis. Opt Eng 21:213427. https://doi.org/10.1117/12.7972925
19. Sutton MA, Wolters WJ, Peters WH, Ranson WF, McNeill SR (1983) Determination of displacements using an improved digital correlation method. Image Vis Comput 1:133–139. https://doi.org/10.1016/0262-8856(83)90064-1
20. Chu TC, Ranson WF, Sutton MA, Peters WH (1985) Applications of digital image correlation techniques to experimental mechanics. Exp Mech 25:232–244. https://doi.org/10.1007/BF02325092
21. Sutton MA, Mingqi C, Peters WH, Chao YJ, McNeill SR (1986) Application of an optimized digital correlation method to planar deformation analysis. Image Vis Comput 4:143–150. https://doi.org/10.1016/0262-8856(86)90057-0
22. Choi S, Shah SP (1997) Measurement of deformations on concrete subjected to compression using image correlation. Exp Mech 37:307–313. https://doi.org/10.1007/BF02317423
23. Shah SPSP, Choi S (1999) Nondestructive techniques for studying fracture processes in concrete. Int J Fract 98:351–359. https://doi.org/10.1023/A:1018620008780
24. Carpiuc-Prisacari A, Poncelet M, Kazymyrenko K, Leclerc H, Hild F (2017) A complex mixed-mode crack propagation test performed with a 6-axis testing machine and full-field measurements. Eng Fract Mech 176:1–22. https://doi.org/10.1016/j.engfracmech.2017.01.013
25. Rajan S, Sutton MA, Li N, Rizos D, Caicedo J, Bartelmo S, Lasprilla S (2017) Experimental determination of transfer length in pre-stressed concrete using 3D-DIC. In: Advancement of optical methods in experimental mechanics, pp 107–113. https://doi.org/10.1007/978-3-319-41600-7_12

26. Roux S, Réthoré J, Hild F (2009) Digital image correlation and fracture: an advanced technique for estimating stress intensity factors of 2D and 3D cracks. J Phys D Appl Phys 42:214004. https://doi.org/10.1088/0022-3727/42/21/214004
27. Li J, Xie X, Yang G, Zhang B, Siebert T, Yang L (2017) Whole-field thickness strain measurement using multiple camera digital image correlation system. Opt Lasers Eng 90:19–25. https://doi.org/10.1016/j.optlaseng.2016.09.012
28. Sutton MA, Turner JL, Chao YJ, Bruck HA, Chae TL (1992) Experimental investigations of three-dimensional effects near a crack tip using computer vision. Int J Fract 53:201–228. https://doi.org/10.1007/BF00017337
29. Schmidt T, Tyson J, Galanulis K (2003) Full-field dynamic displacement and strain measurement-specific examples using advanced 3D image correlation. Photogrammetry Part II. Exp Tech 27:22–26. https://doi.org/10.1111/j.1747-1567.2003.tb00118.x
30. Sun Z, Lyons JS, McNeill SR (1997) Measuring microscopic deformations with digital image correlation. Opt Lasers Eng 27:409–428. https://doi.org/10.1016/S0143-8166(96)00041-3
31. Helm JD, McNeill SR, Sutton MA (1996) Improved three-dimensional image correlation for surface displacement measurement. Opt Eng 35:1911. https://doi.org/10.1117/1.600624
32. Orteu JJ (2009) 3-D computer vision in experimental mechanics. Opt Lasers Eng 47:282–291. https://doi.org/10.1016/j.optlaseng.2007.11.009
33. Kanazawa D, Yoneyama S, Ushijima K, Naito K, Chinzei S (2017) Evaluating stress triaxiality and fracture strain of steel sheet using stereovision. In: Advancement of optical methods in experimental mechanics, pp 271–277. https://doi.org/10.1007/978-3-319-41600-7_35
34. Melenka GW, Carey JP (2017) Experimental analysis of diamond and regular tubular braided composites using three-dimensional digital image correlation. J Compos Mater 51:3887–3907. https://doi.org/10.1177/0021998317695418
35. Cortese L, Genovese K, Nalli F, Rossi M (2017) Inverse material characterization from 360-Deg DIC measurements on steel samples. In: Advancement of optical methods in experimental mechanics, pp 225–231. https://doi.org/10.1007/978-3-319-41600-7_30
36. Valeri G, Koohbor B, Kidane A, Sutton MA, Schreier H (2017) Investigating the tensile response of materials at high temperature using DIC. In: Advancement of optical methods in experimental mechanics, pp 77–82. https://doi.org/10.1007/978-3-319-41600-7_9
37. Brabec M, Lagana R, Milch J, Tippner J, Sebera V (2017) Utilization of digital image correlation in determining of both longitudinal shear moduli of wood at single torsion test. Wood Sci Technol 51:29–45. https://doi.org/10.1007/s00226-016-0848-7
38. Lee C, Peters WH, Sutton MA, Chao YJ (1987) A study of plastic zone formation by digital image processing. Int J Plast 3:129–142
39. Aydilek AH, Guler M, Edil TB (2004) Use of image analysis in determination of strain distribution during geosynthetic tensile testing. J Comput Civ Eng 18:65–74. https://doi.org/10.1061/(ASCE)0887-3801(2004)18:1(65)
40. Mott PH, Twigg JN, Roland DF, Schrader HS, Pathak JA, Roland CM (2007) High-speed tensile test instrument. Rev Sci Instrum 78:45105. https://doi.org/10.1063/1.2719643
41. Tekieli M, De Santis S, de Felice G, Kwiecie A, Roscini F (2017) Application of digital image correlation to composite reinforcements testing. Compos Struct 160:670–688. https://doi.org/10.1016/j.compstruct.2016.10.096
42. Wang YH, Jiang JH, Wanintrudal C, Du C, Zhou D, Smith LM, Yang LX (2010) Whole field sheet-metal tensile test using digital image correlation. Exp Tech 34:54–59. https://doi.org/10.1111/j.1747-1567.2009.00483.x
43. Sutton MA, Yan J, Deng X, Cheng C-S, Zavattieri P (2007) Three-dimensional digital image correlation to quantify deformation and crack-opening displacement in ductile aluminum under mixed-mode I/III loading. Opt Eng 46:51003. https://doi.org/10.1117/1.2741279

44. Kang J, Ososkov Y, Embury JD, Wilkinson DS (2007) Digital image correlation studies for microscopic strain distribution and damage in dual phase steels. Scr Mater 56:999–1002. https://doi.org/10.1016/j.scriptamat.2007.01.031
45. Hunter F, Biver S, Fuqua P (2007) Light science & magic, 3rd edn. Focal Press, Oxford, Reino Unido
46. de Oliveira Baldner F, Costa PB, Gomes JFS, do E Santo Filho DM, Leta FR (2013) Determining camera parameters for round glassware measurements. J Phys Conf Ser 575:7–10. https://doi.org/10.1088/1742-6596/575/1/012021
47. Choi D, Thorpe JL, Hanna RB (1991) Image analysis to measure strain in wood and paper. Wood Sci Technol 25:251–262. https://doi.org/10.1007/BF00225465
48. de Oliveira Baldner F, Costa PB, Gomes JFS, Gomes S, Leta FR (2014) Analysis of the exposure time in cameras on a machine vision system for fluids metrology. In: Proceedings of the 3rd international congress mechanical metrology, Gramado, p 4
49. Han G, Sutton MA, Chao YJ (1994) A study of stationary crack-tip deformation fields in thin sheets by computer vision. Exp Mech 34:125–140. https://doi.org/10.1007/BF02325709
50. Roux S, Hild F (2006) Stress intensity factor measurements from digital image correlation: post-processing and integrated approaches. Int J Fract 140:141–157. https://doi.org/10.1007/s10704-006-6631-2
51. Corr D, Accardi M, Graham-Brady L, Shah S (2007) Digital image correlation analysis of interfacial debonding properties and fracture behavior in concrete. Eng Fract Mech 74:109–121. https://doi.org/10.1016/j.engfracmech.2006.01.035
52. Caduff D, Van Mier JGM (2010) Analysis of compressive fracture of three different concretes by means of 3D-digital image correlation and vacuum impregnation. Cem. Concr. Compos. 32:281–290. https://doi.org/10.1016/j.cemconcomp.2010.01.003
53. Luo P-F, Chao YJ, Sutton MA (1994) Application of stereo vision to three-dimensional deformation analyses in fracture experiments. Opt Eng 33:981. https://doi.org/10.1117/12.160877
54. Hung P-C, Voloshin AS (2003) In-plane strain measurement by digital image correlation. J Brazilian Soc Mech Sci Eng 25:215–221. https://doi.org/10.1590/S1678-58782003000300001
55. Lagattu F, Brillaud J, Lafarie-Frenot M-C (2004) High strain gradient measurements by using digital image correlation technique. Mater Charact 53:17–28. https://doi.org/10.1016/j.matchar.2004.07.009
56. Sutton MA, Turner JL, Bruck HA, Chae TA (1991) Full-field representation of discretely sampled surface deformation for displacement and strain analysis. Exp Mech 31:168–177. https://doi.org/10.1007/BF02327571
57. Lyons JS, Liu J, Sutton MA (1996) High-temperature deformation measurements using digital-image correlation. Exp Mech 36:64–70. https://doi.org/10.1007/BF02328699
58. Liu J, Sutton MA, Lyons J, Deng X (1998) Experimental investigation of near crack tip creep deformation in alloy 800 at 650 °C. Int J Fract 91:233–268
59. Sutton MA, Helm JD, Boone ML (2001) Experimental study of crack growth in thin sheet 2024-T3 aluminum under tension-torsion loading. Int J Fract 109:285–301. https://doi.org/10.1023/A:1011014917851
60. Vanlanduit S, Vanherzeele J, Longo R, Guillaume P (2009) A digital image correlation method for fatigue test experiments. Opt Lasers Eng 47:371–378. https://doi.org/10.1016/j.optlaseng.2008.03.016
61. Lagattu F, Bridier F, Villechaise P, Brillaud J (2006) In-plane strain measurements on a microscopic scale by coupling digital image correlation and an in situ SEM technique. Mater Charact 56:10–18. https://doi.org/10.1016/j.matchar.2005.08.004
62. Abanto-Bueno J, Lambros J (2002) Investigation of crack growth in functionally graded materials using digital image correlation. Eng Fract Mech 69:1695–1711. https://doi.org/10.1016/S0013-7944(02)00058-9

63. Tao G, Xia Z (2005) A non-contact real-time strain measurement and control system for multiaxial cyclic/fatigue tests of polymer materials by digital image correlation method. Polym. Test. 24:844–855. https://doi.org/10.1016/j.polymertesting.2005.06.013
64. Lava P, Van Paepegem W, Coppieters S, De Baere I, Wang Y, Debruyne D (2013) Impact of lens distortions on strain measurements obtained with 2D digital image correlation. Opt Lasers Eng 51:576–584. https://doi.org/10.1016/j.optlaseng.2012.12.009
65. Mott PH, Twigg JN, Roland DF, Schrader HS, Pathak JA, Roland CM (2007) High-speed tensile test instrument. Rev Sci Instrum 78:045105. https://doi.org/10.1063/1.2719643

CFD—A Powerful Visualization Tool in Turbomachinery Applications

Galina Ilieva

Abstract Scientific visualization is an important tool in advancing the state-of-the-art in various applications involving fluid flow modelling. Many research works have shown to obtain reliable physical visualization and correct analysis of the entire complex aerodynamic nature in rotating machinery, a number of studies related to geometry, discretization, numerical set up and solution convergence are needed. Current research work results in a logical sequence for modelling of 3D viscous, compressible and turbulent flow in turbine stages with moving and twisted rotor blades. Approaches to attain high quality grid, physically correct numerical modelling and overcome convergence problems, are established. A complex turbine stage with application in industry is under research. Initially, the geometry modelling is carried out in Gambit, with established approaches to achieve high-quality grid; numerical modelling and calculations are fulfilled in Fluent with the help of additionally implemented user defined codes and convergence approaches, established by the author. Numerical modelling features and various approaches to obtain solution convergence for 3D compressible, viscous and turbulent flow, through rotating machines, are under consideration and discussed in this work. The elaborated methodology was applied for research of boundary layer development; radial gap effects on flow aerodynamics in turbine stages; erosion effects over turbine blades working in wet steam; roughness influence over turbine blade surface; to attain higher efficiency performance, etc. In addition, the methodology is applicable for research of turbomachines and their exploitation in nominal and variable operating regimes, further modernization and reconstruction.

Keywords CFD modelling · Convergence · Turbine stage · Visualization

G. Ilieva (✉)
Heat Turbomachines Lab, Shipbuilding Faculty, Technical University of Varna, Studentska 1, Varna, Bulgaria
e-mail: galina.ilieva@tu-varna.bg

D. Vucinic et al. (eds.), *Advances in Visualization and Optimization Techniques for Multidisciplinary Research*, Lecture Notes in Mechanical Engineering, https://doi.org/10.1007/978-981-13-9806-3_10

1 Introduction

Nowadays, Computational Fluid Dynamics (CFD) numerical modelling and simulations are an important source for scientists in their research practice and reveal a new and detailed world of knowledge. CFD and its scientific visualization methods give the possibility to analyse the same problem in a variety of approaches. In addition, it can reveal reasons for many complex problems.

Here comes the question: why scientific visualization and its application is so important tool in the contemporary research? Scientific visualization is a great asset in the area of engineering research and helps researchers in their work. In addition, it is very important to:

- arrive to ways to improve scientific visualization approaches, providing innovative and better ways of displaying data;
- promote and improve software development platform for industrial applications;
- reveal interesting and important phenomena in those regions where measurements are impossible or very difficult to perform;
- improve the educational process at schools and especially at universities;
- validate numerical and especially experimentally obtained results.

It has well known that the CFD simulation procedure consists of the following stages:

- geometry modelling and grid generation;
- numerical set-up;
- solution procedure with application of various approaches to get to convergence;
- analysis of obtained results;
- verification and validation.

Geometry modelling and grid generation relate to the process of creation of a 1D, 2D or 3D model of the object under consideration and its discretization. Visualization shows geometry and its discretization, as well as it helps to control the process of geometry modelling.

A number of procedures to choose solver type, boundary conditions, and solution options relate to the process of numerical set up. User has to apply the most appropriate approaches to obtain convergence and physically correct results. In the process of analysis of nonlinear and transient problems, as well as to monitor the simulation procedure, obtained residuals and convergence, visualization is applied. It can rapidly help us to pay attention to flow instabilities, divergence in the solution progress and take actions to resolve the problem.

The scientific visualization is exclusively helpful for researchers to tune the solution parameters, maintain and optimize both the accuracy and stability of the computational procedure in progress. The iteration procedures and solution are concerned on various quantitative and qualitative parameters, aerodynamic effects, instabilities, processes interaction, etc. Having a solution, the researcher can obtain more parameters, applying the computed data; plot graphics for analysis and comparisons, in order to represent, discuss and validate the simulation results. The scientific visualization can help and accelerate the CFD modelling in a highly interactive manner, help users to compare various approaches and variants of analysis and leads to a deep understanding of the modelling process, simulation and obtained results. Scientific visualization has been an important tool in advancing the state-of-the-art in various applications involving fluid flow modelling.

One of the various and helpful application of the CFD modelling and visualization is found to be in the area of rotating machinery. In order to arrive to a correct logical sequence, as previous experience has shown, a number of research works is applicable. Last mentioned is targeted to: creation of a geometry model, adequate to future mesh procedures that have already been established or already known as appropriate and to attain a logical scheme for the numerical set up, as well as to understand what convergence approaches could be applied by user to overcome divergence problems and to obtain physically real results.

Modelling, research and analysis of aerodynamic performance of 3D compressible, viscous and turbulent flow, through complex turbine stages, requires the fulfilment of a geometry model and its high quality mesh, in order to obtain physically correct prediction of fluid parameters. In a comparison with many software products, for geometry modelling, Gambit-Turbo permits the user to construct turbo geometries in a faster way. However, instead of all implemented advanced options, the mesh procedure can result in a grid with left-handed elements, low quality elements, negative volumes and others, that are a prerequisite to physically incorrect results for flow parameters distribution or lead to solution divergence. The accuracy of a CFD solution depends on the achieved grid quality.

Another very important thing that must be taken into account is that before a new simulation is being started, one has to think carefully of what it is that should be predicted and what is the physical or computational (mathematical) phenomena that could affect the obtained results. Numerical modelling and problem solution for various geometries, design and off-design conditions, requires a number of very specific modelling and mainly convergence approaches, among others, to be estimated with taking into account of all aerodynamic particularities. Specific convergence approaches are crucial to arrive to a physically correct solution.

This research deals with a methodology for geometry and numerical modelling, research and analysis of flow parameters distribution in a 3D turbine stage with twisted rotor blade. In more details, it aims at the development, verification and implementation of a methodology for geometry and numerical modelling of 3D turbine cascades of complex geometry, discretized with high quality elements, after the application of a set of various logical procedures, developed to eliminate the presence of bad elements and solve all mesh problems and challenges. Furthermore, this research aims to establish a logical sequence for numerical modelling with its specific convergence approaches, for the purposes of successful modelling and research of aerodynamic and specific flow features of fluid through turbine stages.

The manuscript consists of three main sub-chapters, related to geometry modelling, numerical set-up and correct approaches to attain solution convergence and physically correct results.

2 Prerequisites to Proper Physical Results and Scientific Visualization

2.1 What Is the State-of-the-Art?

The level of complexity and various aerodynamic features, caused by 3D and unsteady effects, real physical fluid properties (compressibility, turbulence effects and viscosity) and specific stator and rotor blade geometries, characterize the flow and its behaviour in turbomachines. All the aforementioned contribute to mechanical and thermal stresses, secondary flows and leaks, among others, which significantly affect the reliability and efficiency of turbines. Precise modelling, research and analysis of fluid flow characteristics is important and very relevant task of the modern theory of thermal turbomachines.

The solution procedure for a system of nonlinear, partial differential equations (PDEs) with derivatives on the three coordinates x, y, z and time t relates to the determination of fluid parameters, considering the 3D flow aspect. An adequate approach is to solve the Reynolds-Averaged Navier-Stokes equations (RANS) system [1–7] with appropriate turbulent model [3, 8–15]. RANS equations could be solved with 2 main solvers—Segregated Solver [4, 12, 15–18] and Coupled (Implicit or Explicit) solver [19, 20].

Flow equations can be discretized with First Order Upwind (FOU) [16, 21], Second Order Upwind (SOU) [3, 26], Semi-Implicit Method for Pressure-Linked Equations (SIMPLE) [17, 18], etc.

For accurate prediction of complex aerodynamic phenomena as boundary layer and its separation, effects of rotation, secondary flows and others, it is often necessary to apply specific model, even to implement additional terms to the main turbulence model. In [3, 4, 6, 15, 22–24] the Standard k-ε model is applied, but it only gives general picture of how turbulence affects aerodynamics [3, 25, 26]. Reynolds stress turbulent model (RSM) is appropriate to model effects of additional vortices, shear stress effects over fluid particles, and rotational features [10]. Shear Stress Model (SST) and Spallart-Allmaras (SA) [3, 15, 26] are also applicable in turbomachinery. Shear Stress turbulence model predicts location of transition; provides good agreement for numerical and test cases [27] and has good behaviour in adverse pressure gradients and separating flow. In [28, 29] is studied the capability of Large-Eddy Simulation (LES) to obtain turbulent flow specifics in turbine configurations.

A contemporary approach to stator-rotor interaction modelling is the"Mixing Planes" model [25, 30].

An appropriate set of boundary conditions is total and static pressure, static temperature and flow direction at turbine stage inlet and static pressure and temperature at turbine stage outlet, as is shown in [12, 15, 19, 31–33].

An actual and important problem, related to the process of numerical modelling and solving of 3D flow in turbine stages, is to take into account various geometry and aerodynamic features; to achieve convergence of the numerical procedure and acquire physically correct aerodynamics of flow through turbine channels. This will contribute for detailed research of turbine aggregates, under various exploitation conditions, also to consider specific criteria and approaches to efficiency increase and modernization of turbine aggregates.

This research targets the possibility to establish a numerical modelling procedure, approaches to overcome divergence problems and to obtain solution convergence for a 3D compressible, viscous and turbulent flow through rotating machines.

3 Geometry Model and Approaches to High Quality Elements Grid. Numerical Modelling

Modelling, research and analysis of aerodynamic performance of 3D compressible, viscous and turbulent flow, through complex stages, requires modelling of the geometry with high quality mesh to obtain physically correct prediction of fluid parameters distribution. In a comparison with many software products for geometry modelling, Gambit-Turbo permits users to construct turbo geometries in a faster way. However, instead of all implemented advanced options, the mesh procedure could results in a grid with left-handed elements, low quality elements, negative volumes and others, that are a prerequisite to physically incorrect results for flow parameters distribution or solution divergence. The accuracy of a CFD solution depends on the achieved grid quality.

Various approaches for geometry modelling and grid generation, ways to obtain elements of high quality with positive volumes, have been under research and are under discussion in this paper. The methodology and established techniques to attain high quality grid, are implemented into practice, for the purposes of the geometry modelling of a wide range of turbine and compressor blades of complex geometry.

A 3D turbine cascade with twisted rotor blade is under consideration in this study. The stage, Fig. 1, is the 4th one of a double-flow low pressure turbine, working at a nuclear power plant. The configuration consists of a stator with affixed 113 blades and rotor with 127 twisted blades. Rotor blades rotate counter clockwise with 1500 rpm, extracting work from the fluid as it flows along the inter-blade channels.

After a procedure of implementation of a logical sequence and specific approaches, to meet requirements for high quality grid, the 3D model is under construction in Gambit.

It is very important users to understand that both logic and order in which all lines, surfaces, sub-domains and volumes, are constructed, is crucial for the upcoming mesh procedures, grid consistency and attained so far 2D or 3D elements with very high quality.

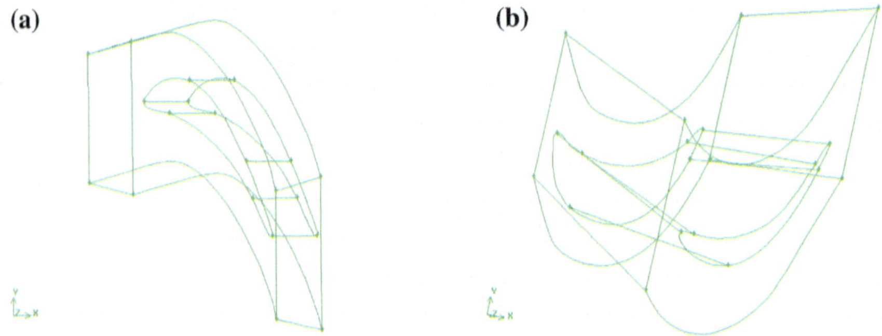

Fig. 1 Volumes formed around stator (**a**) and rotor blades (**b**)

The following sequence is advisable for 3D modelling of stator and rotor blades:

1. Introduction of the key points' coordinates to construct hub and shroud sections for stator and rotor blades as presented in [34, 35].
2. Application of a set-up to obtain leading, trailing and side contours; determination of inlet and outlet geometry angles, based on lines already built.

In the process of the turbo profile definition, the order in which blade tip vertices are under selection is crucial. It is very important to keep the following rule: blade tip vertices must be always select in an order going from hub to shroud sections and in a direction from inlet to trailing edges.

3. Both the stator and rotor sub-volumes, which build the main volumes of the stage under consideration, have to be built, as well as main turbo zones (hub, casing, pressure and suction side domains) must be defined. All required turbo volume characteristics are determined by the turbo profile and by specification of the blades number, tip clearance (if any) and number of span-wise sections.
4. In this step, boundary layer with enough thickness and distributed in several rows of elements, must be modelled, by application of appropriate options such as "growing factor" value and others (Fig. 2).

Sizing functions are important for application as they define the size of boundary layer cells. However, they cannot contribute to the required directional control (anisotropy), so one possible approach is to build elements in direction from field boundaries and form "viscous" layers, Fig. 3.

It is clear that velocity and temperature gradients, in a direction normal to the streamed walls, are much larger than the gradients parallel to the wall. This is the reason to implement "inflation layer control". It creates thin elements that can capture the normal gradient with minimal elements; use inflation layers to visualize velocity and temperature gradients near the no-slip walls.

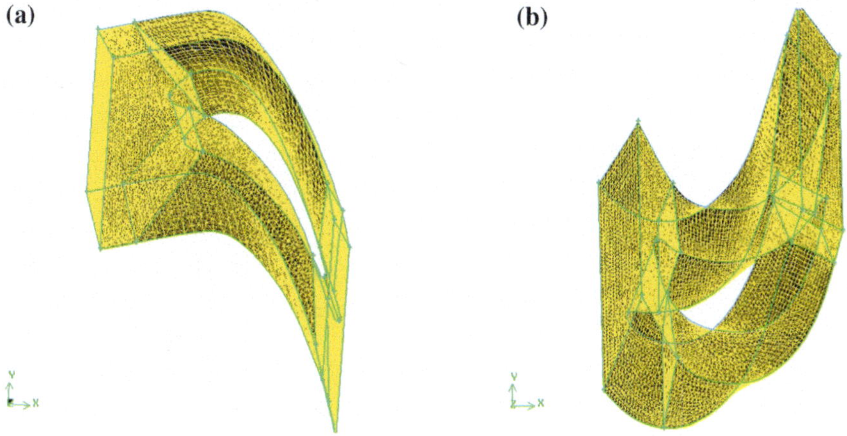

Fig. 2 Discretization of stator (**a**) and rotor (**b**) blade volumes

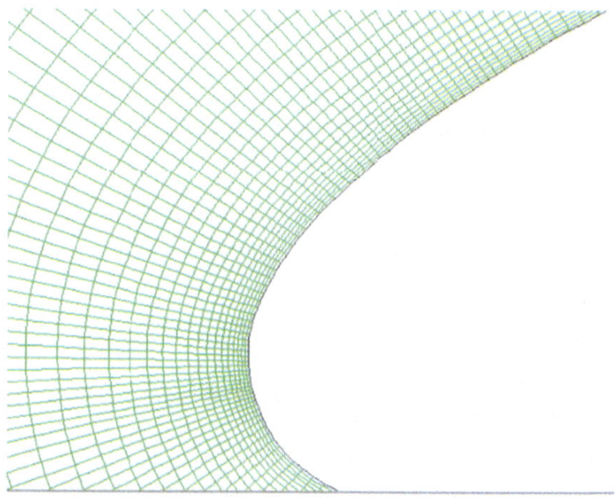

Fig. 3 Viscous layer formation around the leading edge of a stator blade

5. Discretization of hub section and after is the application of Cooper's scheme to create needed 3D grid around turbine blades, Fig. 2a, b. For elements with higher quality were implemented various approaches [35, 36].
6. Elements quality check, Fig. 4a, b.
7. Definition of boundary zones, in this case: "wall", "pressure-inlet", "pressure-outlet", "fluid" "periodic", Fig. 5a, b.
8. Merge of stator and rotor main volumes, Fig. 6.

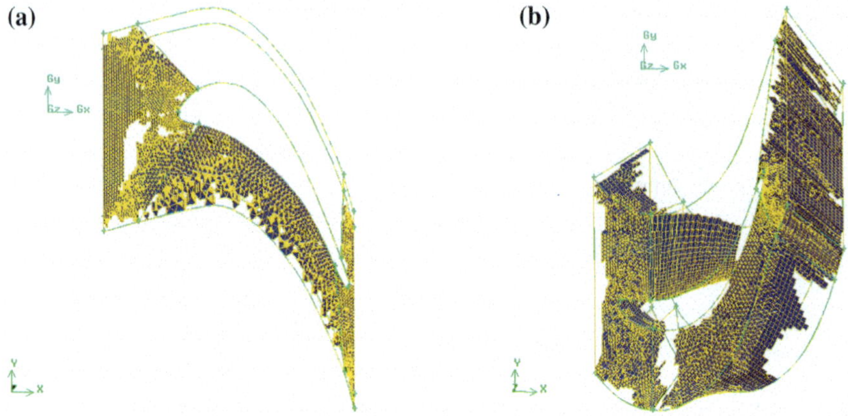

Fig. 4 Element quality check for stator (**a**) and rotor (**b**) volumes

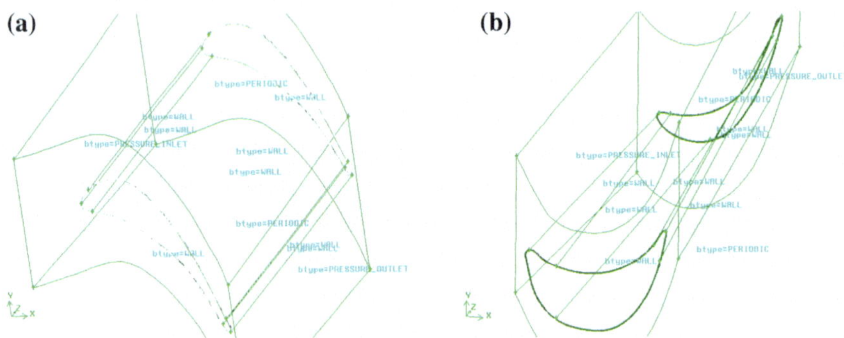

Fig. 5 Defined boundary zones for stator (**a**) and rotor (**b**) turbo volume

3.1 Specific Features, Problems and Solutions

1. Discretization of turbo volumes with grid consisted of high quality elements is very important in the geometry modelling. Gambit gives opportunity for grid quality check. Level of "equiangular skew" for grid elements, in the range of 0 to 1, and presence of negative volumes can be determined. The highest advisable value for 3D grid elements quality check is between 0.6–0.7 [34]. Presence of quality over 0.7 in the resulting mesh can greatly affect the accuracy and robustness of the CFD solution [34, 36]. However, current research found that advisable value of quality factor is 0.3–0.5. A value of 0.7 can provide elements with level of skewness, leading to some solution instabilities, especially in case of vortex structures, boundary layer separation, unsteady effects, among others to appear in the flow [35, 36].

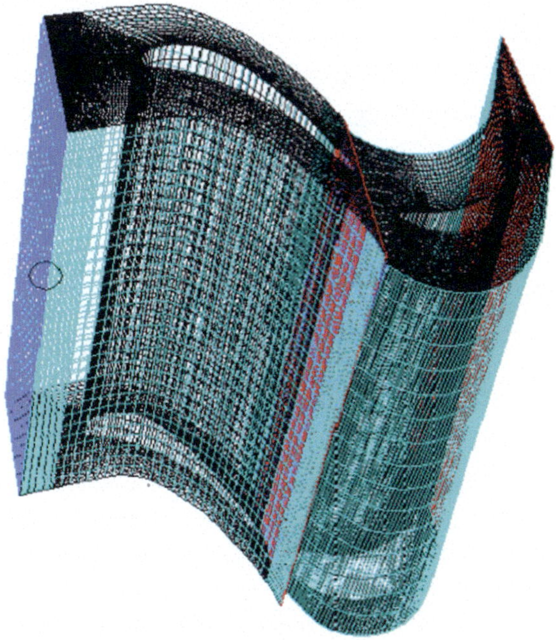

Fig. 6 Merged stator and rotor blade volumes

After a number of research approaches was found that the high level of skewness in downstream direction is a reason for large changes of the unknown parameters (a jump in gradients) and robust iterative procedure. Bad elements (elements of low quality) could be avoided by: mesh refinement around already existing nodes and/or elements; re-mesh in the problematic areas; application of "double sided" mesh option; change in values for "internal ratio" and other options; change of elements type and/or change in the chosen scheme for flow mesh procedures. Other established approaches and many examples of solved serious problems are discussed in details in [36].

2. Zones that contain low quality elements, negative volumes and elements with left-handedness, especially in case of applied Cooper's discretization scheme, are concentrated close to the leading and trailing edges, in radial direction. This is a problem provoked of the blade twistiness ratio, Fig. 7a, b.

For instance, current studies show that the increase of "internal ratio" values, for lines, which describe the leading and trailing edges, and the application of Cooper's scheme, for flow discretization, lead to presence of negative volumes. Research works demonstrate the presence of overlapping elements, in radial direction. It is a result of an inappropriate choice for mesh factors and their values, applied in the process of finite element areas coping, in direction from hub to shroud section, in case of Cooper's method for 3D discretization is applied.

As it is clear after a number of research works an optimal value for "internal count" option to discretize each line, describing profile contours, must be set

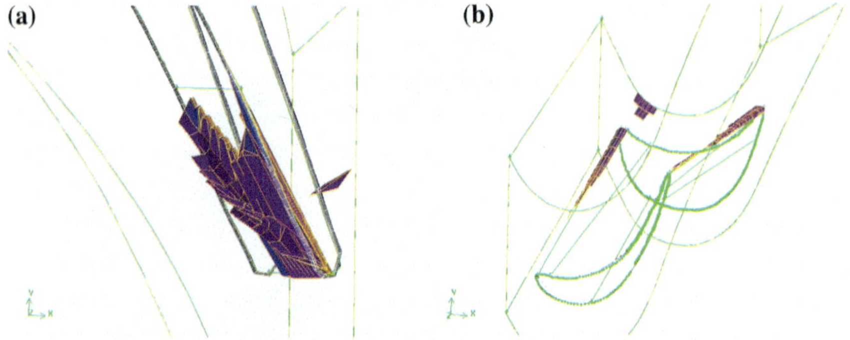

Fig. 7 Elements with negative volume, formed at stator trailing edge (a) and at rotor leading and trailing edges (b), in radial direction

up. What will be that value depends on many factors such as blade's size, consistency of the boundary layer elements and chosen scheme for its creation, curvature of blade walls, etc.

Number of finite elements, applied for approximation of leading, trailing and side contours of blade sections, along the blade height, has to be defined very precisely. Profile contours could be remeshed with high quality elements by selection of the "successive ratio" option, from "Mesh Edges" panel, for grading those nodes along the leading and trailing edges, Fig. 8a, b. In Fig. 9a is shown problematic mesh at the trailing edge of the rotor blade of hub section; Fig. 9b presents the resulting negative elements, along the blade height, after application of Cooper's scheme for discretization.

Research clearly shows that for grading of nodes, distributed along the trailing and leading edges (close to the end vertices of the central camber line), value of the "successive ratio" option, could increase significantly. It is advisable, in the process

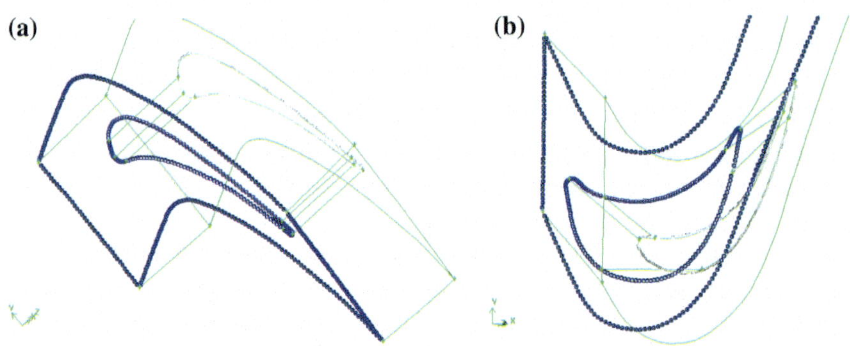

Fig. 8 Approximation of leading, trailing and sideward edges of stator (a) and rotor (b) blades; sections at hub of blades under consideration

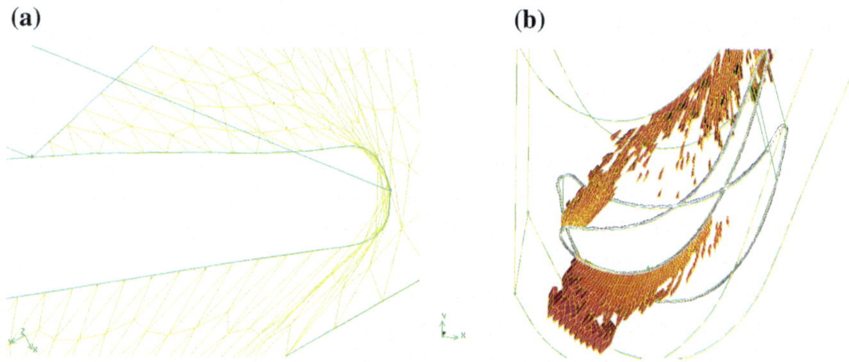

Fig. 9 Highly skewed elements for stator (**a**) and rotor blades (**b**)

of turbo volume discretization, to define higher value for "internal size" option, in order to escape presence of skewed elements in radial direction, Fig. 10.

A complex method to overcome all problems related to the presence of negative volumes is advisable. For instance, modification of already applied discretization schemes; mutually connected values for "internal size" and "successive ratio" options; appropriate type and number of finite elements, which approximate stator and rotor volumes, etc.

For different blade geometries, various individual approaches to attain high quality grid, are available. For instance, flow domain approximation, is very sensible to the blade geometry. For leading and trailing edges, the presence of highly skewed elements is a prerequisite to worse mesh quality, Fig. 11. In this case, the

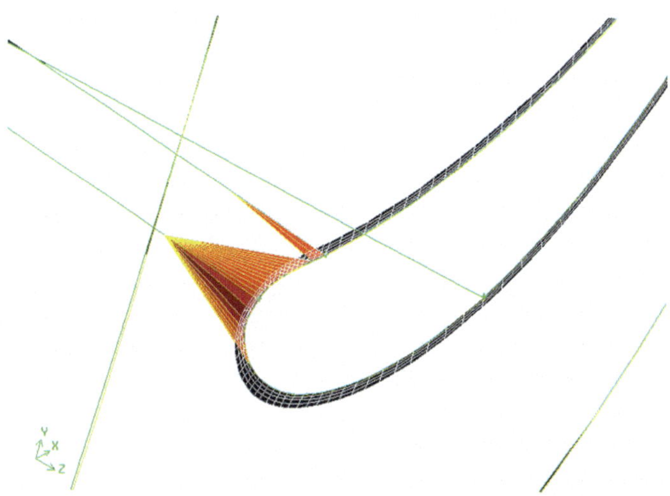

Fig. 10 Highly skewed elements in radial direction

number of elements with bad quality can decrease with increase in the level of discretization along the blade contours. The tendency is to gain a grid with good level of commensurability for edges; it could be achieved after a detailed grid analyses.

As a result of the extensive modelling procedures, the following approach to escape from presence of skewed elements, formed along hub and shroud sections, is advisable:

- Individual discretization of lines, describing hub and shroud sections with appropriate values for "internal size" and "successive ratio" options.
- Discretization of hub section after application of an appropriate set of options by type and value.
- Quality check of the resulting mesh.
- Discretization of volumes.
- Elements quality check.
- The default values for options—"grading" and "max faces", must be changed in agreement with the case under research and according to conformities that were found, in order to obtain high quality grid.
- Before reading of *.msh file in Fluent, boundary types to all turbo domains, must be specified. Boundary zones are of the type: "wall", "pressure-inlet", "pressure-outlet", "periodic", in this case of research.
- Sideward faces must be selected in precise sequence and meshed in a "linked" mode for a correct definition of periodic boundary zones. This leads to correct sign and value of the defined periodic boundaries of rotational type.
- Merge of turbo volumes after their geometry modelling and discretization.

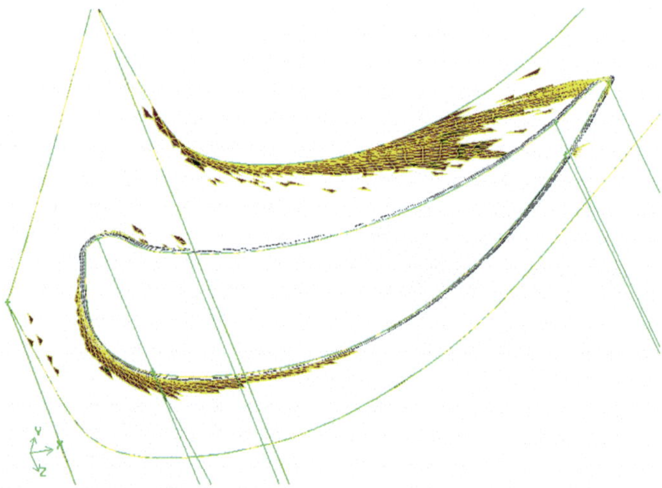

Fig. 11 Highly skewed elements in hub section of rotor blade

3.2 Numerical Modelling Features, Specifies in Our Case of Research of 3D Flow in a Turbine Stage

The procedure is described with main steps in [35].

The fluid is saturated steam, according to the official exploitation documentation.

Boundary zones are "Pressure Inlet", "Pressure Outlet", "Periodic" and "Wall". Boundary conditions are total and static pressure, static temperature, flow direction at turbine stage inlet, static pressure and temperature at stage outlet and turbulence parameters. Boundary conditions values were preliminary calculated in the process of 2D modelling by control sections in radial direction [37].

The chosen solver is Coupled Solver with imposed initial value for the Courant-Friedrich-Levy (CFL) number of 5. During the solution procedure, CFL number has been gradually decreased as follows: 2, 1.5, 1, 0.75, 0.5, 0.25. Options for residuals smoothing - "Iteration smoothing" and "Smooth factor", were set to 1 and 0.1.

Initially, solution was obtained with implemented First Order Upwind convective scheme. In order to obtain more physically correct solution, next the discretization scheme is changed to Second Order Upwind.

Values obtained for the unknown variables, after every one of the iterations, should become closer and closer and thus solution is converged. Due to the non-linear equations set, unsteady solution, separation and other aerodynamic effects, the solution is unstable. That is why to remove the steep oscillation of residuals a set of relaxation factors is applied. For all under-relaxation factors, calculations were performed with initial values equal to 1 and decreased gradually to 0.1 or 0.5.

Scaled residuals in the range of 10^{-4} for all the unknown parameters are advisable to achieve solution convergence; the exceptions are energy and turbulence parameters—equal to 10^{-6}.

All relaxation factors and discretization schemes, as well as the approaches to overcome various divergence problems and access convergence, are described and discussed in more details in [35, 37].

Numerical results together with detailed verification and validation are present in [35].

4 Convergence Problems and Approaches to Their Solution

During the iteration process, by check of residuals, statistics of forces acting on blades and check of the mass flow rates through stages under consideration, convergence is been monitored. Solution is going to stop when each of the unknown parameters meets its specified convergence criteria. In case of non-convergence, various relaxation factors must be changed; specific approaches and schemes must be applied [35]. Approaches to attain solution convergence relate to a strong control of elements' quality, fluid flow properties, turbulence model, boundary conditions, discretization schemes, etc. During the numerical modelling and solution procedure are being settled specific approaches, as discussed below.

- After acceptance of mesh file in Fluent and grid quality check, Coupled Implicit solver is applied. Research on how variations in the type of the solver could influence the flow parameters distribution, is performed. In case of application of Segregated Solver. The level of pulsations of residuals increases, even though the relaxation factors are decreased to a value of 0.1. In the iterative procedure, after the first iterations, the so-called "reverse flows" are reported on [19]. The observed "reverse flows" must gradually decrease, until they disappear in the flow volumes. This is a guarantee for non-admission of solution errors and faster convergence of the iteration procedure. In case of chosen Segregated Solver, in the beginning of the iteration procedure, convergence is not achievable.
- In case of initially imposed more complicated turbulence model, the following approach to get to convergence is advisable: start of the simulation with Standard k-ε turbulent model until solution is converged. Next, the obtained solution serves as an initial solution for further computations with more specific turbulence model.
- If turbulence residuals are going to divergence, one must suspend them from the iteration procedure. In a close proximity to the expected point of convergence, solution for turbulent parameters will run again. The aforementioned approach is applicable as well as in case of divergence for the energy equation, Fig. 12.
- Research works have shown that the points number in the"Mixing Planes" zone, in which momentum, energy and mass exchange are realized, must be more than the default number of 10. This helps in achieving better modelling of stator-rotor

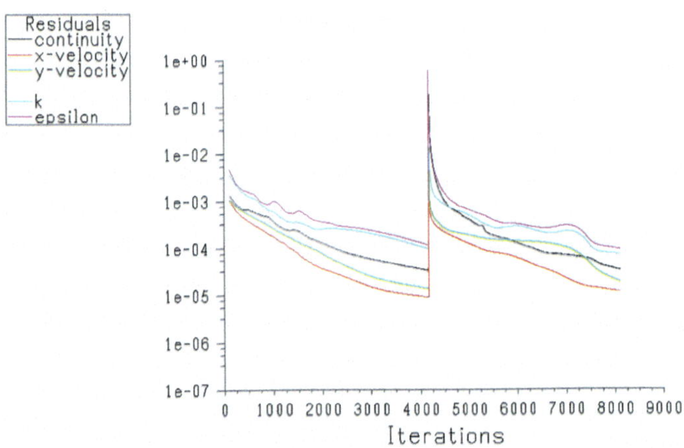

Fig. 12 Suspend of energy equation during the iteration procedure

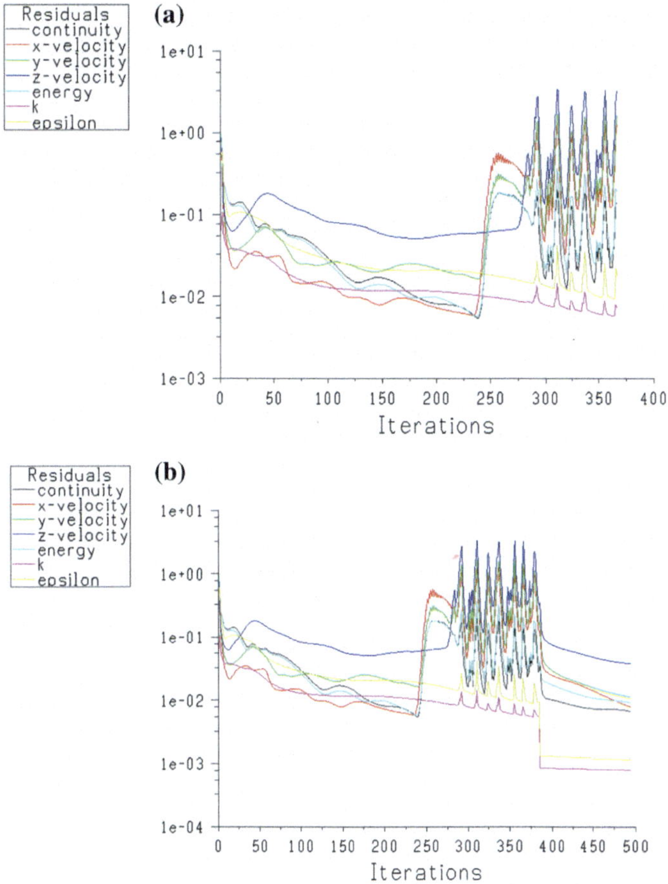

Fig. 13 Strong variations in residuals progress (**a**), progress to convergence after CFL decrease to value of 0.1 (**b**)

interaction, its influence on flow behaviour and for the solution accuracy. The interaction between stator and rotor blades is defined of type "rotational".

- Relaxation factors must be set initially to a value of 0.5. The initially applied discretization scheme for every equation is First Order Upwind; Courant-Friedrich-Levy (CFL) number is set to a value of 2. During the iteration procedure, residuals' variation gradually vanishes until solution is converged, Fig. 13a, b.

For the purposes of initial equations discretization, is applied the First Order Upwind scheme, followed by a significant change in residual values. Furthermore, almost negligible convergence of the iterative process, even after reduction in the value of Courant-Friedrih-Levy number from 2 to 1.0, is observed. Next, after nearly 300 iterations, the SOU scheme for equations discretization is set. In this case, fast convergence of the solution procedure follows the initially reported harsh changes in residual. Effects of viscosity, compressibility and turbulence, affect the residuals for continuity and energy equations, Fig. 13.

- Rotation of rotor blades provokes formation of vortices and vortex structures, secondary flows and thus boundary conditions, imposed at the outlet, are changed. Possible convergence problems are solved after application of a specific approach. The approach is as follows: (a) calculations to determine flow parameters distribution in case of stationary rotor (b) set up of a small value for the rotational speed (c) after attained convergence, a gradual increase of rotor angular velocity must be imposed. In this case, each global iteration set will start based on the previously obtained solution. This process will continue until is imposed full rotational and solution is obtained.
- Minimum CFL value that could be set is 0.10. Figure 14 shows the progress of residuals to convergence after decrease of CFL to 0.10. Figure 15 shows the impossibility to get convergence when low CFL value is set at the beginning of the iteration procedure. In more details, in case of observed large peaks for the residuals of the unknown parameters, there is a significant change in values of all unknown flow parameters. In order to decrease the variation in residuals' values and to control the iteration procedure, a decrease of CFL value from 1 to 0.10, is advisable. Iteration procedure continues with a value for CFL equal to 0.10. During the iteration procedure, physical properties remain constant and characterize the flow behaviour; rotor blades are stationary and the Standard k-ε model is set to reveal turbulence effects. Any unaccounted flow characteristic leads to additional convergence problems, thus change in already defined relaxation factors, is advised. In some cases, even application of new approaches for convergence, is not useful. This has a major impact on the time needed to obtain solution and leads to a lack of convergence, Fig. 15.

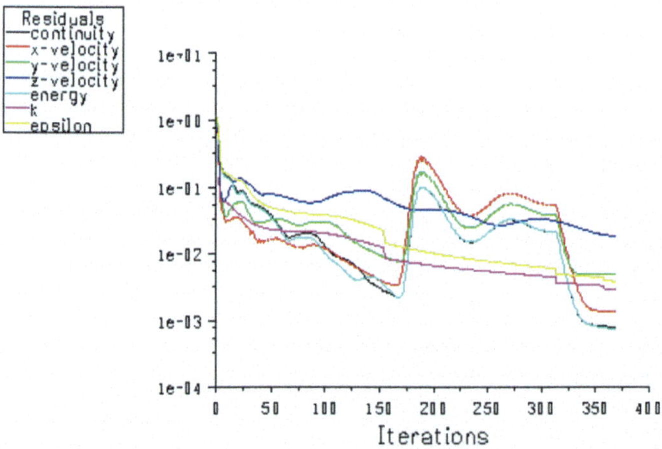

Fig. 14 Progress to convergence in case of decreased value of CFL to 0.10

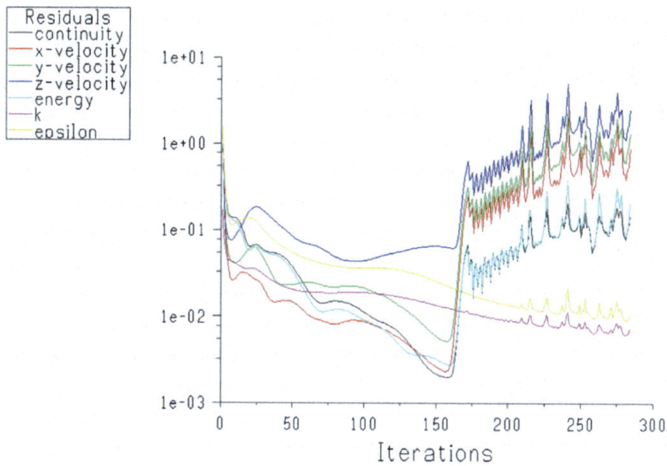

Fig. 15 Impossible convergence in case of imposed lower value of CFL and application of "Residual Smoothing" option

- An additional approach to solution convergence is to apply Residual Smoothing option for all unknown parameters, Fig. 16. Smoothing factor for flow parameters is in the range of 0.5–0.1 for every one or every five iterations. A number of studies prove that the continuous growth of residuals is avoidable by further choice of the Residual Smoothing option from the Control Solution panel. Periodical smoothing and faster convergence in case of set "Residual Smoothing" and Control Solution options could be attain (Fig. 17).

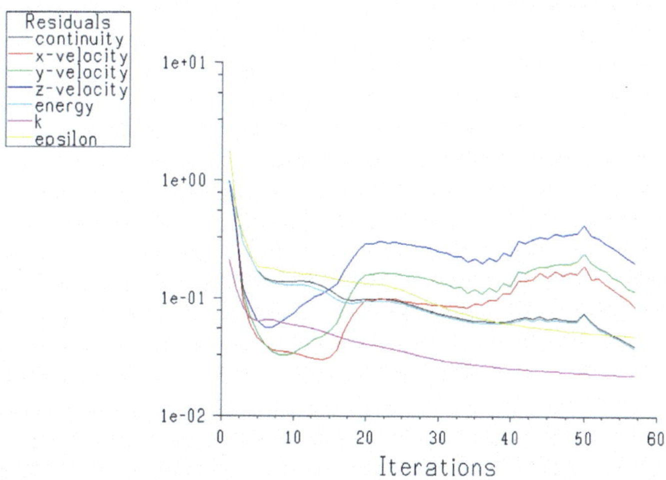

Fig. 16 Progress to convergence in case of application of option "Residual Smoothing"

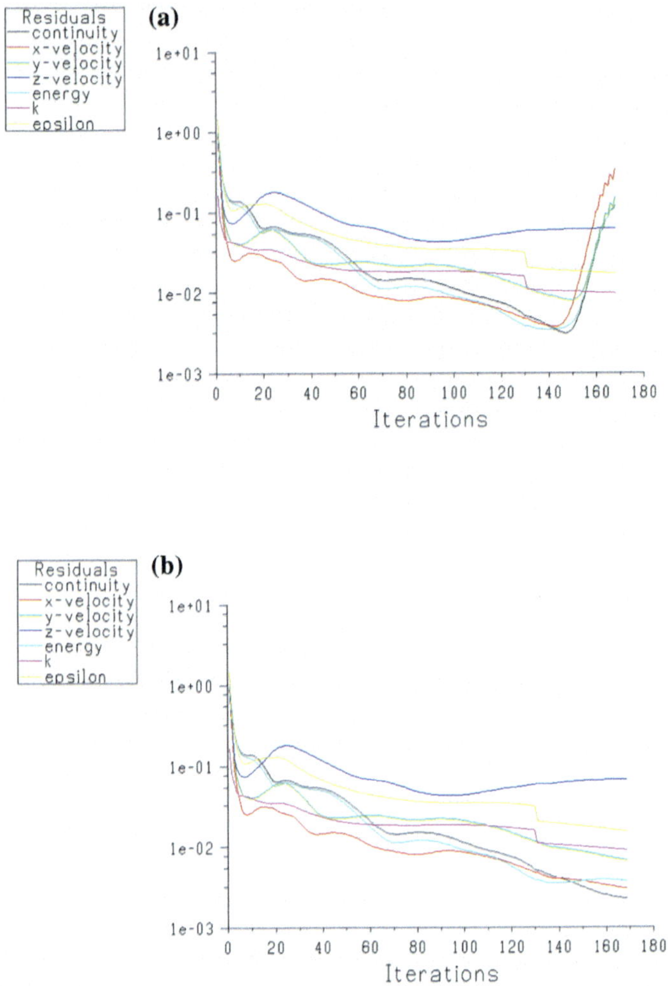

Fig. 17 Change in progress of residuals in case of strong flow perturbations (**a**); residuals progress to convergence in case of preliminarily decrease in CFL value from 1 to 0.75 (**b**)

- An approach to solution convergence is the initial decrease of residuals' values, Fig. 18a, b. Impossibility for convergence is typical in case of unchanged values for the relaxation factors during the entire solution procedure, Fig. 19.

In the solution procedure, the established and recommended scheme to smooth the observed harsh changes for the residuals is as described in the following lines. Before the iteration, after which strong changes in residuals for continuity and energy equations are observed, followed by the impossibility to return to

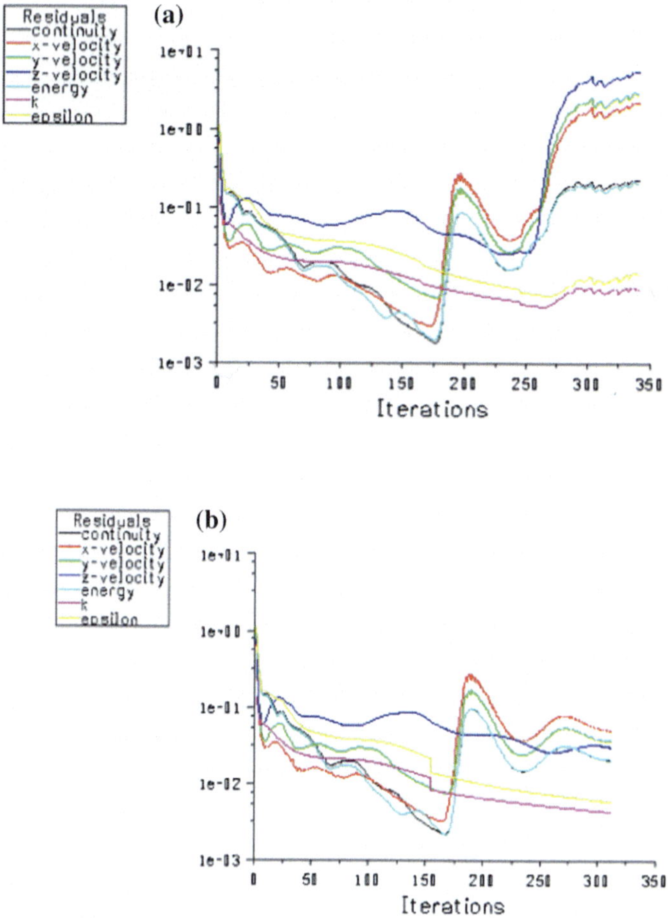

Fig. 18 Variations in the progress of residuals and impossibility for solution convergence (**a**); progress to convergence in case of preliminarily decrease for relaxation factors (**b**)

convergence through described approaches above, a decrease for values of options "Residual Smoothing" and "Smoothing Factor" is required, to carry out the numerical procedure.

After application of the described approach, a slight increase in residuals for two consecutive iterations is observed. Furthermore, residuals could arrive to the solution point after application of specific convergence factors. For example, at the beginning of the computational procedure, with applied SOU discretization scheme and before iteration number 78—relaxation factors of 0.5 and CFL value equal to 2, "Residual Smoothing" equal to 10 and "Smoothing Factor" equal to 0.50 must be defined. Before iteration number 145, to avoid the problem of significant variations

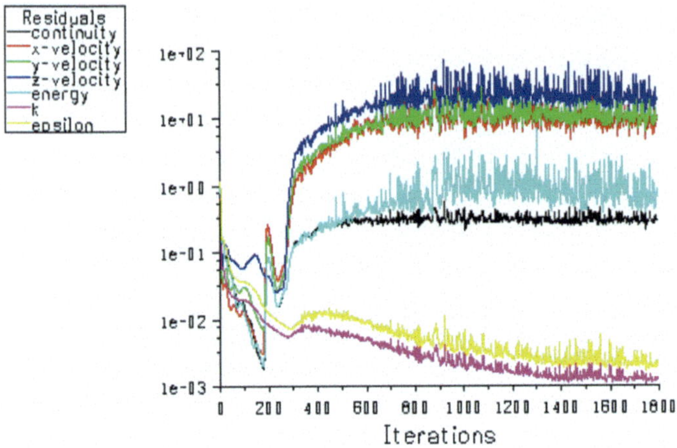

Fig. 19 Impossibility to attain solution convergence in case of applied unchangeable relaxation factors at the beginning of the iteration procedure

in the residuals and their divergence, the CFL number is decreased from 1 to 0.75, Fig. 17a, b. Before iteration number 245, the observed harsh change in residuals, can be avoided by a decrease of CFL number to a value of 0.50, followed by a slight decrease of "Residual Smoothing" to 5.00 and decrease in the value of "Smoothing Factor" to 0.30, Fig. 18a.

Figure 20 shows an example of residuals divergence in case of initially chosen SOU discretization scheme and imposed very low initial value of the Courant number (CFL). In order to model the distribution of turbulence parameters and to

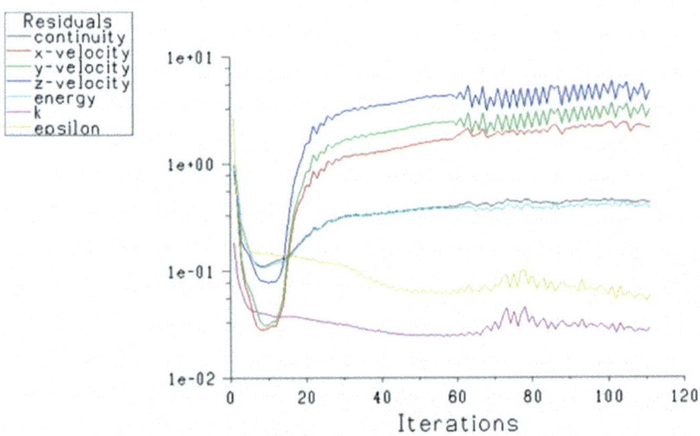

Fig. 20 Divergence in case of initially imposed SOU discretization scheme and low value for CFL number

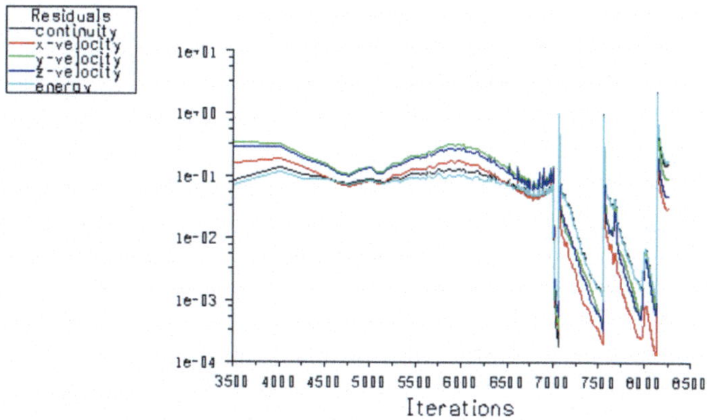

Fig. 21 Process to get solution convergence in case of switched off turbulence equations during the iteration procedure

get to physically correct results, again for calculation in the iteration procedure, turbulence equations are selected, by clicking the check-marked box beside "Turbulence", Fig. 21.

In addition, the approach is applicable in case of problems with convergence for the energy equation.

5 Conclusion

Because of the performed global research, is established a logical scheme with specific approaches to perform 3D geometry and numerical modelling of turbine stages. The methodology provides approaches to attain high quality grid and to surpass convergence problems. This is very important for correct and detailed visualization and understanding of important aerodynamic problems.

The established methodology is applied to model and analyse: static pressure distribution along streamed profiles; aerodynamics and flow features at stator and rotor outlets; radial gap and its influence on flow parameters distribution; both wet-steam conditions, roughness and their impact over blade surfaces; erosion effects over turbine blades; analysis of boundary layer growth along streamed surfaces and prediction of its separation to substitute expensive experiments.

The logical sequence and its implemented approaches are very important and considered as state-of-the-art to attain: detailed visualization and flow parameters distribution and evaluation of energy efficiency; study on thermal stresses; to fulfil stress and modal analysis in turbo aggregates and their elements; research on exploitation in nominal and variable operating regimes; modernization and reconstruction of turbomachines.

Acknowledgements The author acknowledges the precious support and advices of prof. R. Yossifov from the TU-Varna.

References

1. Fridh J (2002) Experience from through-flow calculations with SCAM-T for a two-stage axial turbine—the streamline curvature method-Version 1, Technical Report January 2002 SNEA project P12457-2
2. Oksuz O (2002) Aerodynamic optimization of turbine cascades using an euler/boundary-layer solver coupled genetic algorithm. Doctoral Thesis, 2002
3. Aldo Bonfiglioli M (2006) Sergio campobasso, computing three-dimentional turbomachinery flows with an implicit fluctuation splitting scheme. In: Conference on modeling fluid flow (CMFF'06), the 13th international conference on fluid flow technologies, Budapest, Hungary, 6–9 Sept 2006
4. Craft TJ, Launder BE. Suga K (1995) A non-linear eddy-viscosity model including sensitivity to stress anisotropy. In: Proceedings of tenth symposium on turbulent shear flows, 23-19–23-24
5. Hughes TJR, France LP, Hulbert GM (1989) A new finite element formulation for fluid dynamics the Galerkin/least-squares method for advective-diffusive equations. Comput Methods Appl Mech Eng 73:173–189
6. Buntić Ogor I, Dietze S, Ruprecht A (2005) Numerical simulation of the flow in Turbine-99 draft tube. In: Proceedings of the third IAHR/ERCOFTAC workshop on draft tube flow, Porjus, Sweden, 8–9 Dec 2005
7. Menter FR (1993) Zonal two equation k-w turbulence models for aerodynamic flows. In: AIAA Paper #93-2906, 24th fluid dynamics conference, July 1993
8. GRAPE 2-D Grid Generator for Turbomachinery, User's Manual and Documentation, Version 104, 24 Oct 1997
9. Wilkinson DH (1968) A numerical solution of the analysis and design problems for the flow past one or more aerofoils or cascades. Aeronautical Research Council Report and Memorandum, No 3545
10. Wilkinson DH (1970) Stability, convergence and accuracy of two-dimensional streamline curvature method using quasi-orthogonals. In: IMechE thermodynamics and fluid dynamics convention, paper 35
11. Oyama A, Liou M-S, Obayashi S (2002) Transonic axial-flow blade shape optimization using evolutionary algorithm and three-dimensional Navier-Stokes Solver, AIAA 2002-5642
12. Zobeiri A, Kueny J-L, Farhat M, Avellan F (2006) Pump-turbine rotor-stator interactions in generating mode: pressure fluctuation in distributor channel. In: 23rd IAHR Symposium, Yokohama, Oct 2006
13. Ameri AA (2001) Heat transfer and flow on the blade tip of a gas turbine equipped with a mean-camberline strip. NASA/CR-2001-210764 2001-GT-0156
14. Innovative Turbulence Modeling (2005) SST model in ANSYS-CFX, Technical Brief, ANSYS-CFX, 2005
15. Fernandez-Castaneda J, Corral R (2001) Surface mesh generation by means of Steiner triangulations, Industria de Turbo Propulsores. AIAA J 39(1)
16. Sonoda T, Arima T, Olhofer M et al (2006) A study of advanced high-load transonic turbine airfoils. Journal of Turbomach 128
17. Wilkinson DH (1972) Calculation of blade-to-blade flow in a turbomachine by streamline curvature. Aeronautical Research Council Report and Memorandum, No 3704, 1972
18. Yang Hong, Nuernberger Dirk, Kersken Hans-Peter (2002) Toward excellence in turbomachinery computational fluid dynamics: a hybrid structured-unstructured reynolds averaged Navier-Stokes solver. J Turbomach 128:390–402

19. Kachel CE, Denton JD (2004) Experimental and numerical investigation of the unsteady surface pressure in a three stage model of an axial high pressure turbine, Paper No. GT2004-53626; ASME Turbo Expo 2004: Power for Land, Sea, and Air, vol 5: Turbo Expo 2004, Parts A and B, Vienna, Austria, 14–17 June 2004
20. Holley BM, Becz S, Langston LS (2006) Measurement and calculation of turbine cascade endwall pressure and shear stress. J Turbomach 128
21. Schaffarczyk AP, Pawlak M, Richert F (2001) New model for calculating intensities of turbulence in the wake of wind-turbines. In: DEWEK lecturers notes
22. Laumert B, Matensson H, Fransson TH (2000) Investigation of the flowfield in the transonic VKI BRITE EURAM turbine stage with 3D steady and unsteady N-S computations. ASME paper, 2000-GT-0433
23. Uzol O, Brzozowski D, Chow YC et al (2007) A database of PIV measurements within a turbomachinery stage and sample comparisons with unsteady RANS. J Turbulence 8(10)
24. Xie D, Yu X, Li W, et al (2010) Numerical simulation of water droplets deposition on the last-stage stationary blade of steam turbine. Energy Eng 2:248–253. https://doi.org/10.4236/epe.2010.24036 Published Online November 2010 (http://www.SciRP.org/journal/epe)
25. Yan J, Gregory-Smith D (2000) CFD simulations of 3-dimensional flow in turbomachinery applications. Presented at turbomachinery flow prediction VIII ERCOFTAC WorkshopLac Clusaz, France, Mar 2000
26. Ruprecht A, Bauer C et al (1999) Unsteady forces on the blading of an axial turbine caused by stator-rotor interaction. In: Iahr W.G The behaviour of hydraulic machinery under steady oscillatory conditions. www.ihs.uni-stuttgard.de
27. Хауторн У. Р., Аэродинамика турбин и компрессоров, Изд. „Машиностроение", М., 1968 (in Russian)
28. Novak RA et al (1976) A nearly three-dimensional interablade computing system for turbomachinery. ASME papers 76-FE-19 and 76-FE-20, parts I and II, 1976
29. Сироткин Я. А., Г. Ю. Степанов, Установившееся осесимметричное вихревое течение невязкой жидкости в многоступенчатых турбомашинах, Изв. АН СССР, МЖГ № 6, 1981г (in Russian)
30. Pasinato HD, Squires KD et al (2004) Assessment of Reynolds-averaged turbulence models for prediction of the flow and heat transfer in an inlet Vane-Endwall Passage. J Fluids Eng 126(305)
31. Doi H, Alonso JJ (2002) Fluid/structure coupled aeroelastic computations for transonic flows in turbomachinery. In: Proceedings of ASME turbo expo 2002, Amsterdam, The Netherlands, 3–6 June 2002
32. Gier J, Stubert B, Brouillet B, de Vito L (2002) Interaction of shroud leakage flow and main flow in a three-stage LP turbine. In: Proceedings of ASME Turbo Expo 2002, Amsterdam, The Netherlands, 3–6 June 2002
33. GAMBIT 2 Tutorial Guide, December (2001)
34. Ansys – Fluent: GAMBIT User's Guide http://aerojet.engr.ucdavis.edu/gambithelp/html/users_guide/ugtoc.htm
35. Ilieva G (2009) Modeling, research and analisys of 3D real flow in turbine stages with complex geometry. Thesis, TU-Varna
36. Ilieva G (2015) Numerical modeling and research of 3D turbine stage, engineering applications of computational fluid dynamics. Springer advanced structured materials series, vol 4, pp 103–126
37. Ilieva GI, Iosifov RD (2005) Geometry modeling features for 3D turbine cascade with twisted rotor blade in GAMBIT. Acta universitatis Pontica Euxinus 2(5):7–12

Application of Simulation for Emulating and Visualizing Nanotechnology

Ahmed S. Khan

Abstract Phenomenal interest in new and emerging technologies like nanotechnology has led to an increased demand for graduates who have sound knowledge of fundamental principles underlying the design and operation of complex systems. However, incorporation of such learning processes, including state-of-art technological tools and equipment, requires considerable time and financial resources. Keeping curricula and labs current with the rapid change of technology pose another challenge for academia. Visualization plays an important role in the modern scientific process. In an era where the means of generating data through natural inquiry or simulation have vastly exceeded the power of direct analysis, tools to interpret and explore the data are crucial to the continued pursuit of knowledge (Patterson in Nanoscale visualization: transfer function exploration, 2003, [1]). Visualization of processes can be achieved by using simulation tools. In addition to cost savings, simulation offers a number of other advantages: (a) allowing the user to modify system parameters and observe the outcomes without any harmful side effects (b) eliminating component or equipment faults that affect outcomes (c) supporting users progress at their own pace in discovery and understanding of concepts and issues, and (d) enhancing the presentation of "dry" concepts by integrating theory and practice (Cherner et al. in Use of adaptable simulation-based virtual laboratories for teaching alternative energy and energy conservation in engineering & technology programs, 2011, [2]). Simulation has another major advantage in that a simulation is task- and learner-neutral; it models an object/system construction and operation or learning situation. Within simulation functionality, there are no restrictions on the student's actions. Realistic simulations visualize processes occurring in the devices and enable students to observe the physical processes at different levels (from macroscopic to subatomic), analyse constraints between physical parameters, compare actual and virtual data, and much more. The manuscript presents a literature review of simulation and visualization, and discusses applications of various online

A. S. Khan (✉)
College of Engineering & Information Sciences, DeVry University,
Addison, IL 60101, USA
e-mail: Dr.a.s.khan@ieee.org

D. Vucinic et al. (eds.), *Advances in Visualization and Optimization Techniques for Multidisciplinary Research*, Lecture Notes in Mechanical Engineering, https://doi.org/10.1007/978-981-13-9806-3_11

simulation tools, laboratory instruments, and remote access instruments used for visualizing nanotechnology devices and processes.

Keywords Nanotechnology · Visualization · Simulation · Remote access instruments · Computer-based instruction (CBI) · Feedback · Scaffolding · Cognitive apprenticeship · Sequential design · Parallel design · Mixed design · Atomic force microscope (AFM) · Scanning electron microscope (SEM) · NanoHub · Remotely accessible instruments for nanotechnology (RAIN) · Energy dispersive spectroscopy (EDS) · Fourier transform infrared spectroscopy (FTIR) · X-Ray fluorescence (XFR) · Nanosensors · Nanophotonics · Micro electro mechanical systems (MEMS)

1 Introduction

> The purpose of computing is insight, not numbers.
>
> —R. W. Hamming

Simulations are recognized as an efficient and effective way of teaching complex and dynamic engineering systems. A simulation-based teaching environment enables students to acquire experience and consider their previous results [3]. Simulation has been shown to be effective in improving teaching and learning of various subjects [4]. By reducing practical learning time for students, and for schools and programs, simulation reduces costs for practice oriented educational methodology.

In constructing a simulation, guidelines provide an appropriate model: a physical or mathematical abstraction of a real world process, device, or concept. Because simulation is concerned with modelling of real-world problems, simulation in engineering usually refers to the process of representing the dynamic behaviour of a "real" system by an idealized, more manageable model-system implemented through computation via a simulator [5].

The advantages of using engineering and technology simulation-based training also include reducing the gap between the learning environment and the "real" environment, and making available training "real world" situations that are difficult to simulate in a hands-on lab. Traditionally for teaching technology-based courses, laboratory experiments were offered using a hands-on approach.

In spite of the advantages of simulations, hands-on labs remain tremendously important in the engineering and technology curricula, which is based on Dewey's experiential learning theory. The basic premise of this theory is that students learn as a result of doing or experiencing things in the world, and learning occurs when mental activity is suffused with physical activity [6]. The professional success of an engineer or a technologist is directly related to her/his ability to transfer knowledge gained in the academic environment to real-world situations. Acquisition of manipulative skills is only possible through the use of real instruments and real experimental data. Therefore, to enhance student learning, the engineering and technology curricula must integrate the effective characteristics of both simulations and hands-on lab activities.

A primary objective of today's educator is to prepare students for the world of tomorrow. Pogrow indicated that if students are to be competitive in the years to come, faculty need to be able to provide their students with the cognitive strategies that will enable them to think critically, make decisions, and solve problems [7].

Furthermore the definition of "hands-on" has changed drastically in the last couple of decades. Today, design means endless rounds of modelling and simulation. Design engineers have effectively become computer programmers. Engineers who build systems, whether chips or boards, seem to be doing less actual design of circuits and more assembly of prepackaged components. Still, students need to have good hands-on experience to understand the system behaviour and to troubleshoot system or equipment in case of a problem. Design models are the "soft technologies" that influence and activate the thought processes of the learners rather than the "hard technology" of the computer [8]. The proper use of soft technologies (simulation software) as a tool can improve student performance in gaining hands-on experience. In traditional education, the educator is responsible for the students' learning. Educators typically lecture to students who take notes and then memorize and recall the material to perform well on examinations [9]. This type of learning environment is not appropriate for college students who bring life skills and increased reasoning ability to the classroom. In such a situation, it may be appropriate for students to take responsibility for their own education. One method of transferring the responsibility from the educator to the student is through guided discovery. Discovery is "in its essence a matter of rearranging or transforming evidence in such a way that one is enabled to go beyond the evidence so reassembled to additional new insights" [10].

Guided discovery was developed by Dr. Charles E. Wales at the Center for Guided Design at West Virginia University. Guided discovery has been found to be an effective learning method that stimulates group interaction and is challenging enough to force students to use resources beyond what are available in the classroom [11]. Menn evaluated the impact of different instructional media on student retention of subject matter [12]. It was found that students remember only 10% of what they read; 20% of what they hear; 30% if they see visuals related to what they are hearing; 50% if they watch someone do something while explaining it; but almost 90% if they do the job themselves even if only as a simulation. In other words, guided discovery through labs and computer simulations that are properly designed and implemented could revolutionize education.

2 Importance of Feedback in Simulation

The use of feedback is a critically important attribute in computer-based instruction (CBI) such as multimedia simulations, as it promotes learning by providing students with information about their responses [13]. Especially when it comes to novice learners, research has demonstrated that novices do not learn as well when they are placed in unguided training environments [14]. Novices need to be given some degree of guidance when learning new information, especially those involving complex tasks. The content of the feedback should help the novice develop accurate knowledge structures and build schema in order to better learn the information and eventually become an expert [15]. Therefore, feedbacks, being an essential part of a guided discovery-based learning platform such as simulation, deserve serious attention by the instructional designers.

3 Tools for Simulation and Visualization at Nanoscale

The use of simulation and visualization tools emulating nanotechnology can provide instructional opportunities in many modes: on campus or on-site, online and hybrid or blended (onsite plus online). By using laboratory instruments, online simulation and remote access instruments, nanoscale devices and processes can be visualized. Realistic simulations enable learners to:

1. Observe the physical process insightfully at different levels of detail.
2. Analyse the constraints between relevant parameters (relationships).
3. Push these parameters beyond normally allowed values to simulate infrequent operating conditions or casualty situations (cause and effect).
4. Run what-if scenarios (minimizing risk).
5. Acquire data from virtual experiments in a theory-to -practice approach (narrowing the understanding gap between theory and practice).

Recent studies suggest that students should be first be exposed to theoretical concepts in the simulation environment and then required to perform a hands-on-activity. This approach ensures progressive and sequenced learning in the form of scaffolding, an aspect of cognitive apprenticeship. Furthermore, to enhance student learning, the instructional design should consider three approaches. The first is to use simulation-based experiment, followed by hands-on activity or experiment in the second phase (sequential design). The second approach is simultaneous use of simulation and hands-on activities (parallel design). The approach is to use simulation and hands-on activity in an alternating mode (mixed design). All three approaches support a combinational approach or hybrid instructional delivery for promoting student learning [16].

Visualization plays an important role in understanding of nanoscale processes. Visualization of processes can be achieved by using laboratory instruments, online simulation and remote access instruments. Two most commonly used resources for simulation and visualization at nanoscale are:

- Remote Accessible Instruments for Nanotechnology (RAIN)
- NoanoHUB: www.nanohub.org.

4 Remote Accessible Instruments for Nanotechnology (RAIN)

The Nanotechnology Applications and Career Knowledge (NACK) has developed the Remotely Accessible Instruments for Nanotechnology (RAIN) initiative [17]. RAIN program allows students to access and control microscopes and analytical tools, to look at nanosized materials from the ease of classrooms, or home computers, all across the country. Students can control the following tools (see Fig. 1) over the Internet from 19 centres in real-time [18]:

- Atomic Force Microscope (AFM)
- Optical Microscope
- Confocal Microscope
- Scanning Electron Microscope (SEM)
- Energy Dispersive Spectroscopy (EDS)
- Profilometer
- Ultraviolet–visible Spectrophotometer
- Molecular Analyser
- Fourier Transform Infrared Spectroscopy (FTIR)
- X-ray fluorescence (XRF)
- Fabrication Tools.

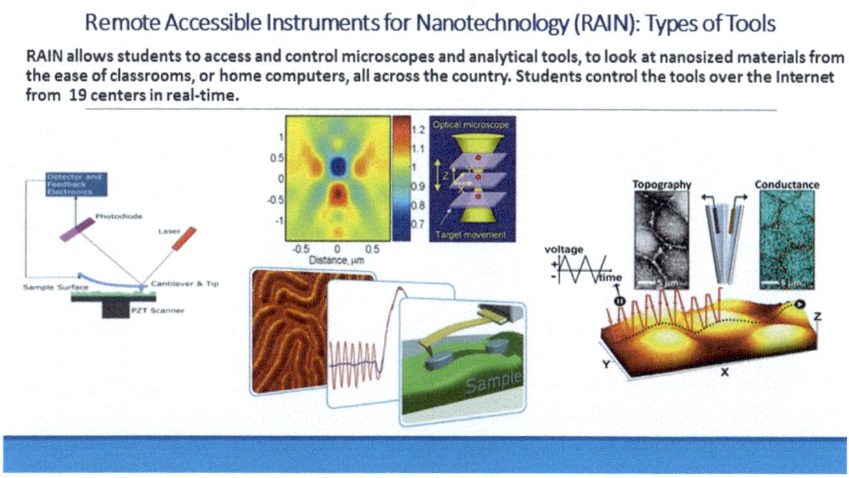

Fig. 1 Remote accessible instruments for nanotechnology (RAIN)

5 NanoHUB.Org

NanoHUB, computing cloud located at Purdue University, offers more than 500 simulation tools and 5500 resources to more than 1.4 million global users (see Fig. 2). Simulation tools appear to run as applets in the browser window powered by the computing cloud of Purdue University. Any user with a nanoHUB username and password can run simulations on nanoHUB. Anyone can register online at nanoHUB.org. Registration is quick, easy, and completely free. NanoHUB is operated by the Network for Computational Nanotechnology (NCN), whose mission is to enable the use of modelling and simulation in the advancement of nanoscience and nanotechnology. Most of the tools on nanoHUB are derived from NCN's contributions to research in the nano areas of electronics, mechanics, bio, photonics, and materials, primarily through leveraged funding [19]. One of the most commonly used simulation tool at NanoHub is Crystal Viewer (see Figs. 3 and 4). The Crystal Viewer simulation tool allows:

(a) viewing all materials which have periodical structure
(b) building crystal structure even not exists in nature.

RAIN and NanoHub simulation and visualization resources can be used to teach the following areas:

- Nanosensors
- Nanoelectronics
- Nanobiosensors
- Nanophotonics
- Nanotransistors
- Thermal Energy

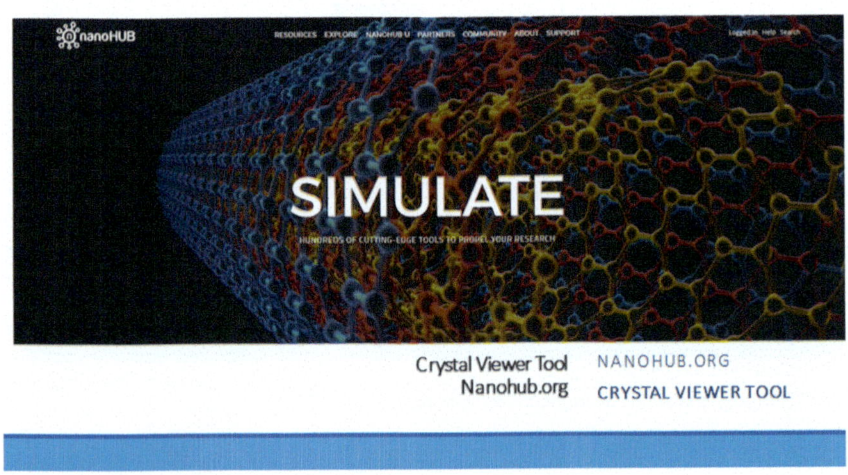

Fig. 2 NanoHub portal

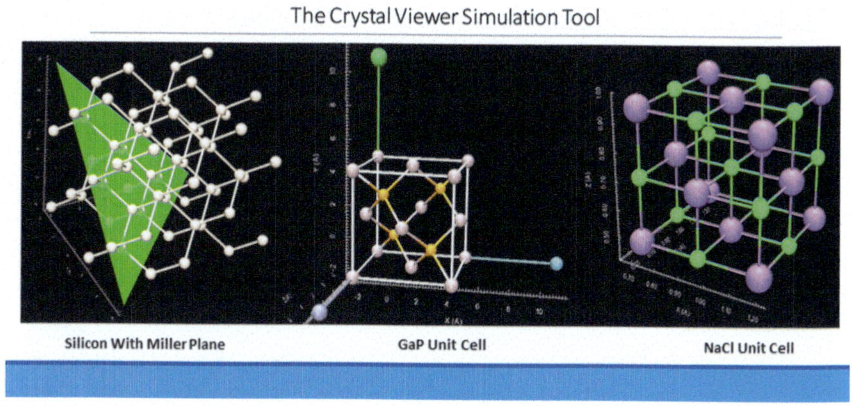

Fig. 3 The crystal viewer simulation tool

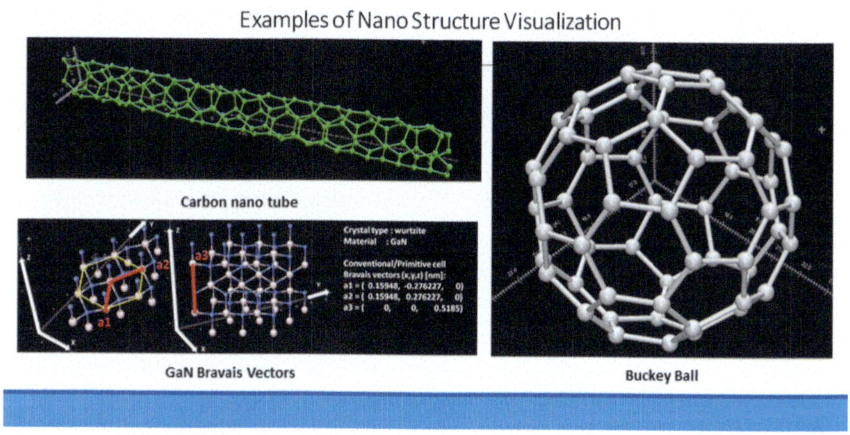

Fig. 4 Examples of nano structure visualization using the crystal viewer

- Thermal Resistance
- Semiconductor Manufacturing
- Material Science
- STEM
- Cellular design
- Thermal Engineering
- Thermal Imaging
- Micro electro mechanical systems (MEMS).

6 Limitations of Simulation and Visualization Tools

Today, simulation and Visualization plays an important role in the modern scientific processes; the ways of generating data through simulation have vastly exceeded the power of direct analysis; tools to interpret and explore the data are crucial to the scientific inquiry. The field of visualization aims to bridge the gap between raw data and natural human perception through visual cues. One critical open problem in the field deals with how to create a meaningful transfer function [20]. So the users of simulations and visualization software tools ought to be aware of their limitations. Discussing the limitations, Henry Petroski observes [21]:

> All software begins with some fundamental assumptions that translate into fundamental limitations, but these are not always displayed prominently in advertisements. Indeed, some of the limitations may be equally unknown to the vendor and to the customer. Perhaps the most damaging limitation is that software can be misused or used inappropriately by an inexperienced or overconfident engineer.

Regarding how to use a simulation program, Constantin Bulucea states [22]:

> The basic difference between an ordinary TCAD user and an true technology designer is that the former is relaxed, accepting on faith the program's results, the latter is concerned and busy checking them in sufficient depth to satisfy himself that the software developer did not make dangerous assumptions. It takes years of training in good schools, followed by hands-on design practice to develop this capability. It cannot be acquired with short courses, or with miracle push-button simulation tools that absolve the engineer of understanding in detail what he is doing.

Discussing the importance of the model, Ronald Rohrer observes [23]:

> Many members of the Spice generation merely hack away at design. They guess at circuit values, run a simulation, and then guess at changes before they run the simulation again… and again… and again. Designers need an ability to create a simple and correct model to describe a complicated situation - designing on the back of an envelope. The back of the envelope has become the back of a cathode ray tube, and intuition has gone on vacation.

Advising to stand up to a computer, Eugene Fergusson expounds [24]:

> The use of sophisticated computer simulation tools is a growing component of modern engineering practice. These tools are unavoidably based on numerous assumptions and approximations, many of which are not apparent to the user and may not be fully understood by the software developer. But even in the face of these inherent uncertainties, computer simulation tools can be a powerful aid to the engineer. Engineers need to develop an ability to derive insight and understanding from simulations. They must be able to "stand up to a computer" and reject or modify the results of a computer-design when dictated to do so by engineering judgement.

7 Conclusion

The use of sophisticated simulation tools to visualize nanoscale processes is a growing component of modern nanotechnology engineering practice. In this day and age, simulations have vastly exceeded the power of direct analysis, tools to interpret and explore the data are crucial to gain new understanding and knowledge at nanoscale. Visualization of nanoscale devices and processes bridges the gap between raw data and natural human perception through visual signs. Application of simulation and visualization resources such as RAIN and NanoHub enhance teaching/learning of nanoscale phenomenon, and supports development of Nanotechnology workforce. NanoHUB is an excellent simulation tool for introducing and analysing nanotechnology phenomena at all educational levels at no cost. RAIN also provides free online remote access to real world Nanotechnology tools.

References

1. Patterson S (2003) Nanoscale visualization: transfer function exploration. Available online at: https://nanohub.org/resources/829/download/2003_suri_patterson_abs.pdf
2. Cherner Y, Khan AS, Karim A, Mullett GJ (2011) Use of adaptable simulation-based virtual laboratories for teaching alternative energy and energy conservation in engineering & technology programs. Paper presented at 2011 ASEE Annual Conference & Exposition, Vancouver, BC, Canada
3. Nahvi M (1996) Dynamics of student-computer interaction in a simulation environment: Reflections on curricular issues. In: Proceedings of the IEEE Frontiers in Education, USA, pp 1383–1386
4. Hsieh S, Hsieh PY (2004) Integrating virtual learning system for programmable logic Controller. J Eng Educ 93(2):169–178
5. Sevgi L, Goknar IC (2004) An intelligent balance on engineering education. IEEE Potentials 23(4):40–43
6. Dewey J (1938) Democracy and education. Macmillan, New York
7. Pogrow S (1994) Students who just don't understand. Educ Leadership 52(3):62–66
8. Jonassen DH (1991) Evaluating constructivist learning. Educ Technology 31(9):28–33
9. Leutner D (1993) Guided discovery learning with computer-based simulation games. Learn Instr 3(2):113–132
10. Bruner J (1965) Readings in the psychology of cognition. Holt, Rinehart & Winston, New York
11. Veenman MV, Elshout J, Busato V (1994) Metacognitive mediation in learning with Computer-based simulations. Comput Hum Behav 101(1):93–106
12. Menn D (1993) Multimedia in education. PC World, M52-M60
13. Clariana RB, Ross SM, Morrison GR (1991) The effects of different feedback strategies using computer-administered multiple-choice questions as instruction. Educ Tech Res Dev 39(2):5–17
14. Institute for Creative Technologies. (2009) Intelligent guided experiential learning: tutoring for practice, retrieved from http://ict.usc.edu/projects

15. Cuevas HM, Fiore SM, Bowers CA, Salas E (2004) Fostering constructive cognitive and metacognitive activity in computer-based complex task training environments. Comput Hum Behav 20(2):225–241
16. Khan Ahmed S (2012) Nanotechnology: ethical and social implications. CRC Press, Boca Raton, FL, pp 313–317
17. Nanotechnology Applications and Career Knowledge. Available online at: http://nano4me.org/remoteaccess_resources/newsletter/newsletter_august.html
18. Remote Accessible Instruments for Nanotechnology (RAIN). Available online: http://nano4me.org/remoteaccess
19. Nanohub.org. Available online at: https://nanohub.org/about/simulate
20. Patterson S (2003) Nanoscale visualization: transfer function exploration. Available online at: https://nanohub.org/resources/829/download/2003_suri_patterson_abs.pdf
21. Petroski H (1994) Failed promises. Am Sci 82(1):6–9
22. Bulucea C (1998) Process and device simulation in the era of multi-million transistor VLSI—a technology developer's view. IEEE Workshop on Simulation and Characterization, Mexico City, 7–8 Sept 1998
23. Rohrer R (1990) Taking circuits seriously. In: IEEE circuits and devices, 19 July
24. Fergusson E (1993) Engineering in the mind's eye. MIT Press, Cambridge

Hybrid Optimization Method and Algorithms for Monochrome Images Tone Approximation with Implementation

Albert Aghajanyan, Rudolf Neydorf and Dean Vučinić

Abstract The chapter considers the Monochrome Multi-tone Images (MMI) Tone Approximation (TA) problem. The TA procedure consists in reducing the image single-color tones palette size by replacing the original tones values with the approximated ones. The main problem is the selection of the appropriate approximation tones; in other words, there is a need to define the optimal palette. To provide optimal TA for monochrome images a hybrid algorithm is developed and implies a 2-stage MMI processing. In the first stage the modified evolutionary-genetic algorithm is used. The main goal of the first stage is reducing the search area for the optimal approximation palette. In the second stage, the simple, but effective deterministic algorithm scans the nearest neighbourhood of the suboptimal solution, which was found in the first stage. The scanning of the nearest neighbourhood guarantees that the found extreme approximation palette is fulfilling the optimization criterion and that it is sub-optimized in respect to the total TA processing time.

1 Sub-optimization of Monochrome Images by Modified Evolutionary Approach

1.1 Mathematical Modelling of Tools and Processes of Monochrome Images Approximation

The Technical Sight Systems (TSS) are applied for autonomous objects navigation in space. In general, TSS-s use a recognition technique for tracking their location

A. Aghajanyan · R. Neydorf (✉)
Hardware and Software Engineering Department, Don State Technical University (DSTU), Rostov-on-Don, Russia
e-mail: rudolf.neydorf.44@mail.ru

D. Vučinić
Vesalius College (VeCo), Vrije Universiteit Brussel (VUB), Pleinlaan 2, 1050 Brussels, Belgium
e-mail: dean.vucinic@vub.be

D. Vucinic et al. (eds.), *Advances in Visualization and Optimization Techniques for Multidisciplinary Research*, Lecture Notes in Mechanical Engineering, https://doi.org/10.1007/978-981-13-9806-3_12

and predicting their safe route, which are extensively based on the acquired Monochrome Multitone Images (MMI).

It is well known that the acquired digital image can be created in the raster or vector graphics form. The raster graphics considers an image as a matrix of pixels, each defining a certain colour, and used in the majority of such applications. The vector graphics uses the geometric primitives for displaying an image, and this type of graphics is applied only in special applications, as example, are the applications in Computer-Aided Design (CAD). In this work, the raster graphics is considered in all its complexity details associated with the real images processing. Moreover and well known for its flexibility, the raster graphics serves multiple purposes in many software applications.

One of the main properties of MMI is Tones Palette (TP) applied for displaying graphic forms. TP is an integer vector of tones used for colouring the image. Firstly, the TP display capabilities depend on the TP's size or "length", secondly, on the TP tones values.

The standard scales are based on the equal distribution of the tones values, within their value range (minimum to maximum). For example, the standard 256-tones palette is represented with vector: $(0, 1, 2, \ldots 255)^T$. The special purpose scales (with approximated values) are characterized with a non-equal distribution of tones. For example, the optimal 8-tone TP, which is the approximation of the original 256-tones image (Fig. 1a), has the following tones distribution: $(33, 62, 81, 96, 111, 132, 151, 241)^T$.

In Fig. 1b the TP approximation is shown, and in Fig. 1c, d—brightness frequency (BF) of the original and the approximated images are shown too. The BF structure, shown in Fig. 1d, is the approximated optimal tones distribution obtained from the pixels brightness, which has a non-uniform distribution characterizing the original image, see Fig. 1c.

When comparing Fig. 1a, b, the visual conclusion seems quite easy, and intuitively pointing that the use of the original big-size palette is not needed for supporting the obstacles recognition tasks, where the boundaries identification are computed from the brightness intensities. The important problem, to be solved in the MMI images processing, can be formulated from the received TSS's photosensitive matrix, which is applied to convert the MMI images to its smaller form based on the approximated palette, which is expected to simplify the pattern recognition tasks, as smaller in size.

However, it is obvious that various algorithms can provide such needed transformation and the difference between them is related to their implementation complexity and approximation quality. Thus, the possible question is—how to develop the optimal or suboptimal algorithms that reduce the original MMI palette size. The 3 possible requirements to be fulfilled are:

1. Highest accuracy, in other words to find the most nearest approximated image to the original one, which enables the robust obstacles identification;
2. Minimal computing time of such approximation, as imposed for real-time obstacles identification;
3. Leverage between algorithm quality and its complexity.

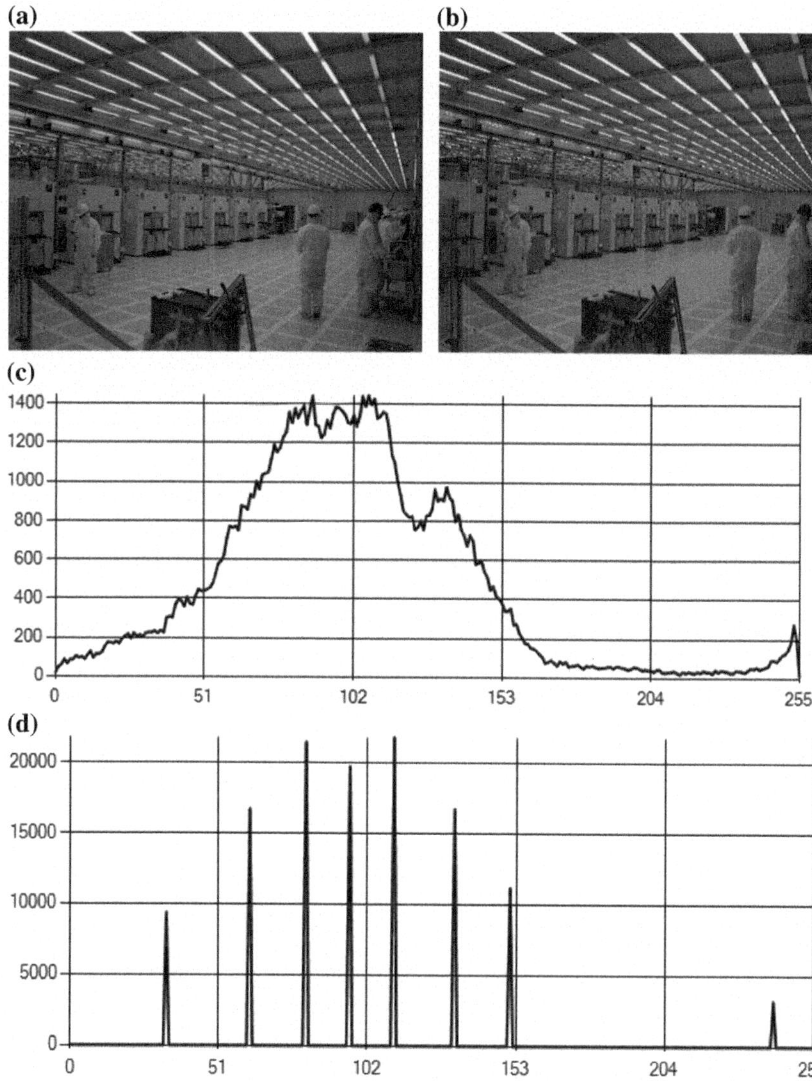

Fig. 1 The original image with 265 tones palette (**a**) and the approximated one with 8 tones palette (**b**); **c** brightness frequency of original image; **d** brightness frequency of approximated image

The first problem requires developing the replacement rule for the transformation of the original TP pixels to their approximated TP pixels values, in order to fulfil the highest quality of approximation. Obviously, it is necessary to develop the Mathematical Model (MM) for such quality estimation. In addition, the pixels replacement rule should provide the effective replacements values among their possible variations. Thus, there is a need to develop the MM algorithm to optimise the image TP approximation.

However, from the known algorithms and their related theories, the conclusion is that there exists only one rule that can provide the required global optimization, which is the NP-full algorithm, characterized with its exponential complexity. The related complexity and the computational time for solving such problem, even for small web images is a serious challenge for the state-of-art high-performance computers. In TSS, much weaker processors are used, and this is why the global MMI approximation optimization is unreasonable.

The second problem is obvious. There is a need to create a simple rule, which allows selecting one tone from the optimise approximation Palette (AP) in order to substitute with it some different, but nearby tones in the Basic Palette (BP). In this approach, the approximation quality depends on the selected correcting rule. It is important to note that in such approach, it is impossible to control the approximation quality.

The third problem represents the most effective solution. There is a need to develop the sub-optimization algorithm for the MMI approximation quality according to its complexity limits. In other words, the computing time should be adapted to the application case, in order to be efficient. Thus, there is a clear reason to solve such search problem to find the conditional extreme fulfilling the imposed quality criteria.

1.1.1 Mathematical Model of MMI

The raster image is represented with the point array, which is called pixels (p). The pixels that form an image are ordered by rows i and columns j, as follows:

$$P = \{p_{ij} | i \in [1, r]; j \in [1, c]\}. \quad (1)$$

It is obvious that the array P is the rectangular matrix with r rows and c columns (i and j define certain row and column index):

$$P[i,j] = \begin{pmatrix} p_{11} & \cdots & p_{1i} & \cdots & p_{1r} & \cdots & p_{1c} \\ \vdots & \ddots & \vdots & \vdots & \vdots & \vdots & \vdots \\ p_{i1} & \cdots & p_{ii} & \cdots & p_{ir} & \cdots & p_{ic} \\ \vdots & \vdots & \vdots & \ddots & \vdots & \vdots & \vdots \\ p_{r1} & \cdots & p_{ri} & \cdots & p_{rr} & \cdots & p_{rc} \end{pmatrix}, \tag{2}$$

where case $c > r$ is shown for the «widescreen image».

The MMI pixels have the same colour, but different brightness, called tones. Any MMI is characterized by the image palette—ordered tones array (s):

$$B^S = \{b^s | s \in [1, S]\} = \left(b^1, \ldots b^s, \ldots b^S\right). \tag{3}$$

where S is the palette size.

The recognition algorithm applies ordinary MMI image with 256 tones, which are variations of grey colour values (between 0—absolute black, and 255—absolute white), where each pixel is defined with 8 bits ($2^8 = 256$).[1]

Thus, beside the raster image pixels coordinates (i and j), which are shown in formula (2), there is the brightness defining the pixel tone. The brightness is defined by the s index number according to formula (3). Consequently, the full MM raster image should be defined with 3-dimensional matrix, as follows:

$$P^S[i,j,s] = \begin{pmatrix} p_{11s} & \cdots & p_{1is} & \cdots & p_{1rs} & \cdots & p_{1cs} \\ \vdots & \ddots & \vdots & \vdots & \vdots & \vdots & \vdots \\ p_{i1s} & \cdots & p_{iis} & \cdots & p_{irs} & \cdots & p_{ics} \\ \vdots & \vdots & \vdots & \ddots & \vdots & \vdots & \vdots \\ p_{r1s} & \cdots & p_{ris} & \cdots & p_{rrs} & \cdots & p_{rcs} \end{pmatrix}, s = \overline{1, S}. \tag{4s}$$

The quantity characteristic of the MMI's model (4s) is the pixel brightness and it is more appropriate to use a matrix representing the MM image, where each element is the pixel brightness. The down indexes i, j, as in (2) and (4s), show the coordinates of pixel, and upper index s shows its value in the palette vector. So the MMI matrix is as follows:

$$B^S[i,j,s] = \begin{pmatrix} b^s_{11} & \cdots & b^s_{1i} & \cdots & b^s_{1r} & \cdots & b^s_{1c} \\ \vdots & \ddots & \vdots & \vdots & \vdots & \vdots & \vdots \\ b^s_{i1} & \cdots & b^s_{ii} & \cdots & b^s_{ir} & \cdots & b^s_{ic} \\ \vdots & \vdots & \vdots & \ddots & \vdots & \vdots & \vdots \\ b^s_{r1} & \cdots & b^s_{ri} & \cdots & b^s_{rr} & \cdots & b^s_{rc} \end{pmatrix}, s = \overline{1, S}. \tag{5}$$

[1]The standard RGB model allocates 24 bits for pixel, 8 bits for each color channel. In monochrome image used only gray color vector from 0 to 255 that require only 8 bits.

The forms (4s) and (5) are equal, because the both matrix elements are digital codes of the respective pixel brightness. Both forms represent the 2-dimensional arrays cuts of the related 3-dimensional matrix, which the MM digital image. The only difference is in the indices positions, and the form (5) is easier to interpret. The form (4), depending on the MMI processing task, allows formulating the MM digital image in 3 forms, represented as arrays cuts of the 3-dimensional image matrix, according to the different pixels characteristics.

For example, arrays of image's MM cuts with its rows is defined as follows:

$$P^S[i,j,s] = \begin{pmatrix} p_{i11} & \cdots & p_{i1s} & \cdots & p_{i1c} & \cdots & p_{i1S} \\ \vdots & \ddots & \vdots & \vdots & \vdots & \vdots & \vdots \\ p_{ij1} & \cdots & p_{ijs} & \cdots & p_{ijc} & \cdots & p_{ijS} \\ \vdots & \vdots & \vdots & \ddots & \vdots & \vdots & \vdots \\ p_{ic1} & \cdots & p_{ics} & \cdots & p_{icc} & \cdots & p_{icS} \end{pmatrix}, i = \overline{1,r}, \tag{4i}$$

showing the case (4i) $S > c$.

For MM cuts the columns are defined as follows:

$$P^S[i,j,s] = \begin{pmatrix} p_{1j1} & \cdots & p_{1js} & \cdots & p_{1jc} & \cdots & p_{1jS} \\ \vdots & \ddots & \vdots & \vdots & \vdots & \vdots & \vdots \\ p_{ij1} & \cdots & p_{ijs} & \cdots & p_{ijc} & \cdots & p_{ijS} \\ \vdots & \vdots & \vdots & \ddots & \vdots & \vdots & \vdots \\ p_{rj1} & \cdots & p_{rjs} & \cdots & p_{rjc} & \cdots & p_{rjS} \end{pmatrix}, j = \overline{1,c}, \tag{4j}$$

showing the case (4i) S > r.

1.1.2 Approximation of MMI

The raster image approximation problem is the reduction of the image palette size. In other words, there is a need to decrease the number of tones, which are characterizing the image. The pixels brightness values in the Original Monochrome Multitone Image (OMMI) is defined with the Original Palette (OP) as:

$$B^{S^0} = \{b_s^0 | s \in [1, S^0]\} = (b_1^0, \ldots b_s^0, \ldots b_{S^0}^0), \tag{6}$$

and replaced with the brightness values from the Approximation Palette (AP), which size is $S^a < S^0$:

$$B^{S^a} = \{b_s^a | s \in [1, S^a]\} = (b_1^a, \ldots b_s^a, \ldots b_{S^a}^a). \tag{7}$$

The original size of OP S^0 depends on many factors, but primarily is related to the amount of the image details. The approximation palette S^a size can be different and only depends on the approximated MMI application field. For example, the MS Office Paint editor uses 3 types of colour palettes: (1) 24-bit image $(S = 2^{24})$, (2) 256-colors image $(S = 256)$ and (3) 16-colors image $(S = 16)$. They are all defined with formula (3).

The replacement process of the image pixels palette (6) with pixels from the reduced size palette (7) is called the MMI tone approximation, which transforms OMMI to the Approximated Monochrome Multitone Image (AMMI). It is expected that such operation will potentially simplify the future technology of the image processing. From many examples, we mention: the obstacle recognition for autonomous navigation objects, the detection of objects having certain marks and the formation of simplified objects shapes for intelligent systems.

This proposed approximation can be realized by significantly different algorithms. However, the underlying basis of all such algorithms is the same, and consists in changing the pixel tone from OP (6) to some AP (7). In addition, in OP $\left(B^{S^0}\right)$ some sub-arrays of pixels $B_s^{S^0}$, show the set of pixels that will be covered by one B^{S^a} from:

$$B_s^{S^0} \subset B^{S^0} : \forall s \in [1, S^a] \rightarrow \bigcup_{s=1}^{S^a} B_s^{S^0} = B^{S^0} \& \bigcap_{s=1}^{S^a} B_s^{S^0} = \emptyset. \tag{8}$$

The properties in (8) of $B_s^{S^0}$, $s = \overline{1, S^0}$ subarrays assume including all the OP pixels for its set and constraining them that one pixel does not enter in 2 sub-arrays simultaneously. In addition, taking into the consideration the properties in (8) that all the OMMI pixels are used in the approximation, thus, it is necessary to create the OP pixels replacement rules on all the AP—B^{S^a} pixels. And these rules are the basis of this proposed approximation algorithm.

There is a need to pay attention to count the sub-arrays $B_s^{S^0}$, which according to (8) is equal to size of AP—S^a. The reason for that is the need to use all the pixels in AP $b_s^a \in B^{S^a}$. In this way the replacement rule of any pixel $b_s^0 \in B_s^{S^0}$ to one pixel of AP $b_s^a \in B^{S^a}$ is defined as:

$$\forall s : b_s^0 \in B_s^{S^0} \rightarrow b_s^0 \cong b_s^a \in B^{S^a}. \tag{9}$$

Therefore, only 2 factors define the variation and effectiveness of the approximation algorithm that converts OMMI to AMMI:

1. OP structure dividing the sub-arrays $B_s^{S^0}$, in other words, how many and which pixels enter in a sub-array, where all elements change their value to b_s^a;
2. b_s^a values, where each separate value equal the one in the sub-array $B^{S^0} \ni b_s^0 = b_s^a$, elements, because the palette values are always natural numbers.

This approach provides the quality and effective solution to the MMI approximation problem, as required in solving the proposed optimization problem.

Therefore, there is a need for the approximation quantity estimation. As the result, the problem of the quality estimation appears, and can be solved by using different approaches and mathematical tools.

The result of processing an image that is described in MM (4s) with any approximation algorithm is AMMI, which MM's structure is different, as all pixels change their tone value from OP (6) to AP (7) having smaller size. If the OMMI's palette model (6) is:

$$P^{S^0}\left[i,j,s^0\right] = \begin{pmatrix} p_{11s^0} & \cdots & p_{1is^0} & \cdots & p_{1rs^0} & \cdots & p_{1cs^0} \\ \vdots & \ddots & \vdots & \vdots & \vdots & \vdots & \vdots \\ p_{i1s^0} & \cdots & p_{iis^0} & \cdots & p_{irs^0} & \cdots & p_{ics^0} \\ \vdots & \vdots & \vdots & \ddots & \vdots & \vdots & \vdots \\ p_{r1s^0} & \cdots & p_{ris^0} & \cdots & p_{rrs^0} & \cdots & p_{rcs^0} \end{pmatrix}, s^0 = \overline{1, S^0}, \quad (10)$$

than the AMMI's palette model (7) is:

$$P^{S^a}\left[i,j,s^a\right] = \begin{pmatrix} p_{11a} & \cdots & p_{1is^a} & \cdots & p_{1rs^a} & \cdots & p_{1cs^a} \\ \vdots & \ddots & \vdots & \vdots & \vdots & \vdots & \vdots \\ p_{i1s^a} & \cdots & p_{iis^a} & \cdots & p_{irs^a} & \cdots & p_{ics^a} \\ \vdots & \vdots & \vdots & \ddots & \vdots & \vdots & \vdots \\ p_{r1s^a} & \cdots & p_{ris^a} & \cdots & p_{rrs^a} & \cdots & p_{rcs^a} \end{pmatrix}, s^a = \overline{1, S^a}, \quad (11)$$

1.1.3 Monochrome Image Approximation Quality Estimation

The approximation rules should be estimated by criteria, that shows the algorithm effectiveness. In order to take into account the huge criteria variations of the approximation algorithms, it is necessary to define the conceptual and mathematical models for OP and AP, and covering the whole approximation process.

In the general case, the quality of the approximation is described by a certain criteria. The most common criteria is the Root-Mean-Square Deviation (RMSD) of pixels brightness between AMMI and OMMI or Least Square Deviation (LSD), which criteria is defined as follows:

$$\Delta_2 = \frac{1}{\mathrm{n}} \sum_{\mathrm{i}=1}^{\mathrm{n}} \frac{1}{\mathrm{m}} \sum_{\mathrm{j}=1}^{\mathrm{m}} \left(\mathrm{p}_{\mathrm{ij}}^{a} - \mathrm{p}_{\mathrm{ij}}^{s}\right)^2 \quad (12)$$

where $\mathrm{p}_{\mathrm{ij}}^{a}$, $\mathrm{p}_{\mathrm{ij}}^{s}$,—elements of $r \times c$ matrix AMMI—$\mathrm{P}^{a}[\mathrm{i,j}]$ and OMMI—$\mathrm{P}^{s}[\mathrm{i,j}]$; $r \times c = N^{p}$—number of image pixels.

However, investigations show that RMSD's estimation has significant error due to its square influence of the OP tones approximated deviations with small amount.

These deviations are few, but their amplitude is significant, and the effect of increasing their value when squaring prevails over influence to estimation of the more numerous, but smaller deviations.

For this reason, the proposed criteria is another one, found more appropriate, as the criteria for comparing the original with the approximated image, which is the Least Module of Deviation (LMD), as follows:

$$\Delta_m = \frac{1}{n}\sum_{i=1}^{n}\frac{1}{m}\sum_{j=1}^{m}\left|\mathrm{p}_{\mathrm{ij}}^{a} - \mathrm{p}_{\mathrm{ij}}^{s}\right|. \tag{13}$$

and, which is experimentally validated.

1.1.4 Conclusion

The consideration of the tone approximation problem makes clearly relevant the development of this proposed method, which provides high quality results within minimal processing time. The proposed Monochrome Multitone Image Approximation is expected to simplify and bring to a new level the future image processing technology. The following chapters are devoted to modelling, development and experimental investigation of this proposed method.

1.2 Experimental Research of Modified Gene-Chromosome Model for Optimal Tone Approximation of MMI

The problem of sub-optimum approximation of MMI by a palette with reduced amount of tones, called Support Palette (SP), is solved. The SP palette tones are defined with the images analysis, which arise in related scientific topics as: technical sight, recognition of images, etc. In this work the research objective was to assess the opportunity of using efficiently such analysis, by applying the Genetic Algorithms (GA) for sub-optimum approximation of MMI, considering the original big size tones palette. The proposed approximation consists in replacing the original MMI pixels with the approximated pixels from a smaller size tone palette. This procedure is of importance in the synthetic vision approach, where image recognition procedures are expected to define the main contours within the image. The developed method reduces the amount of tones used to display an image, which approximation approach is presented in this section. In order to solve it, 2 alternative problems are considered: (1) minimization of losses in such image transformation, and (2) minimization of the SP size (for example, to simplify the image recognition process). The approximated MMI quality is defined as the mean square deviation of pixels brightness (original to approximated). The chromosome in GA is SP, where tones are represented as genes. Such approximations are resulting from the mutational variation of the MMI palette tones, within gene alleles, which are formed by applying the original palette tones. The palette is iteratively changing from generation to generation, where the reduction of the stop risks is done on the local extreme. This fact increases the available search opportunities, as provided by the multi-point crossing-over algorithm, which parameters are able to mutate during such evolution process. In addition, to demonstrate the result of this work, appropriate software has been developed, with an easy-to-use user interface, enabling to show the achieved highly efficient processing of the investigated algorithm. The presented solutions are validated with the photo examples of several technical objects, on which the sub-optimum method has been applied.

1.2.1 Introduction

The Monochrome Multitone Image (MMI), called also the grayscale image, is an array of different brightness pixels of the same tone, which forms a picture. Quite often, the monochrome images are used for processing the technical sight, where detecting the orientation of areas forming objects is used in the recognition process. There are huge class of MMI's approximations problems, which have significant difference in their realization and direction. For example, image compression algorithms with minimal losses, image filtering or pre-processing for recognition algorithms, image converting to format of some devices with a limited number of supported tones, and others [1–4]. However, many of them use a method to lower the palette size, which motivated the authors to research this transformation, as in their knowledge it was not addressed till today. Important to mention is that this step of image processing has some underlying rules and thus envisaged to be useful in different problems of image analysis. Therefore, the considered problem of MMI's approximation is the mapping of an image to a smaller palette, and in such way simplifies the pattern recognition tasks, image defects detection, and image transformation, for printing equipment.

The proposed approximation consists in replacing the Basic Palette (BP) pixels of the Original MMI (OMMI) with the Approximated Palette (AP) pixels, which contain less number of tones. This operation could be achieved by different algorithms, having in common a pixels mapping from some original array $P_{parent} = \{p_i | i = \overline{1, n_p}\}$ of pixels, where n_p—size of BP, to some single pixel $p_{appr} \in \{p_j | j = \overline{1, n_a}\}$, where n_a—size of AP. The rules of this mapping have to follow some exact criteria, which evaluate the effectiveness of the approximated image. In case of big variations of the criteria and related approximation methods, it is necessary to specify the appropriate conceptual and mathematical models.

The quality of the MMI approximation depends on the number of AP tones, the values of AP tones, which are together called the Support Palette (SP) and the SP tones are covering the range of OMMI tones (Fig. 2).

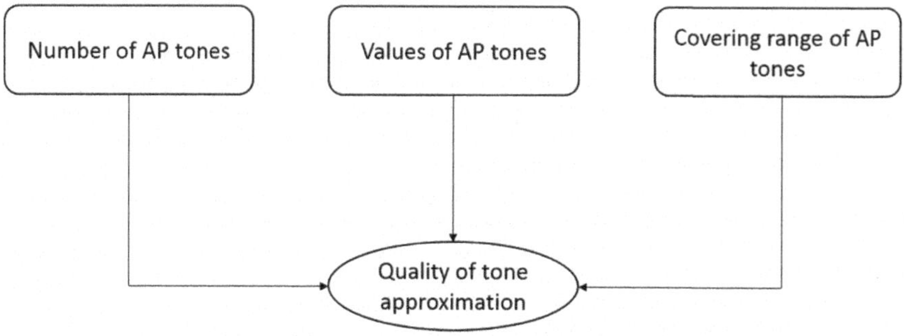

Fig. 2 The influence on tone approximation quality

One of the most important factors is to correctly choose the SP tones and their values, which define SP. Often, the size of SP (number of tones) is the user prerogative, because it depends only on the applying sphere, which has its own allowable error.

In the previous research [5], the SP covering range is defined by dividing BP in equal parts, where the number of parts and the number of SP tones is the same. Such homogeneous distribution was found quite effective to initialize this research.

However, the MMI Frequency Diagram of Brightness (FDB) is quite rare to be found equal (Fig. 3).

And for that reason, the strategy of maximal measured calculation of pixels number for each range was applied to define the SP covering range in this presented modified algorithm, which consists of the following steps:

1. determine the count of all tones N,
2. find average count of the pixels N_c according the size of SP,
3. scan DFB by separating ranges, which have the full pixels number N_i of BP, and the minimal deviation from N_c.

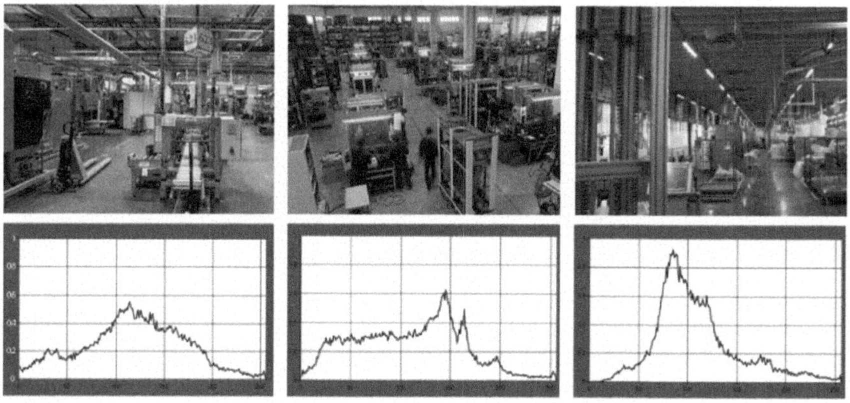

Fig. 3 Industrial workshops photos and their respective FDB

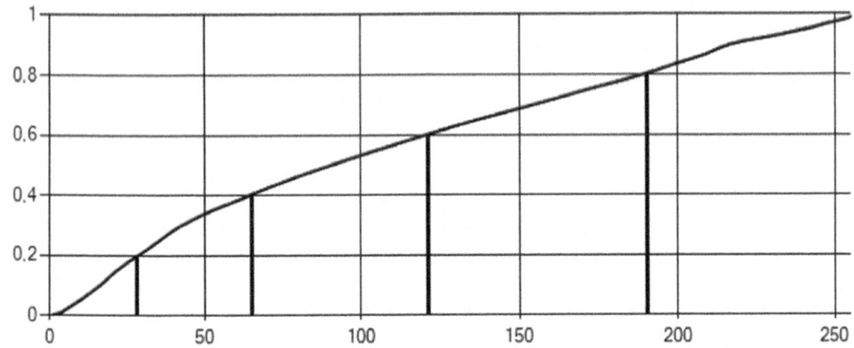

Fig. 4 The integral diagram of the Workshop-1 photo

In this way, the zones, which have smaller number of tones, receive the larger range, because their contribution to the MMI's correctness is less significant. The example of this strategy is present through the integral (accumulated) diagram of brightness, which is analogue of the integral probability distribution (Fig. 4).

The above diagram shows in the relative scale the accumulated number of pixels for increasing brightness, where on Y-axis (number of pixels), shows the strategy of ranges partition. However, the follow up research confirmed a good computational efficiency of this model in defining the covering range.

1.2.2 Image Approximation With Genetic Algorithms

The research decision was to use Genetic Algorithm (GA) [6–9] to optimize MMI's approximation, which modifications have been developed for the solvable problem.

The GA's modification considers the individual to be AMMI and the chromosome to be SP. The genes are represented as separate elements of SP. The BP tones, which enter in a separate gene are covering the range as gene alleles, which contain the genome of each gene. In the beginning, the chromosome model considers the like bitmap (matrix) of AMMI, but influenced directly by the genetic operators on the matrix, resulting in a small change. The new gene-chromosome structure simulates correctly the evolutionarily genetic field, because the chromosome contains the whole genetic information about each individual, so there no need to push the matrix, which is only used for the chromosome estimation. The algorithm creates the matrix of individuals according to the genetic information contained in each chromosome, which is evaluated by the optimization criteria. The simplified scheme of gene-chromosome structure is present on Fig. 5.

The first step of the algorithm is the creation of an initial population, which is provided by the random variation of the SP pixels tones within the defined ranges limits, which allow creating different chromosomes in the developing population. The individuals of the initial population are estimated according the criteria (12) or

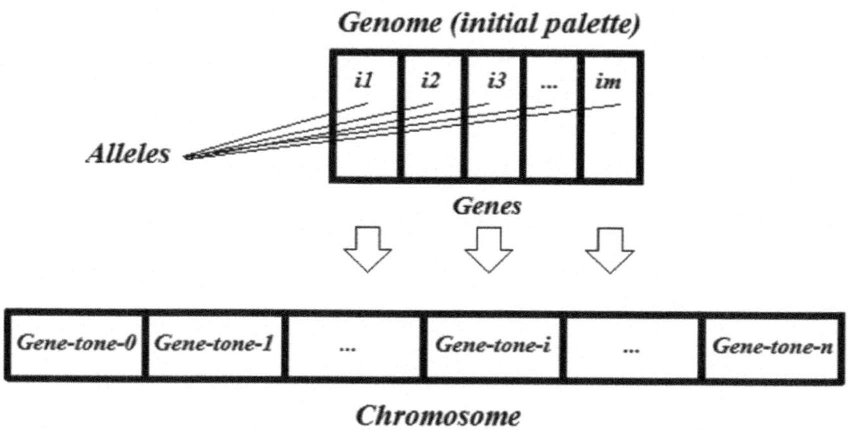

Fig. 5 Scheme of gene-chromosome structure

(13). The evaluation results are used to perform «roulette» selection of the population. After selecting the changed population under the influence of genetic operators «crossover» and «mutation», the new generation is formed. The convergence of the algorithm is attainment for the last generation, which is quite simple and sufficient for motivating the start of the current research.

Current research is aimed to learn the influence of GA's probability factors on the output result. Already in the previous report [5] the sub-optimal values of quantity factors from the standpoint of result/time has been found out, but because of the algorithm modification, and quite predictable behaviour of the quantity factors influence on the chosen values, is estimated to be sufficient to validate the algorithm's performance.

1.2.3 Image Approximation Based on Parametric Sub-optimization of Genetic Algorithm

For the consider problem probabilistic component of modified GA consist of 4 following parameters:

1. amplitude of initial SP—A_{sp};
2. crossover probability. Chromosomes recombined the genes with probability p_{cross};
3. mutation probability. Each gene of chromosome changes with probability p_{mut};
4. amplitude of mutation—A_g. In the case of mutation, the gene changes in range of defined amplitude.

which are necessary to be defined to perform the Full Factorial Experiment (FFE) with $n = 4$. The values and their level of variation were chosen based on the experience acquired in the previous research on heuristic's sphere, and shown in Table 1.

Table 1 The structure of experiment

Factor	x_1			x_2			x_3			x_4		
	Variation of initial SP			Crossover probability			Mutation probability			Amplitude of mutation		
Level	Max	Mid	Min	Max	Mid	Min	Max	Mid	Min	Max	Mid	Min
Value	3	2	1	60	40	20	60	40	20	3	2	1
Symbol	+	0	−	+	0	−	+	0	−	+	0	−

The performed simulations consist of 3 experiments, which have different input OMMI for performing image approximation, based on the industrial workshop pictures, called «Workshop-1», «Workshop-2» and «Workshop-3», where the resolution of these 3 photos is 280 × 210 pixels (Fig. 2).

Each experiment consists of 16 tests, where every test is repeated 30 times. The experiment's structure is presented in Table 2.

In the performed experiment evaluation of the graphic-approximation modification (GAM) of GA quality is made according to the following main parameters:

- y1—the best individual according criteria (13) in 30 parallel tests of FFE line;
- y2—average value of criteria (13) among the best individuals of FFE line;
- y3—dispersion of the best individuals of FFE line;
- y4—RMS of the best individuals of FFE line.

The above Table 2 structure of the performed experiment contains the output results for the workshop №1 approximation, and under the same scheme the experiments are performed for workshop №2 and №3. The purpose of these experiments is to find out the influence of the FDB form to the GA parameters. The results of all the 3 experiments are done in accordance to the main criteria (13) presented in Table 3.

The FFE analysis shown in Tables 2 and 3 put in evidence that the best approximation (absolutely minimal deviation from OMMI), almost for all the 3 quite different images, was obtained for T9 parameters (see Table 3, row 11). This is quite a strong confirmation of the positive influence of parameters around ***T9***, which almost exclude the probability of statistical error. In the point ***T9*** high level has only $\boldsymbol{A_{sp}}$ value. That was predicted, because this parameter response directly to the genetic variety of the population, which according to the genetic point of view is always considered as one of the key factors of the successful population evolution.

Other parameters of ***T9*** are directed to the minimal side, and close in the accuracy to the ***T9*** results are the results characterized by the minimal value of parameters $\boldsymbol{p_{cr}}$, $\boldsymbol{p_g}$ and $\boldsymbol{A_g}$. For the performed experiments, when considering the observed vector for these 3 parameters, it is required to perform additional experiment with the center of plan that will be close to ***T9***. This will provide the possibility to find sub-optimal parameters of the needed probabilistic factors.

Table 2 Matrix and results of optimization experiment of the workshop №1 approximation from 2^8 pallet to 2^4

№		0.	1.	2.	3.	4.	5.	6.	7.	8.
Factor	X1	0	–	–	–	–	–	–	–	–
	X2	0	–	–	+	+	–	–	+	+
	X3	0	–	+	–	+	–	+	–	+
	X4	0	–	–	–	–	+	+	+	+
Response	Y1	0.015288	0.015249	0.015268	0.015249	0.015275	0.015263	0.015393	0.015263	0.015383
	Y2	0.015348	0.015257	0.015318	0.015260	0.015314	0.015303	0.015554	0.01530	0.015564
	Y3	1.02E−09	3.85E−11	8.15E−10	1.08E−10	4.73E−10	5.2E−10	5.22E−09	4.43E−10	6.06E−09
	Y4	3.20E−05	6.21E−06	2.85E−05	1.04E−05	2.17E−05	2.28E−05	7.22E−05	2.10E−05	7.78E−05
№		0.	9.	10.	11.	12.	13.	14.	15.	16.
Factor	X1	0	+	+	+	+	+	+	+	+
	X2	0	–	–	+	+	–	–	+	+
	X3	0	–	+	–	+	–	+	–	+
	X4	0	–	–	–	–	+	+	+	+
Response	Y1	0.015288	*0.015248*	0.015276	0.015248	0.015273	0.015258	0.015386	0.015260	0.015403
	Y2	0.015348	*0.0152587*	0.015314	0.0152588	0.015322	0.015305	0.015563	0.015298	0.015522
	Y3	1.02E−09	*8.62E−11*	6.00E−10	7.61E−11	5.77E−10	4.41E−10	8.08E−09	4.27E−10	6.25E−09
	Y4	3.20E−05	*9.28E−06*	2.45E−05	8.72E−06	2.40E−05	2.10E−05	8.99E−05	2.06E−05	7.90E−05

Table 3 Results of experiments (minimal deviation from OMMI)

№ Test	y_1 (Workshop-1)	y_1 (Workshop-2)	y_1 (Workshop-3)
T0	0.015288	0.015322	0.013876
T1	0.015249	0.015277	0.01382426
T2	0.015268	0.015302	0.013825
T3	0.015249	0.015277	0.013820
T4	0.015275	0.015317	0.013831
T5	0.015263	0.015278	0.013823
T6	**0.0153930**	**0.0154540**	**0.014112**
T7	0.0152630	0.0152740	0.01382452
T8	**0.015383**	**0.015412**	**0.014370**
T9	*0.0152480*	*0.0152650*	*0.01382406*
T10	0.0152760	0.0152810	0.013842
T11	0.0152480	0.0152750	0.013821
T12	0.0152730	0.0153000	0.013826
T13	0.0152580	0.0152800	0.013827
T14	**0.015386**	**0.015473**	**0.014148**
T15	0.0152600	0.0152800	0.013828
T16	**0.0154030**	**0.0154800**	**0.014522**

For all the 3 different images, the worsts results were obtained exactly in the same points of the experiment (***T6***, ***T8***, ***T14*** and ***T16***), where parameters $\boldsymbol{p_{mut}}$ and $\boldsymbol{A_g}$ are on their highest level. Therefore, this confirms the requirement to perform the search of the sub-optimal values of these factors in their respective minimal value areas. In addition, they influence the stability of the designed algorithm, because these parameters give the same influence in respect to different images, avoiding absolutely random results.

For achieving a more reasonable assessment on the considered influence for the optimization of approximation process of workshop №1 we have obtained the mathematical model of the regression equation for y ($\boldsymbol{x_1}$, $\boldsymbol{x_2}$, $\boldsymbol{x_3}$, $\boldsymbol{x_4}$):

$$\begin{aligned} y_{comp}(x_1, x_2, x_3, x_4) = {} & 0.0152934 + 0.00000056 \cdot x_1 + 0.00000081 \cdot x_2 \\ & + 0.00003869 \cdot x_3 + 0.00003269 \cdot x_4 + 0.00000119 \cdot x_1 \cdot x_2 \\ & + 0.00000181 \cdot x_1 \cdot x_3 + 0.00000006 \cdot x_1 \cdot x_4 \\ & + 0.00000056 \cdot x_2 \cdot x_3 + 0.00000031 \cdot x_2 \cdot x_4 \\ & + 0.00002644 \cdot x_3 \cdot x_4 + 0.00000094 \cdot x_1 \cdot x_2 \cdot x_3 \\ & + 0.00000244 x_1 \cdot x_2 \cdot x_4 + 0.00000081 x_1 \cdot x_3 \cdot x_4 \\ & + 0.00000006 x_2 \cdot x_3 \cdot x_4 + 0.00000219 x_1 \cdot x_2 \cdot x_3 \cdot x_4 \end{aligned} \quad (14)$$

The presented equation is from the experiment with the OMMI workshop №1 input. The regression equations of workshop №2 and workshop №3 have shown the same results. According to the regression equation the largest influence was resulting from factors x_3 and x_4, which is the confirmation of the early made conclusions.

Equation (13) received for estimation the curvature of response surface in space of setting parameters and to find minimum RMS area of AMMI from IMMI. It is allow finding out anti-gradient direction (shortest path to minimum) taking partial derivatives from (13) according to setting parameters x_i. Three additional steps according to anti-gradient direction allow identifying the optimal area of the probabilistic factors, giving the best RMS in whole experiments around 3.05E−06. For that reason, the decision was to perform an additional experiment around that area.

By considering the influence of $\boldsymbol{A_{sp}}$ in this additional experiment, this parameter will be fixed. In this way, the plan of the additional experiment will be presented as FFE with $n = 3$. The structure of this FFE is shown in Table 4.

This experiment is analogue to the previous one, except the numbers of tests are 9. In light of almost equality of previous approximation results for all the 3 images, we perform only one experiment with input OMMI workshop №1.

The results analysis of this additional experiment show that it is impossible to find the best point of the experiment with criteria y_1, because most of them have equal accuracy result, which are estimated by the optimization criteria (13) as 0.015248. But according to criteria y_2, y_3 and y_4, the best point of the experiment is ***T8*** (see Table 5, column 10). The analysis made over all the experiments for OMMI workshop №1 has shown that there are no more accurate results found, thus this one is considered as the extreme candidate, at least for the defined covering range of tones from SP.

By considering the found candidate for the extreme it is required to add the new criteria for the GA's search quality evaluation as y_5—the count % of results where $y_1 = 0.0152485$. The new criteria also show that the best point of experiment is ***T8***.

Based on the results of the experiment the made decision was to calculated the anti-gradient for the point ***T8*** and perform 2 steps for factors x_2 and x_3 according to the received vector. The result of the calculation is step with −1 for factor x_2 and 0 for x_3.

The first step, according to y_5 *has* repeated the result of ***T8***, but the second step has improved the percentage of extreme to 19% (see Table 6, column 6).

Table 4 The FFE structure with $n = 3$

Factor	x_1			x_2			x_3		
	Crossover probability			Mutation probability			Mutation amplitude		
Level	Max	Mid	Min	Max	Mid	Min	Max	Mid	Min
Value	20	15	10	20	15	10	2	1	1
Symbol	+	0	−	+	0	−	+	0	−

Table 5 Output results of additional experiment

№		0	1	2	3	4	5	6	7	8
Factor	X1	0	+	−	+	−	+	−	+	−
	X2	0	+	+	−	−	+	+	−	−
	X3	0	+	+	+	+	−	−	−	−
Response	Y1	0.015248	0.015256	0.015249	0.0152485	0.0152485	0.0152485	0.0152485	0.0152485	*0.0152485*
	Y2	0.015252	0.01527	0.01527	0.015253	0.015252	0.015259	0.015256	0.015250	*0.015250*
	Y3	3.07E−11	2.75E−10	6.37E−10	2.77E−11	2.56E−11	7.05E−11	5.16E−11	1.01E−11	*4.51E−12*
	Y4	5.54E−06	1.66E−05	2.52E−05	5.26E−06	5.06E−06	8.39E−06	7.18E−06	3.18E−06	*2.12E−06*
	Y5	0.20	0.00	0.00	0.13	0.27	0.03	0.03	0.50	*0.57*

Table 6 Results of additional steps

№ Step	y_1	y_2	y_3	y_4	y_5
*S*1	0.015248	0.0152507	1.57E−11	3.96E−06	0.57
*S*2	0.015248	0.0152494	9.08E−12	3.01E−06	0.76

As mentioned before, every experiment is repeated 30 times. By analysing the results of the implemented «steep descent» is the proposed algorithm for finding the optimal (or possibly sub-optimal) solution of OMMI approximation problem. Now, we can calculate the minimal number of parallel GA's runs with the found optimal parameters, which guarantee at least one optimal result with an acceptable probability (for example, three sigma).

The calculation of the required number of algorithm runs (every of them is independent) is defined according to the following formula:

$$N = \frac{\ln(1 - p_d)}{\ln(1 - p_o)} = \frac{\ln(1 - 0.9973)}{\ln(1 - 0.76)} \approx 4.1 \tag{15}$$

where p_d—is desired probability to find the extreme, p_o—probability to find extreme in the single run.

The number of parallel runs should be rounded up to a greater value, for example in our case $N = 5$, and thus provides a high probability $P = 0.999204$.

For the verification 50 experiments have been conducted, where each of them was repeated 5 times. As expected, at least one optimum was found in all thirty. Of whole number of repetitions (250) was negative 0.336%. So the actual probability of obtaining a positive result after 5 repetitions for this experiment was 0.9957.

1.2.4 Conclusion

The proposed conceptual and mathematical model of the GA's modification has proved to be an extremely effective tool (when ranges are fixed) for the optimization of the MMI approximation process.

It is suggested to modify the model by including the number of variable settings to better cover the ranges of each separate gene.

There is an idea that such modification of GA will enable to have a solution when performing the absolutely optimization process of AMMI, i.e. fulfilling the minimal deviation from OMMI.

1.3 Parametrical Optimization of Evolutionary-Genetic Algorithm for Improving Monochrome Images Tone Approximation

The section considers the problem of suboptimal approximation of the monochrome multitone image. The proposed approximation consists in replacing the original image tone palette with a reduced size tone palette. The suboptimum approximation is based on the evolutionarily genetic algorithm. The algorithm provides the sub-optimal selection of tones for the new palette and its covering range. The weight-dividing strategy of the original monochrome multitone image's frequency diagram of brightness is used to define the initial tones of the new palette and its covering range. These numerical vectors, considered as chromosomes, define the approximated image created by 2 respective chromosomes. The standard genetic operators of mutation is crossed over with the selection strategy, which provides an effective approximation optimization according to the criteria of the least square deviation between pixels of their original tones, when related to the new palette tones.

The developed algorithm can be applied to the wide class of problems. Examples are, the pattern recognition tasks, image defects detection, and image transformation, for printing equipment. The report illustrates the image approximation of the on board electronic circuit with the sub-optimization of the specific algorithm probabilistic parameters.

1.3.1 Introduction

The transformation of the graphic information for a pattern recognition task can be supported by the tone approximation created from the separate image's parts. This becomes important when there is a need to determine the boundaries of the image elements. Such problem can be considered as a separate process present in the image transformation. The goal is to reduce the picture palette size, which turns out to simplify further the image transformation. It is important to mention that we find this problem when developing systems requiring technical sight.

The previous research [10] has proposed and investigated the idea of using the modified Evolutionarily Genetic Algorithm (EGA) for solving and the optimization problem of the Monochrome Multitone Images (MMI) approximation. In those investigations, the performed experiments confirmed the effectiveness of this approach.

The effectiveness of EGA is largely related to the correct structure of the evolutionarily genetic model, which integrates to the subject area. The development of the algorithm's gene-chromosome structure is the fundamental part found in the integration process.

In the EGA works, an alternative solution to the problem is referenced as the individual or the chromosome. However, the natural analogue, the chromosome stores the whole genetic information to create the new individual, i.e. the individual is the final product of its respective parameters, which are contained within the chromosome. Therefore, the both terms identified in EGA are correct.

1.3.2 Problem Formulation

The purpose of the current research is the development and study of the effective algorithm for sub-optimal MMI approximation, which could be used in the recognition tasks.

The most important issues that have to be solved in order to achieve such goal are:

- To develop a specialized Gene-Chromosome Structure (GCS) of EGA, which is orientated to the image approximation problem;
- To statistically investigate the GCS parameters settings in order to solve the optimization problem in image approximation;
- To design the criteria/strategy of solving the optimization task, which aims to receive the statistically predictable result and possibility to plan the experiment for image processing;
- To perform the parametrical optimization of the whole approximation algorithm, based on the common criteria and to develop the algorithm for the experiment plan when solving the approximation tasks.

1.3.3 Gene-Chromosome Model of Image Approximation

The proposed gene-chromosome structure considers the chromosome in the form of 2 chains. The X-chromosome is 1st chain, which contains SR, and the Y-chromosome is 2nd chain, which contains information about SP. The gene is a separate element combined from X or Y chains.

The main stages of the algorithm include the following steps:

1. Creation of an initial population, which is provided by the random variation of the initial SP and SR genes, and within the defined ranges limits;
2. Individuals of the initial population are estimated according the optimization criteria (1) for further selection with the «roulette» method;
3. Population is undergoing the influence of the genetic operators, «crossing over» and «mutation», which form the new generation;
4. Convergence of the algorithm is attained for the last generation.

1.3.4 Experimental Research and Sub-optimization of Evolutionary-Genetic Algorithm

The algorithm has large number of parameters to set and the full research of all factors could not be described in this short report. For that reason, the description includes the planning and processing of the Full Factorial Experiment (FFE) with the following factors: probability of SP and SR crossing over and probability of SP and SR mutation.

The research includes 2 experiments with different images as input. The 2 images, representing the printed circuit board, are called «Circuit-1» and «Circuit-2», with the resolution of 350 × 250 pixels (Fig. 6).

Each experiment consists of 16 tests, where every test is repeated 50 times. Their selection is based on the experience acquired from the previous research, defining values and levels in their factors variation, as shown in Table 7.

The experiment evaluation of EGA quality is made according to the following parameters: y_1—the best individual according criteria (2) in 50 parallel tests of FFE line, y_2—average value of criteria (2) among the best individuals of FFE lines. To determine the best point of each experiment we need to add another 2 parameters, which show parts of sub-optimal results among 50 repeated tests. Parameter y_3 shows the percentage of individuals with fitness less than 0.0249, y_4—less than 0.0247.

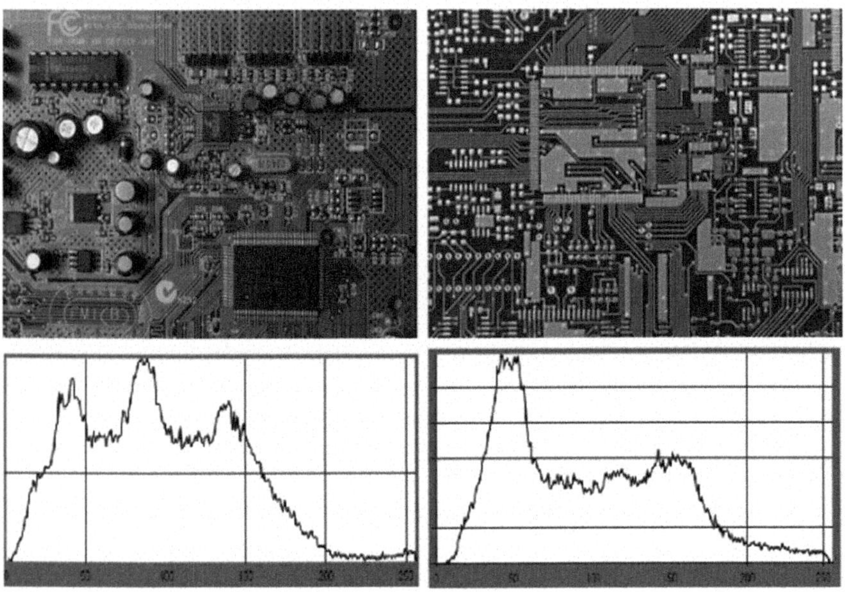

Fig. 6 2 printed circuit board photos and their respective FDB

Table 7 Structure of experiment

		–	0	+
x_1	Probability of SP crossing over	3	5	7
x_2	Probability of SR crossing over	3	5	7
x_3	Probability of SP mutation	9	12	15
x_4	Probability of SR mutation	9	12	15

Table 8 Results of experiment

№ Test	y_1	y_2	y_3	y_4
0 (0000)	0.024678	0.024928	62	2
1 (− − − −)	0.024703	0.024941	46	0
2 (− − + −)	0.024756	0.02498	36	0
3 (− + − −)	0.024723	0.024953	38	0
4 (− + + −)	0.024665	0.024962	46	2
5 (− − − +)	0.024698	0.024916	56	2
6 (− − + +)	0.024678	0.024961	48	2
7 (− + − +)	0.024723	0.024945	42	0
8 (− + + +)	0.024749	0.024966	42	0
9 (+ − − −)	0.024726	0.024903	62	0
10 (+ − + −)	0.024718	0.024943	46	0
11 (+ + − −)	0.024700	0.024936	48	0
12 (+ + + −)	0.024726	0.024959	40	0
13 (+ − − +)	0.024744	0.024912	48	0
14 (+ − + +)	0.024687	0.024908	52	4
15 (+ + − +)	0.024688	0.024897	62	2
16 (+ + + +)	0.024729	0.024943	48	0

The results of the performed experiment with the input OMMI «Circuit-1» is shown in Table 8.

The results of the separate tests are quite close to each other. For that reason, there is no specific way to determine the significance of this factor by performing a simple analysis. The quality analysis shows that according to y_1 parameter the best point of experiment is test 4 and according to y_2 the best is test 15. Based on the results of all 4 parameters the best point of experiment is definitely test 15. However, the tests 0 and 14 are quite close to the results of test 15. Such small difference between the tests is not contributing to build confidence through the quantitative evaluations.

The attempt to describe the influence of the selected factors, by the regression equation, shows that only b0 coefficient exceed the prescribed level of significance. There are 2 reasons behind this conclusion: (1) the area of the factors variation (the selected values estimated as sub-optimal in previous researches), and (2) limited number of parallel test, which is not enough in order to estimate correctly their influence factors.

1.3.5 Additional Experiment for the EGA Sub-optimization

Due to the impossibility to solve the problem of EGA's optimization with standard procedures when planning the experiment, the decision was taken to continue the research by testing the best points of the performed plan. The test 15 was repeated 20 times. It allows us to find out that the sub-optimal individuals are quite rare in the output results. The analysis shows that one of the reasons of such rare evolution of individual's fitness with level less than 0.0247 is due to the small difference in the genetic material. The difference between individuals is provided from the basic population's initialization step. In this way, we change the level of initial SP and SR variation.

For the current research, the value of the initial SP and SR variation is fixed on [−5; 5]. The decision is made to perform 3 experiments with the variation [−6; 6], [−7; 7] and [−8; 8].

This performed additional experiment fully confirms the idea that the small difference in the genetic material received by the individual with high fitness is influencing such results. The initial variation level of SP and SR [−7; 7] shows that the quality of created individuals was rose-sharply (Fig. 7).

These results confirmed that the genetic difference of the population represents an important part in the evolution of individuals, which can be associated with some limits of the evolutionarily searching. The small genetic difference leads the population to de-generation, and on another hand, the huge genetic difference will increase the scale of search. It is obvious that too large scale in searching will also give a negative effect, because it will reduce the probability of finding the extreme. Therefore, an important issue in setting the EGA's level is coming from the genetic material difference.

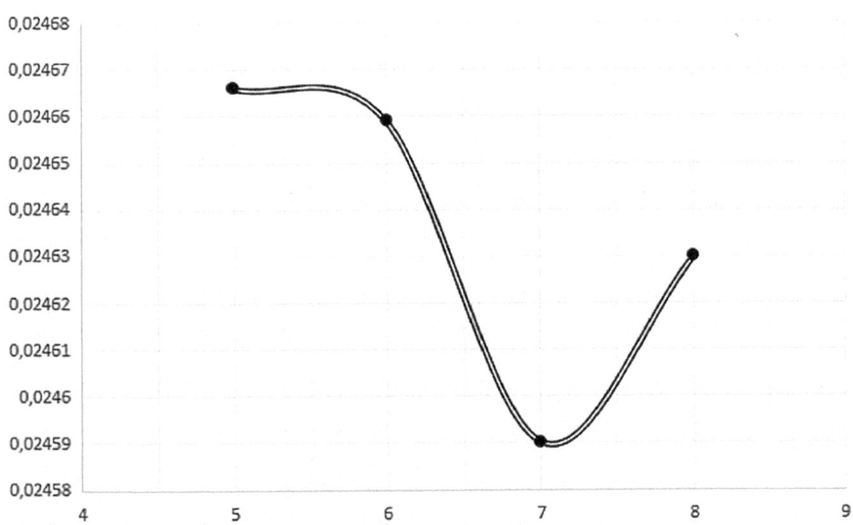

Fig. 7 The influence of the initial variation level on the output result

1.3.6 Conclusion

The main research results are:

(1) The developed double chromosome model of the approximated image representation has increased the searching capabilities of the algorithm on 3.2% for «Circuit-1» and 6.9% for «Circuit-2» compared with previous single chromosome model [10].
(2) The appearance frequency, found in the sub-optimal results, allow to calculate, on their basis, the necessary number of required approximations that will guarantee the sub-optimal result with a given confidence level.
(3) The found area of the sub-optimal EGA's settings provides the best results to the MMI approximation.
(4) The scientific novelty of the work consists in the new mathematical and numerical model of the image representation approximation within the evolutionarily genetic space.

2 Hybrid Methods Increasing Quality and Speed of Monochrome Images Tone Approximation

2.1 Improvements in Quality and Speed of Monochrome Images Tone Approximation by Hybridization of Modified Evolutionary Approach Algorithm Based on Direct Extreme Search

The problem to improve the speed of Monochrome Multitone Image (MMI) approximation, which consists in replacement of original palette with palette that has less number of tones, is considered. Often, for solving such a problem, the heuristic optimization-search algorithms are used. Their drawback is that, they cannot guarantee to find the solution for the given optimization criteria, which motivates to find the approximating palette of the potential extreme. The solution of this problem is to define the deterministic algorithm, that guarantees finding the extreme, through the integration with the heuristic algorithm. Such hybridization takes in consideration the main advantage of these 2 approaches: (1) speed of the heuristic algorithm and (2) accuracy of the deterministic one. The hybrid algorithm applies the heuristic approach to decrease the search area and the deterministic approach guarantees to find extremes within this area. This approach defines the time, as the optimization criteria for the algorithm processing speed. Thus, the proposed hybridization enables to find the bi-optimal solution for the defined MMI approximation problem that provides both, the optimal quality of approximation within the available time frame.

2.1.1 Introduction

The image approximation problem consists in reducing the number of palette of colours with minimal image quality losses [11, 12]. The problem is defined due to the necessity to display a large palette size image on the devices that have allowable palette size limits [13]. In addition, the image approximation can be used to simplify the pattern recognition problems [14, 15], as required for the web application to have the minimal memory footprint. The files size reduction can significantly increase the speed of website download time. In many technical applications (for example, in systems of technical sight [14–16]) it is necessary to solve the approximation of monochrome multitone images (MMI), to be able to obtain the optimal shaded grey scale image, obviously easier to process in respect to the respective colour images. The already performed research in this field has been published in several reports [5, 17] in which the different directions to solve the minima deviation, as the optimization task relating the original image to the approximated one. For example, the possibilities of particle swarm algorithm have been tested to optimize the mentioned approximation process [3, 4], where, the effectiveness and perspectives of using evolutionarily genetic algorithm (EGA) [18–20] have been shown.

However, EGA, as any other heuristic algorithm is characterized with impossibility to estimate the accuracy of computed solution in terms of nearness to the optimum [20]. This is due to the fact that the core of any heuristic algorithm is a random search, despite each heuristic algorithm has some special rules, which tune the random search and make it much more effective. However, investigations aimed at providing and estimating the extreme result of the graphical approximation, which has not been found in the available sources.

2.1.2 Problem Formulation

The solution of the above-mentioned problem requires the development of the MMI palette reduction algorithm and to experimentally validate its usage, as follows:

1. The algorithm has to be able to check any MMI's approximating palette in respect to the specified approximation quality;
2. The hybrid MMI approximation algorithm is combines evolutionarily genetic algorithm and the deterministic algorithm to find local extremes.

2.1.3 Modified EGA's Gene-Chromosome Structure

In the EGA algorithm that aims to optimize the image approximation, every AMMI is considered as an individual that enters the population. Thus, the AP tones can be considered as genes that spawn the image. Therefore, the OP (1) represents as genome $\mathfrak{G}^{\mathcal{P}}$ that EGA operates for optimization:

$$\mathfrak{G}^{\mathcal{P}} = (\mathfrak{g}_i | i = \overline{1, \mathcal{N}}). \tag{16}$$

The construction of each separate individual AMMI by OMMI approximation is provided with scheme that is defined by certain AP, which applies $s < S^0$ tones [18, 19]. This s—vector is considered as chromosome that spawns an individual during the algorithm process, defined with the AP structure represented as s tones—genes $\mathfrak{g}_{ij}$ too:

$$\mathfrak{G}^{\mathcal{S}} = \left(\mathfrak{g}_{ij} | i = \overline{1, \mathcal{N}}, j = \overline{1, \mathcal{S}}\right). \tag{17}$$

Because of earlier set condition the value of s is already defined. For this reason, the quality of the approximation algorithm will depend only from tones—genes $\mathfrak{g}_{ij}$ selection. In this individual spawn process, only the genes from genome $\mathfrak{g}_i \in \mathfrak{G}^{\mathcal{P}}$ participate and enter the chromosome (AP) after applying the genetic operators—mutation and crossing—to the natural selection.

The change of each separate j gene $\mathfrak{g}_{ij}$ in the chromosome is the variation of value i that increase, decrease or leave unchanged the fitness of each individual. The modified EGA algorithm finds the best chromosome (AP) resulting from all the generations. However, the extreme of the solution and its nearness in respect to the absolute optimum, remains unknown, which in this case represents the minimum, as defined in (13).

2.1.4 Theoretical AP Extreme Estimation Prerequisites

It is known that multidimensional function $f(x)$, where x—vector of its arguments, has the minimum in x_0 point of the multidimensional factor space, if exists a point neighborhood in which, for all $x \neq x_0$, the following neighborhood inequality is fulfilled:

$$f(x_0) > f(x + \Delta x), \tag{18}$$

where Δx—randomly increment of x vector within the searched neighbourhood [22].

The formulated feature guarantees the presence of at least one local extreme, which simultaneously determines the necessary conditions for global extreme too. In the context of the optimization problem it checks the space of any solution. In our case, the solution is represented with a chromosome (AP) that is a vector of natural numbers:

$$P_{\mathcal{S}}^{a} = \left(\mathfrak{g}_{i_1 1}, \mathfrak{g}_{i_2 2}, \ldots \mathfrak{g}_{i_j j}, \ldots \mathfrak{g}_{i_{\mathcal{S}-1} \mathcal{S}-1}, \mathfrak{g}_{i_{\mathcal{S}} \mathcal{S}}\right)^T \tag{19}$$

For example, for $s = 8$, such AP vector—P^a, is:

$$P_8^a = (14, 47, 79, 114, 149, 182, 213, 242)^T. \tag{20}$$

The genes discreteness in vectors (19) and (20) creates a topology in the chromosome space, which is a finite array of points generated by the 3-combinatory states: adding 1, −1 and 0 for different variations, which influence all the genes values. For the 8-element vector (20) this operation is illustrated in Table 9.

The central column contains the coordinates of the 8-dimensional vector's space, for which the extreme has to be found. Any one-element combination from each Table 1 row, with the exception of the combinations of the all elements from the central column, represents the nearest vector (20) neighbourhood, which is easy to calculate from the $3^8 = 6561$ combinations. Thus, the chromosome (20) neighbourhood in the factor space of its genes contains 6561 chromosomes (points), which have to be checked. The AP (11) local extreme investigation is a sequential combination of 8 rows elements in Table 1, which results in a new chromosome. This chromosome fitness is estimated according to optimization criteria (13) and it is compared with the fitness of AP (20) that is under checking (central column). If for the entire neighbour investigated AP points the fair condition (18) is satisfied, then the found AP solution is at least the local extreme. If there are some points that did not satisfy the extreme conditions, then there is a possibility to choose a better point and repeat for it checking procedure. Such possibility motivated the creation of a deterministic algorithm, which seeks the real extreme in neighbourhoods of that point.

2.1.5 Hybrid Optimization Algorithm Features

The idea of joining these algorithms consists in the AP sequential processing. First, EGA finds the AP that is supposed to be extreme neighbourhood. It allows decreasing the search area. Second, the deterministic algorithm checks the AP neighbourhood and in case of success, the better fitness AP initiates the checks to find the new better one. This continues until condition (18) is satisfied, as the

Table 9 The neighbourhood of chromosome (20)

$g_{ij} - 1$	g_{ij}	
$g_{ij} + 1$		
13	14	15
46	47	48
78	79	80
113	114	115
148	149	150
181	182	183
212	213	214
241	242	243

extreme hybrid algorithm solution is guaranteed. In addition, it makes sense to choose the appropriate time period during which the extreme is to be found in accordance with the defined optimization criteria.

The EGA processing time depends on the following quantity factors: population size, number of generations, number of repeats. The processing time of the deterministic algorithm depends on the size of the investigated space and number of checking iterations. Therefore, the EGA performance efficiency significantly influences the overall computational time of the hybrid algorithm.

2.1.6 Experimental Investigation

For investigating the performance efficiency of the hybrid algorithm, we need to find appropriate values for the EGA's parameters, in order to quickly calculate the AP suboptimal solution, which will be, further on, minimized by the hybrid algorithm.

In order to research the proposed approach, the investigation on the printed circuit board was represented with the MMI approximation having the resolution of 350 × 250 pixels (Fig. 8). This MMI is defined with 256-tones palette and approximated to the 8-tones palette.

For the parametric setup of the hybrid algorithm there were planned the 4-factor full factorial experiment (FFE) resulting in 16 tests, where each test has been repeated 6 times.

In order to achieve effective solution to the hybrid algorithm approximation, the stage of deterministic extreme search has been provided, which enables to choose the algorithm processing time as the parameter in the setup criteria. For this reason

Fig. 8 MMI of the printed circuit board

the EGA parameters with higher influence on the time have been chosen, as follows: population size K, number of generations M, number of search repetitions N. The chosen factors values are from 10 to 22, where the central level is 16. These values have been selected according to the initial investigations and also based on experience of previous works [5, 18, 19]. The 4th parameter is the initial variation of starting AP δ, that partly determines the size of the search EGA area in this experiment (varied from 0.1 to 0.3).

The best algorithm processing time (351 s.) is a test that had maximal value of K, M, N factors and minimal δ. However, the statistical calculations showed the only the influence of 2 factors: N and K. The realization of the gradient descent procedure in which the best test parameters values receive experimental optimum of hybrid algorithm, for them the processing MMI time was around $\sim$203 s, when $N = 25$, $K = 24$, $M = 22$ and $\delta = 0.1$.

2.1.7 Conclusion

(1) The developed approximated palette-checking algorithm, in addition defined the basis of the deterministic extreme algorithm contributed to significantly reduction the search time and raise the quality of image approximation results, due to its possibility of finding the extreme solution of approximation.
(2) The second significant result of the hybrid-developed algorithm is its bi-optimization characteristics, which allows receiving the extreme result according to the approximation quality and simultaneously guaranteeing minimal processing time.
(3) It makes sense to continue with follow-up investigations of the formulated problem in order to find the general regularity within the EGA setting parameters, which optimize the hybrid algorithm processing time based on the analyses of the image structure, as example, it can be the brightness frequency diagram.

2.2 *Modified Bi-optimal Hybrid Algorithm of Monochrome Images Tone Approximation*

This section investigates the image tones approximation algorithm for the multitone image processing, which applications examples are in Web development, compression algorithms, machine vision etc. It considers the Monochrome Multitone Image (MMI) approximation of the original palette to be replaced by a pallet having significantly less number of tones. For solving such problems, the optimization strategy requires the approximation quality, which maximize the tones reduction deviation between the original and approximated images. In particular, such problems are effectively solved with the heuristic Evolutionarily Genetic

Algorithms (EGA) fulfilling the required accuracy, while computational costs still remain significant. Thus, this research is focusing on the hybrid algorithm that is combining the heuristic algorithm, in order to provide suboptimal approximation quality, and the deterministic Algorithm of Local Discrete Optimization (ALDO) for finding the local extreme. EGA minimizes the local discrete optimization search area and ALDO guarantees to find the extreme within the search area. In conclusion such hybrid algorithmic architecture enables the MMI bi-optimization approximation.

2.2.1 Introduction

The image transformation has a wide range of applications. The examples can be found in the art image processing and in the processing that aimed on simplifying the graphical complexity of images by minimizing their file size and makes their processing faster. Another examples are the recognition algorithms used for finding Objects for Autonomous Navigation (OAN). The goal of this research is to reduce the image palette, which simplifies and makes more efficient the recognition process.

In Technical Sight Systems (TSS) [14], when tracing the safe autonomous navigation route, the encountered objects are recognized, by finding their orientation in space, and respective forms. This objective motivates the usage of the Monochrome Multitone Images (MMI). The term «multitone» is introduced to represent the images characteristic, which are defined with pixels of the same colour, where each of them has different brightness. Often, the image pixels have different tones of grey colour, called improperly «black and white», thus the term «halftone» is found more appropriate to use.

The processing of the real images with huge number of details is focusing on the raster graphics and this research addresses the Raster MMI (RMMI) approximation.

One of the main properties of any MMI is associated with the Tones Palette (TP). TP is defined as an integer vector of the image tones. The TP properties depend on the TP's size or length, and on the respective TP tones values.

The standard scales are based on the equal tones values distribution. For example, the standard 256-tones palette is defined as the vector: $(0, 1, 2,\ldots 255)^T$. The special purpose scales (which include approximation) can be characterized with a non-equal tones distribution and having the different number of tones in the related TP.

The TP size depends on the application domain and the related tones distribution rules depend on the structure of the image tones.

This section describes the algorithm that transform the images with large TP size, where each tone is maximized to its respective nearest original, resulting with the significantly smaller TP size, by applying an appropriate tones reduction algorithm. In addition, the computational efficiency is taken into account in order that the solution is computed as fast as possible.

2.2.2 MMI Approximation Technology

According to already considered information, the OMMI to AMMI approximation is the sequential execution of the following operations:

(1) Chose the size of S^A AP;
(2) Divide the OP on S^A subarrays $B_s^{S^0}$;
(3) Chose S^A tones b_s^A AP from subarray $b_s^0 \in B^{S^0}$;
(4) Replace sequentially OMMI $N = r \cdot c$ pixels that belong to OP with the nearest to them pixels from AP, which result is AMMI;
(5) Evaluate the criteria (13) for AMMI.

To achieve the best possible approximation result there is a need to organize the algorithm in a way that allows reaching the following condition:

$$\Delta_m\left(B^{S^0}, B^{S^A}\right) = \min_{B^{S^a}} \Delta_m. \tag{21}$$

Thus, it is necessary to realize the algorithm 1–5 as an algorithm for the minimization searching criteria (12). The high resolution of graphical objects (size of AP and number of pixels) motivates developing and using the heuristic Evolutionarily-Genetic Algorithm (EGA). The classic EGA modification on solving the approximation problem showed good accuracy [17–20].

It is obvious that the processing time of Extreme Estimation Algorithm (EEA) depends on the AP size. For example, for 8-tone AP it is necessary to produce 6561 calculations of criteria (13) and checking the condition (21). On computer with the Intel quad 3.3 GHz processor these operations require $\sim$23.85 s. But for 16-tone AP the number of combinations will increase to 3^{16} = 43046721, which means that the processing time will be 6561 times longer, requiring $\sim$44 h. Therefore, the application of EEA is only reasonable for small AP sizes. At the same time the algorithm is quite appropriate for STS, which uses small size palettes.

In such case, it is possible to develop and use the Algorithm of Nearest Extreme Finding (ANEF). The algorithm is based on EEA with the difference of having a new cycle starting condition, which will be initiated after finding the first AP that is better according to criteria (13) than the investigated one, but simultaneously it has not satisfied the condition (21). The stop condition ending the algorithm will be when the condition (21) is satisfied for the entire AP neighbourhood. Therefore, ANEF can be also used as EEA.

Figure 3 shows the protocols of the extreme AP searching steps for the image fragment (Fig. 1, displayed by red rectangle) with using ANEF from different starting positions. And as a result it takes different numbers of steps. However, the last part of the route is the same for all the cases, which can be seen according to equal time to the nearest AP search. In addition, the final extreme check step takes the same time (the small difference is explained by the processor background

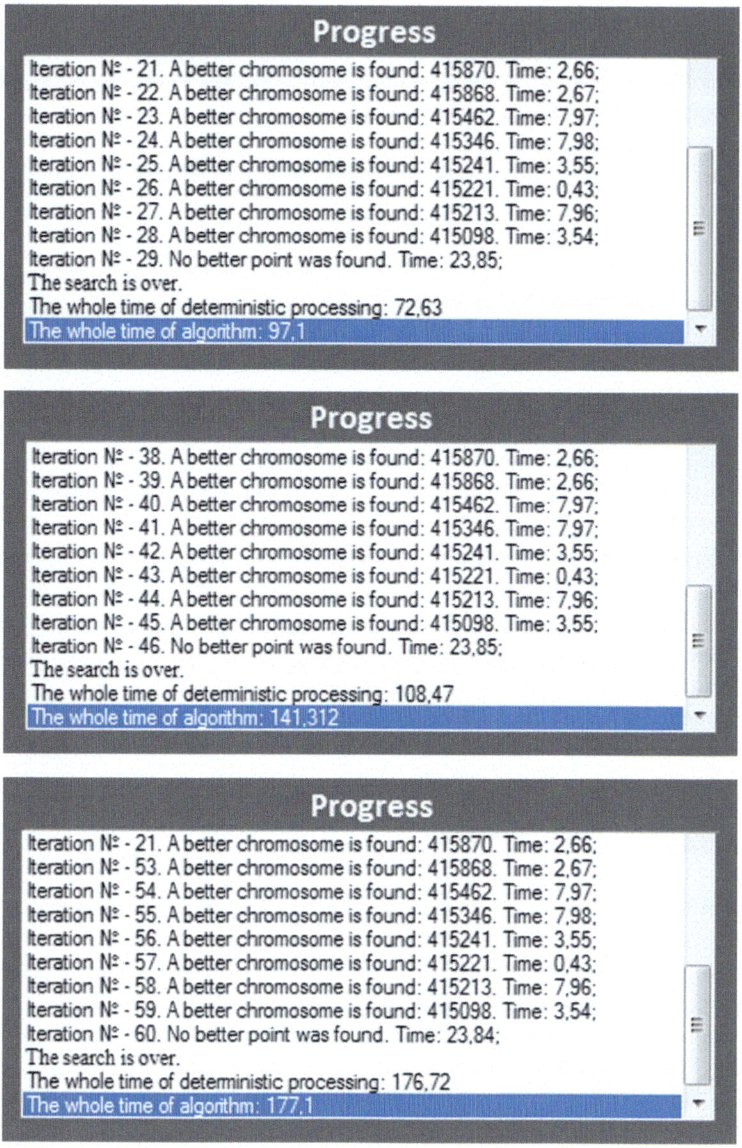

Fig. 9 The ANEF results of the extreme AP search with different starting positions

activity). If the EGA stage is excluded and only ANEF is used as start for the so-called "weighted" distribution [18, 19], then the processing time is 148.7 s.

However, to plan the searching strategy based only on ANEF is not reasonable, because even in this example, it is easy to notice that it depends on the starting position and the computing time can takes 1.5–3 min (as shown in this considered case). But, if the AP is provided with criteria value, for example, 415221 (Fig. 9), then the time to find the optimum solution will decrease to $\sim$36 s.

Therefore, the fair EGA initial setup is to provide a position near to extreme, and to further on apply ANEF, which will check for the solution significantly faster.

For the above-mentioned reasons, it is proposed to use Hybrid Extreme Search Algorithm (HESA), which is based on sequential EGA and ANEF. The results of this research are described below.

2.2.3 Optimization of HESA

For solving the HESA problem, it is necessary to find such EGA parameters that can find the solution in order that entering the extreme area is realized within minimal time. These parameters require that the number of parallel runs of EGA x_1, size of population x_2 and number of generations x_3.

The 2 Full Factor Experiments (FFE) have been planned and implemented. In the first FFE, the factors variation was on the same low level affecting all the factors: $x_1, x_2, x_3 \in \{2, 4, 6\}$. The results of FFE showed minimal average time (60.7 s.), when all x_i factors had high values. The regression coefficients values were negative for all the linear effects $b_1 = -15.3$; $b_2 = -8.3$; $b_3 = -11.6$.

Because the best FFE average time was too long and increment observed for all the 3 factors was expected to give significantly better result. The second FFE was implemented with $x_1, x_2, x_3 \in \{4, 6, 8\}$. And again the best average time (47.0 s.) was obtained, when all factors have their maximal values. The regression coefficients give bigger values for the linear effects $b_1 = -18.1$; $b_2 = -17.6$; $b_3 = -12.9$.

The best result of the second FFE is quite near to the maximum faster time of HAES—23.8 s, which gives the basis for new investigation—to consider gradient descent effect. The best result showed that the following parameters $x_1 = 12$; $x_2 = 12$; $x_3 = 10$, give average time of 40.7 s to obtain the solution. The further EGA's parameters increment increased the computational time, but simultaneously the output result was much nearer to extreme according to criteria (13). The explanation can be found in the EGA processing time. For example, when setting $x_1 = 16$; $x_2 = 20$; $x_3 = 16$, the computational time to find the extreme becomes 54.4 s.

2.2.4 Conclusion

The main research results are:

1. The new developed deterministic Algorithm of Nearest Extreme Finding (ANEF), together with the extreme solution estimation showed to be highly efficient and thus it is found to be a very good tool for solving monochrome multitone image tones approximation problems.

2. ANEF became the basis of the hybrid extreme search algorithm, allowing significantly increasing the performance of MMI TP approximation. It enabled the approximation bi-optimization providing the quality estimation in respect to the required computational time.
3. ANEF is the NP full and thus not appropriate for solving the approximation problem, in case of the large TP size needs, which motivates future studies to investigate the possibilities on how to increase the ANEF computational performance.

2.3 *Adaptive Optimization of Monochrome Images Tone Approximation*

One of the main factors that affects on the quality of tone approximation is a proper definition of approximation palette's (AP) structure. The structure represents tones that the AP contains. Thus, there is appear the optimization problem that consists in correct selection of AP for an image. With the purpose to solve the problem, a hybrid algorithm is used. The developed algorithm combine heuristic and deterministic searching approaches in such a way to guarantee at least suboptimal tone approximation. Also the algorithm make possible to realize two-parametric optimization of the tone approximation procedure.

The current research is devoted to developing and implementing an adaptive system of changing the number of searching iteration of heuristic stage. The evolutionarily-genetic algorithm is used as a heuristic stage in the developed hybrid algorithm. Statistical investigation allowed defining a criterion of iteration's significance during the search, which allows to approximately estimating the nearness of result to extreme zone. A condition of switching heuristic stage to deterministic one devolved based on the criterion of iteration's significance. The condition allows decreasing the total computational time of the hybrid algorithm. In fact, it is also can considered as optimization of the hybrid algorithm computational time, because the deterministic stage guarantees the extreme result. The experimental investigation confirms the effectiveness of developed adaptive scheme of the hybrid algorithm stages switching compared to fixed strategy.

2.3.1 Introduction

The size of brightness palette that produce a Monochrome Multitone Image (MMI) is one of the key parameters that affects on visual quality of the image. Commonly the parameter is called «color depth», but according to the authors opinion the palette's size is more straight term. The modern computational devices allows to process and produce images that represents by 24-bit palette's size, which means operating more than 16 mil colours. However, monochrome images and their tone approximation (reduction of the palette's size) are still demanded in many technical fields. Reduction of the palette's size directly leads to decreasing the image file's size and their visual quality. The optimization of the procedure could allow to minimize the visual quality losses. Besides of effective information compression, the procedure may be useful for simplifying the noises filtration, pattern recognition etc. Reduction of the palette's size makes the borders between image objects more obvious, because of significant brightness changing. Such a result could used by described kinds of processing. The current research investigates the process of the MMI tone approximation. The standard palette of MMI contains and operates 256 tones of grey color, where the tone 0 is a black color and 255 is a white color. The consideration of MMI can give more fundamental results on image's tone approximation in general, because of the clear and comprehensible MMI structure.

2.3.2 Structure and Settings of Hybrid Algorithm for Tone Approximation Optimization

The pilot model of developed Hybrid Optimization Model (HOM) of MMI approximation results showed possibilities to increase quality of tone approximation and computational speed of the problem solving as well [25–27]. Besides, the algorithm provides bi-optimization of tone approximation procedure according described criterions. Bi-optimization provided due the ANES that guarantee sub-optimization, which allow excluding accuracy of tone approximation and considering the total time of hybrid algorithm [28–30] as the only criterion.

The hybrid model consists in sequential searching of AP structure. The first stage is heuristic approach and the second stage is deterministic one. The reliability level of heuristic stage is not high, but compared to ANES, MEGA need much less computational time. This is allow to use MEGA as a tool for decreasing a searching area. The optimization of MEGA parameters provides increasing the effectiveness of that process. One of the most important settings of hybrid algorithm that affect on algorithm's total time is condition of switching the heuristic stage on deterministic one. If MEGA does not reach an extreme zone close enough, the total time of hybrid algorithm sharply rises due to high computational costs of deterministic stage. The excess steps of MEGA increase the total time too, but much less.

The condition of the switch for HOM investigated and fixed on 20 iterations of MEGA. The experimental investigation on wide set of MMIs showed that this condition can considers as suboptimal. Performed investigation on some OMMIs with size 430 on 240 pixels showed that total time of the algorithm takes 90 s under selected condition of switch. During the investigation opened potential possibility to decrease the total time by developing an adaptive algorithm of changing iterations of search in real time. The paper considers results of developing and experimental investigation of managing strategy of MEGA's iterations number that minimizing total time of the HOM. For the purpose designed three-parametric model of adaptive managing of MEGA's iterations number.

For adaptation system three setting parameters are used: number of guaranteed iterations g, number of additional iterations z, and e—constant of effective decreasing of criterion ΔQ during iteration. The scheme of the adaptation (r—index of current searching iteration) represents by following steps:

1. $r < g$—repeat a search iterations of MEGA;
2. $r = g$—calculated $\Delta Q_0 = \overline{\Delta t\left(P_0^a\right)_{g-1}} = \overline{\Delta t\left(P_0^a\right)_g}$;
3. $g = g + z$—switch to additional search iterations;
4. $r < g$—check the condition of ΔQ improvement

$$\overline{\Delta t\left(P_0^a\right)_{g-1}} - \overline{\Delta t\left(P_0^a\right)_g} < e; \tag{22}$$

5. if (22) is not satisfied—back to step 3;
6. if (22) is satisfied—switching to ANES stage.

Adding the parameter g is based on the fact that algorithm MEGA needs some minimal number of searching iterations for reaching to extreme zone. With rise of g also increase the probability of reaching extreme zone and searching time. The MEGA is heuristic algorithm, which negatively affects on reliability factor. The parameter z in this case will help decrease a risk of «looping», because additional iterations give a chance to overcome the zones of local extreme. For finding the appropriate value of criterion e there are performed investigation of images sub-optimal approximation. Analysis of ΔQ_a increments for each images allowed to choose $e \leq 0.0001$.

For investigation the adaptation process of MEGA by changing the number of searching iterations performed two-factorial experiment with following parameters: z = 3, 4 and 5; g = 12, 14 and 16 (where 4 and 14—coordinates of “zero point”). As objects of tone approximation selected exactly the same 6 images that used for sub-optimization of fixed number of MEGA iterations. Obviously, it will allow comparing both approaches. All images will processed by hybrid algorithm for each pair of adaptive system parameters. The results of performed experiment showed in Table 10. The column 1 represent the results of non-adaptive approach with 20 fixed iterations, the columns 2-10 showed the results of performed experiment. In rows «Avg. (average)» and «SD (standard deviation)» are shown the average computational time of HMO among all tested images. All settings, except 4 (90.1 s.), allow decreasing the average total time compared to **1** (89.3 s.). The best setting (**3**) decrease the average total time only on ~6% (83.9 s.).

However, the SD analysis (for 1–15.4 s, for 5–7.7 s) showed his significant reduction. It means that adaptive approach to each image decrease the probability of non-reach to extreme. Thus, the adaptive system allows to improvement the reliability of HOM’s output results.

This conclusion is clearly confirmed by diagram on Fig. 10 that is illustrated below. The diagram columns present the average total search time (the top of 7 grey) and its range (from the grey bottom to top light). The confidence probability

Table 10 Result of experiment

№	1	2	3	4	5	6	7	8	9	10
z	– – –	3			4			5		
g	20	12	14	16	12	14	16	12	14	16
Avg.	89.3	86.9	83.9	90.1	84.3	86.3	85.2	85.9	86.8	87.1
SD	15.4	11.7	10.9	10.1	7.7	10.1	11.4	9.6	11.1	9.2

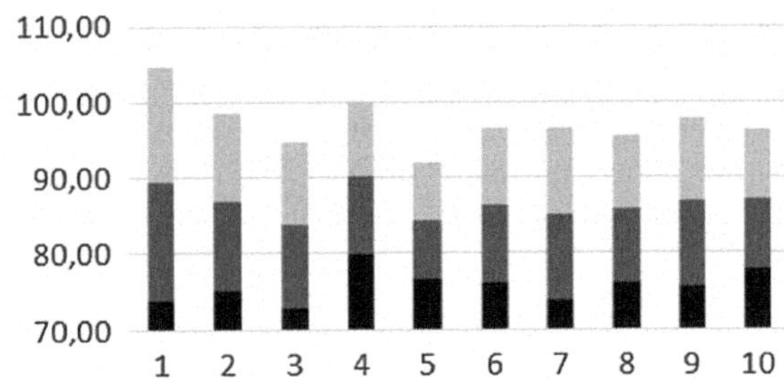

Fig. 10 Dependency of the hybrid algorithm time intervals from MEGA's adaptive system settings

for shown ranges is ~70%. It is easy to notice that the possible maximum time of HOM is significantly decreased.

As quasi-optimal setting of adaptive system for HMO can be considered with $z = 4$ and $g = 12$, which give the possible minimum time interval and the second best average time for the algorithm. However, significant results uncertainty makes reasonable to continue investigating the best values for the parameters, and finding the ways to modify the adaptive system.

In addition, it is necessary to notice that for potential effectiveness in the development of adaptive system needs follow-up investigation of the different values for the e parameter.

2.3.3 Conclusion

The main results of the performed investigation are:

1. Proposal of the hybrid algorithm HOM for monochrome images tone approximation, which combines heuristic Modified Evolutionarily-Genetic Algorithm (MEGA) supporting the initial search for optimal quality area, and the deterministic algorithm for the Nearest Extreme Search (ANES), showing improved performance with obvious potential perspectives;
2. ANES in HOM guarantees the suboptimal or even optimal result according to the required approximation quality, and the MEGA's settings significantly affects the HMO total time, which enables to choose the time as the only criterion for the HMO optimization;

3. Experimental investigation and validation of the integrated adaptive system for managing the MEGA search iterations allow finding the suboptimal settings that provide decreasing ($\sim 6.3\%$) of the processing time, and satisfying the condition in finding extremes, and in addition, reduce the maximum searching time to the confident level;
4. It is found reasonable to continue the investigations in modifying and sub optimizing such adaptive system.

Bibliography

1. Plonka G, Tenorth S, Rosca D (2010) A new hybrid method for image approximation using the easy path wavelet transform, vol 20, p 327
2. Figueras R, Simoncelli E (2007) Statistically driven sparse image approximation 1:461
3. Neydorf R, Derevyankina A (2010) Solving the multiextremal problems with particle swarm method. Vestnik DSTU 4(47)
4. Neydorf R, Derevyankina A (2010) Solving the recognition problems with particle swarm method. Izvestiya SFedU 7(108)
5. Aghajanyan A, Neydorf R (2016) Optimization of monochrome multitone images approximation based on evolutionarily algorithm. Omega Sci 108:11
6. Mitchell M (1999) An introduction to genetic algorithms. Fifth printing, England (162)
7. Luke S (2014) Essentials of metaheuristics
8. Bäck T (1996) Evolutionary algorithms in theory and practice—evolution strategies, evolutionary programming, genetic algorithms. Oxford University Press, New York, Oxford
9. Lewin B (1985) Genes. Wiley, New York (716)
10. Aghajanyan A, Neydorf R (2016) Optimization of monochrome multitone images approximation based on evolutionarily algorithm. Omega Sci 108:111
11. Puzicha J, Held M, Ketterer J (2000) On spatial quantization of color images. IEEE Trans Image Process 9(4):66–82
12. Emre C (2011) Improving the performance of K-Means for color quantization. Image Vis Comput 29:260–271
13. Color quantization. url wikipedia.org/wiki/Color_quantization (date of access: 9.02.2017)
14. Bishop C (2006) Pattern recognition and machine learning. Springer, Heidelberg
15. Duda RO, Hart PE, Stork DG (2001) Pattern classification. Wiley, New York
16. Chirov D, Chertova O, Potapchuk T (2017) Methods of study requirements for the complex robotic vision system. Spiiran Proc 2(51):152–176
17. Aghajanyan A, Neydorf R (2016) Optimal approximation of monochrome multi-tone images using the evolutionarily genetic algorithm. Com-Tech 2016:108–112
18. Neydorf R, Aghajanyan A, Vucinic D (2016) Monochrome multitone image approximation on lowered dimension palette with sub-optimization method based on genetic algorithm. ACE-X 2016, Split (Croatia)
19. Neydorf R, Aghajanyan A, Vucinic D (2016) Monochrome multitone image approximation with low-dimensional palette. In: IEEE East-West Design & Test Symposium (EWDTS)
20. Neydorf RA, Aghajanyan AG (2017) The research of the application possibilities of tones approximation in a technical vision for the autonomous navigation objects. Izvestiya SFEDU, Technical sciences, № 1–2 (186–187), pp 133–145
21. Pierre C, Jean-Philippe R (2006) Stochastic optimization algorithms. Handbook of Research on Nature Inspired Computing for Economics and Management Hershey
22. Vinogradov I (1977) Mathematical encyclopedia

23. Neydorf RA, Aghajanyan AG, Vucinic D (2018) Monochrome multitone image approximation on lowered dimension palette with sub-optimization method based on genetic algorithm. Improved performance of materials. Springer International Publishing, pp 144–154
24. Neydorf RA, Aghajanyan AG, Vucinic D (2016) Monochrome multitone image approximation with low-dimensional palette. In: IEEE East-West Design & Test Symposium (EWDTS)
25. Neydorf RA, Aghajanyan AG, Neydorf AR (2017) Optimization of approximation result of halftone images and assessment of their extremality. Mathematical methods in technic and technology. Saratov: SGTU n. Y.A. Gagarina, vol 1, pp 19–26
26. Neydorf RA, Aghajanyan AG, Vucinic D (2017) A high-speed hybrid algorithm of monochrome multitone images approximation. In: IEEE East-West Design & Test Symposium (EWDTS)
27. Neydorf RA, Aghajanyan AG, Vucinic D (2017) Improved bi-optimal hybrid approximation algorithm for monochrome multitone image processing. In: ADVCOMP 2017, The Eleventh International Conference on Advanced Engineering Computing and Applications in Sciences. IARIA, pp 20–25
28. Schütze O, Hernandez V (2016) The hypervolume based directed search method for multi-objective optimization problems. J Heuristics, vol 22, pp 273–300 (Springer US)
29. Gillette A, Wilson C, George A (2017) Efficient and autonomous processing and classification of images on small spacecraft. In: 2017 IEEE National Aerospace and Electronics Conference (NAECON), pp 135–141
30. Sun JQ, Schütze O (2017) A hybrid evolutionary algorithm and cell mapping method for multi-objective optimization problems. In: 2017 IEEE Symposium Series on Computational Intelligence (SSCI), pp 1–9

The manufacturer's authorised representative in the EU is Springer Nature Customer Service Centre GmbH, Europaplatz 3, 69115 Heidelberg, Germany. If you have any concerns regarding our products, please contact ProductSafety@springernature.com

Printed and bound by CPI Group (UK) Ltd, Croydon, CR0 4YY
24/07/2026
02175923-0007